COMPETITION, COLLUSION, AND GAME THEORY

LESTER G. TELSER
COMPETITION, COLLUSION, AND GAME THEORY

AldineTransaction
A Division of Transaction Publishers
New Brunswick (U.S.A.) and London (U.K.)

Second paperback printing 2009

Library of Congress Catalog Number: 2007012545
ISBN: 978-0-202-30925-5

Printed in the United States of America

Library of Congress Cataloging-in-Publication Data

Telser, Lester G., 1931-
    Competition, collusion, and game theory / Lester G. Telser.
        p. cm.
    Includes bibliographical references and index.
    ISBN 978-0-202-30925-5 (alk. paper)
    1. Competition—Mathematical models. 2. Price—fixing—Mathematical models. 3. Game theory. I. Title.

HB238.T45   2007
338.6'.048015193—dc22   2007012545

TO JOSHUA AND TAMAR

Foreword

For only a little over a decade, economic theorists have been working on a new and fundamental approach to the theory of competition and market structure, an approach inspired by appreciation of the earlier work of Edgeworth and Böhm-Bawerk and making use of the new tools of the theory of games as developed by von Neumann and Morgenstern. This new approach bases itself on the analysis of competitive behavior and its implications for the characteristics of market equilibrium rather than on assumptions about the characteristics of competitive and monopolistic markets. Its central concept is "the theory of the core of the market," and it is concerned, very broadly speaking, with the conditions under which markets will or will not achieve the characteristics of uniform prices and welfare optimality posited by traditional theory. This concern entails, among other things, a shift of emphasis from the prevailing concentration on oligopoly as a type of market structure and on advertising as a weapon of competition to the influence on the outcome of market processes of such factors as number of traders on the two sides of the market, transactions costs, brokerage, the way in which firms form their expectations of future demand, and the costs of collusion.

Professor Lester G. Telser has been at the forefront of the development of the new approach and is one of the few practitioners of it capable of communicating to his fellow economists its theoretical techniques and, more important, its implications for the empirical analysis of market phenomena. This he does both by the construction and analysis of simple economic problems to illustrate the theory and by the presentation of empirical research of his own designed to formulate and test propositions suggested by the new "theory of the core" approach to market analysis. His own introduction provides a sufficiently concise summary of the scope of the book to make it unnecessary for me to attempt a still briefer summary here. Suffice it to say that this is the first book to present a comprehensive synthetic overview of an important new line of development in fundamental economic theory and that I am delighted to be associated with its publication as an Aldine Treatise.

HARRY G. JOHNSON

Acknowledgments

The research presented in this book has received the generous support of the National Science Foundation, initiated with grant GS-365 in 1963 and continued with grant GS-1783 in 1966. This financial aid was indispensable in carrying out the work. It made possible my sojourn as a visiting research fellow at the Cowles Foundation for Research in Economics for the academic year 1964–65 where I first learned about the theory of the core. In 1969–70 I received a Ford Faculty Research Fellowship, which enabled me to finish this book. During this time I worked at the Center for Operations Research and Econometrics at the Catholic University of Louvain, Belgium.

Much of the material in this book has been exposed to my students in the course "Theories of Competition," which I have taught in the economics department at the University of Chicago for the past five years. The comments of the students in this course have taught me a great deal; of particular helpfulness were those of Uri Ben-Zion, Yoram Peles, and Donald Parsons. I have also been fortunate in having the aid of several very able research assistants. E. H. Thornber gave invaluable assistance in carrying out the empirical analysis reported in chapter VII. The work in chapter VIII was done with the assistance of Uri Ben-Zion, Harry Bloch, Josef May, and Stan Horowitz. Carl Berliner helped with appendix 4 of chapter VIII.

I am also very grateful to many of my professional colleagues for their comments and criticisms of various parts of the book. In particular, Lloyd Shapley and Herbert Scarf saved me from some serious errors in chapter II when I presented this material at the Mathematical Social Science Board Seminar on Game Theory at the RAND Corporation, Santa Monica, California, in June 1969. The research in chapter VIII has benefited from the comments by H. Gregg Lewis, Walter Oi, George Stigler, and Leonard Weiss. Zvi Griliches has read and commented on both chapters VII and

VIII. Yoram Barzel read with painstaking care chapters I, V, VII, and VIII. Much of the improvement in the exposition was inspired by the comments of Harry Johnson and Robert Wesner. Charles Cox and Howard Marvel eliminated errors in the final stages by reading the proofs and checking the index. I must assume, however, the sole responsibility for any errors and shortcomings in this book and must absolve all of those mentioned from any blame for these.

One other person has been a constant source of encouragement as well as a sounding board for my ideas—my wife, Sylvia. The burden of editing and preparing the manuscript for publication, of endless checking, and of polishing the style was lightened thanks to her help.

Contents

Introduction

Although the nature of a market and what happens there is surely a proper subject of economic analysis, the student will search the literature in vain for more than passing mention of this fundamental topic prior to 1881 when Edgeworth published his profound analysis of markets. The next important contribution did not appear until a decade later in Böhm-Bawerk's celebrated study of a horse market containing the first rigorous constructions of market supply and demand schedules. This paucity of early analysis is all the more surprising when we recall that in the 1870s economics embarked on its modern rigorous course with the contributions of Jevons, Menger, and Walras. After Böhm-Bawerk a half century passed before the next major contribution, the publication of von Neumann and Morgenstern's *Theory of Games* (1944). But the reviewers found in game theory little of relevance for economics, and it was not until 1959 that Martin Shubik pointed out that Edgeworth's theory could be married to game theory to produce a formidable new approach to the study of competition. This approach is now known as the theory of the core. Nor is this all. In 1838 Cournot developed a mathematical theory of competition generally condemned and misunderstood in most textbooks, which turned out to be the forerunner of the minimax theorem of game theory as applied to nonzero-sum games. This became clear shortly after J. F. Nash in 1950 published his work on equilibrium points, and economists became aware of their connection to Cournot's theory. Rigorous research into competition has been growing lustily only since 1959. This curious tale may engage the attention of the historian of economic thought, but it is not our further concern.

Why is it that prior to the recent studies of competition economists paid so little attention to the foundations of their discipline? One can find much attention given to questions of monopoly, cartel, and competition, but virtually all of this literature takes for granted some of the intrinsic properties of markets and competition without properly understanding them. For example, the common textbook description of competition runs in terms of given prices that the individual is powerless to change. Thus, a firm is said to operate in a competitive market if changes in its rate of sales exert no perceptible influence on the price of the product. Another common description of competition asserts that price equals marginal cost in a competitive market. The former approach often goes on to observe that individuals are powerless to affect prices in a competitive market if there are many traders, so competition results from large numbers. The equality between price and marginal cost raises more complicated problems, as witness Marshall's need to appeal to theories of externalities to explain some phenomena obviously inconsistent with the posited equality. At best these discussions of competition are merely vague and uninformative, and at worst they are positively erroneous. We need a theory that defines competition and then deduces its implications. The theory of the core provides that long sought for theory of competition capable of meeting this need.

It will be helpful to sketch how core theory looks at a competitive market for a single good. Each participant is assumed to start with a given stock of the good and to have preferences describing the terms at which he is willing to change his holdings. He is also assumed to act independently in finding a set of trades that will make him as well-off as possible. Since every trader is activated by the same motive, each is constrained by the conflicting desires of the others. Also, no one is forced to trade, and all are free to make contracts with anyone at mutually acceptable and, therefore, legally enforceable terms. A group of traders of whatever size may also make contracts embodying mutually acceptable terms of exchange. The basic tool of analysis is called "the characteristic function" in game theory. It describes the greatest gain obtainable by a coalition of traders under the most adverse conditions. We imagine that a number of traders combine and attempt to allocate their initial holdings among themselves so as to be as well-off as possible. In these circumstances the group faces the most adverse conditions when the members are confined to trade among themselves. Therefore, the valuation of the optimal allocation of their goods among themselves by the given coalition determines the value of the characteristic function for the coalition. A trader's imputation is the difference between his valuation of his initial holdings and his valuation of his final holdings so that his imputation measures his gains from trade. Since individuals are free to join any coalition and since there is a value of the characteristic function for every coalition, it is assumed that the trader's imputation must be large enough so that no coalition can prevent a proposed allocation of goods by offering him better terms. In other words, the imputations are said to be in the core of the market if they satisfy a system of inequalities. A systematic study of the properties of these inequalities becomes a theory of competition.

In the first two chapters I focus on a single market where one good is exchanged for money. This model is the analogue of partial equilibrium analysis according to Marshall. Consequently, we can learn much about the structure and performance of markets. Some of the questions considered are as follows: When does a competitive equilibrium exist? This is equivalent to finding when a market has a nonempty core. When is there a set of trades capable of implementing the core constraints? When will there be a common price per unit in the market? Under what conditions will there be Pareto optimality? How do transactions costs affect the equilibrium? What is the role of brokers in a market? How do changes in the number of traders affect the equilibrium? How efficient is random contact among the traders? Can there be an equilibrium if trade is confined to coalitions consisting of pairs of traders? In answering many of these questions we shall find that the concept of "a balanced game" is of primary importance.

Core theory can also throw light on problems involving public goods and increasing returns. For example, with the latter it is easy to show the possibility of an empty core, which implies that there is no competitive equilibrium. Therefore, core theory can determine when increasing returns lead to the situation known in the classical literature as a natural monopoly.

The classical theory derives the properties of the market equilibrium in terms of supply and demand schedules, but, as is shown by core theory, this procedure is not always correct even for a competitive market. This leads to the subjects forming the contents of chapters III and IV. Chapter III applies core theory to oligopoly by developing a relation between the core and the Cournot-Nash equilibrium. One of the interesting questions dealt with in this chapter is, When will duopolists collude rather than compete? The answer is related to the famous problem in game theory known as the prisoner's dilemma. Two prisoners suspected of a crime are captured and questioned separately. Each is told that if they both confess then they will receive light sentences, while if one confesses but the other does not then the full penalty of the law will fall on the nonconfessor. If neither confesses, then both must go free because of there being no evidence upon which to base a conviction. The analogy with duopoly is clear. If the duopolists collude, they can divide the monopoly return. If they compete, then at best they can obtain the imputation implied by the Cournot-Nash equilibrium. If they agree to collude but one of them cheats, then the cheater can temporarily obtain a larger return than his share of the maximum, joint monopoly return. When does it pay to collude?

Chapter IV extends the study of the Cournot-Nash equilibrium to the case of n competing firms on the assumption of static demand combined with uncertainty. The uncertainty requires consideration of how the firms choose their current actions in response to their forecasts of future conditions. The problem divides into two parts. In the first the n firms are assumed to make identical goods with constant marginal and average cost. We consider three kinds of forecasting equations. The first, originally proposed by Cournot, asserts that the current level of demand is the best forecast of next period's level. It is shown that the Cournot forecasting equation gives a stable equilibrium for two firms but not for three or more. The second forecasting equation, called "adaptive expectations," makes the forecast a geometrically weighted moving average of past observations. In this case it is shown that there is a stable equilibrium with three or less firms but that with four or more the weights attached to past observations must approach zero with sufficient rapidity, depending on the number of firms, for an equilibrium to exist. The third forecasting equation assumes that each firm uses the arithmetic mean of past observations as its prediction of future values. It is shown that in this case there is a stable equilibrium for any number of firms.

The preceding results pertain when the n firms make perfect substitutes permitting them each an independent choice of quantity but not of price. The second part of the analysis assumes imperfect substitutability among firms' products. If their products are imperfect substitutes and if for any positive rates of sale there is a corresponding set of maximum nonnegative prices, then it is shown that a stable equilibrium exists for Cournot expectations for any number of firms. A fortiori the same is true of adaptive expectations and arithmetic mean expectations.

A central question in this investigation is when competition or collusion will prevail. Since collusion can secure the maximum monopoly return, one might think that it would always prevail. But, then, why is there not a world of monopolies? Fortunately for consumer welfare, collusion has its costs, and it is not true that the participants to a collusive agreement can always obtain a higher net return then they would under competition. The main topic of chapter V is a study of the determinants of the costs of collusion and the conditions under which competition will prevail.

This chapter also studies the nature of competition resulting from product variety, by using a spatial model which assumes a homogeneous product that can be made by an arbitrary number of identical firms whose average production cost is a decreasing function of the output rate. Customers are uniformly distributed along a straight road of finite length so that the demand function relating the rate of purchase to the delivered price is everywhere the same. Unit transport costs are constant with respect to quantity and distance. It is shown that, depending on the level of demand, competition can result in a single firm at the midpoint of the road segment that it serves, implying an efficient spacing of the firms. But there is no marginal cost pricing, and firms compete directly for patronage only at the boundary points of their market territories.

The last topic of chapter V is the question of how the monopoly return for a group of colluding firms is shared among them. There is no fully satisfactory theory of this problem. One can apply core theory, but the results are disappointing. Indeed, there is an embarrassingly large number of imputation theories based on the core which, unfortunately, either give trivial results or for which the core is empty. A trivial result is when the value of the characteristic function is zero for all coalitions except the one including all of the firms, and for this the value of the characteristic function equals the maximum, joint, net monopoly revenue. This chapter concludes with a modest conjecture about the sharing of the collusive return that seems promising because some of its implications correspond with observable phenomena.

Chapter VI continues the analysis of chapter IV by extending and generalizing the results to dynamic demand relations. N firms are assumed to make competing durable goods for which the demand depends on current and expected quasirents. The term quasirent refers to the buyer's evaluation of the cost of the services of one unit of the durable good for one time period. The development of the monopoly and the Cournot-Nash equilibria requires more advanced mathematical techniques in a dynamic setting; hence, this chapter includes some discussion of topics in functional analysis pertinent to the economic application. It is shown that the existence of a monopoly equilibrium implies the existence of a Cournot-Nash equilibrium. Further, for a very general class of forecasting equations embracing all those studied in chapter IV, it is shown that a Cournot-Nash equilibrium exists if there is a monopoly equilibrium. Thus, the existence of a monopoly equilibrium imposes restrictions on the nature of the dynamic demand relations that have several implications. One interesting result bears on the choice between rental and sale of a durable good if there is only one supplier. We often observe that monopolists prefer rental to sale. The analysis shows that sale of a durable can never yield a higher monopoly return than rental when nonnegativity conditions on stocks and rates of sale are taken into account.

The final two chapters are mostly empirical. Chapter VII gives estimates of the demand relation between market shares and prices using monthly data for the leading brands of four advertised consumer products: Minute Maid, Snow Crop, Libby's, and Birds Eye in frozen orange juice concentrate; Chase and Sanborn and Maxwell House in instant coffee and in regular coffee; Folger's and Hills Brothers in regular coffee; and a leading brand in instant mashed potatoes. These results give the own and cross price elasticities from which one can infer limits for the ratios of price to marginal cost. These empirical studies also give evidence for judging the relevance of some of the theory in chapter VI. The first three products were well established during the sample period. It is shown that market shares in them quickly respond to price change and that price differentials among the leading brands tend to remain constant in the long run although there is considerable short-run variation. The persistent long-run price differentials apparently reflect brand cost differentials. A different picture emerges for instant mashed potatoes, a relatively new product during the sample period. The leading brand studied was the innovator here and had a high but falling market share, not so much because it lost absolute sales as because later entrants were able to attract new customers. Consequently, the leading brand share fell as the total market grew. The price policy of the firm producing the leading brand is consistent with the estimates of the demand relation. These estimates show that market share did not respond as quickly to price change as was true for the leading brands of the established goods. Thus the firm producing the leading brand of the new product behaved as if it had a temporary monopoly. Combining the evidence for all of the products gives a single coherent explanation of the competitive relation among the brands.

The last chapter studies the determinants of the returns to manufacturing industries using data for more than 400 four-digit manufacturing industries from the 1963 Census of Manufactures. This chapter contains the evidence bearing on some of the theory in chapter V. It is shown that rates of return are a nonlinearly increasing function of the four-firm concentration ratio. This is demonstrated by stratifying the sample of industries into three strata: $0 \leqslant C < 25$, $25 \leqslant C < 50$, and $50 \leqslant C \leqslant 100$, where C equals the share of industry sales held by the four leading firms. There is about the same number of firms in each stratum. The coefficient relating the rate of return to the concentration ratio rises gently from the first to the second stratum and is sharply higher in the stratum $C \geqslant 50$.

An important problem in this empirical work was to obtain a proper measure of the firm's capital including its intangible as well as its tangible capital. Accounting measures generally include such tangible items as the original cost of plant, equipment, inventory, and land; but there are other components of a firm's true capital contributing to its return although they do not appear on its balance sheet. These include the firm's outlays on research and development, on advertising, and on the training of its employees in firm-specific skills. The latter outlays comprise part of the firm-specific human capital. The empirical work stresses this capital's contribution to the firm's return in two ways: first, by directly relating the firm's return to proxy measures of specific human capital and, second, by indirectly studying the implications of this theory on quit and layoff rates. The chapter contains a careful study of the determinants of labor force turnover using Bureau of Labor Statistics data.

Although all eight chapters provide a unified treatment of the subject, they do fall into three somewhat autonomous groups that may be read independently. Chapters I to III emphasize the core, chapters IV, VI, and VII stress oligopoly, and chapters V and VIII relate industry structure to rates of return and other variables. Chapter V also draws on the analysis in chapter III, and chapter VI generalizes the theory in chapter IV. Chapter VII depends on the theory in chapter VI, and chapter VIII relies on the material in chapter V. Since each chapter gives the appropriate references to the material required from elsewhere in the book, the reader may follow his own interest in choosing the order in which to read the material and still be able to follow the reasoning.

The mathematical level varies among the chapters. The most mathematically advanced is chapter VI followed by chapter II. Chapter I is the most elementary. Chapter II requires understanding of matrix algebra, the theory of linear inequalities, and calculus. Chapters IV and VI make much use of linear difference and differential equations. Some knowledge of advanced calculus and the theory of functions would also be helpful in understanding some parts of chapter VI. It is fair to describe the mathematical level in this book as less demanding than much of modern mathematical economics since little use is made of measure theory or topology. This level is adopted not because one should shun the use of powerful tools on principle but because it is clumsy to use them if less powerful ones are adequate. In fact the approach in chapters I and II is deliberately chosen to avoid the use of measure theory so that certain economic, rather than technical, questions can be placed in the forefront. Similarly, the theory of dynamic monopoly and Cournot-Nash equilibria in chapter VI works in discrete time to avoid some purely technical complications arising in continuous time that have no economic relevance.

A full-scale analysis of advertising is not included in the book, but the interested reader is invited to consult several of my published articles on this subject, which are listed in the References at the end of the book. Perhaps the two most important of these are the ones published in 1962 and in 1964.

Finally, it is necessary to describe the reference system used. Theorems and lemmas are numbered sequentially in each chapter beginning with number one. Each number in cross-reference is preceded by a roman numeral which refers to the chapter in which the theorem or lemma appeared. Thus, theorem II.3 means theorem 3 in chapter II. Equations are numbered sequentially beginning with one in each section and are always given in parentheses. For example, (11) means equation number 11 in the same section, while I.5 (11) would mean equation (11) in chapter I, section 5. Tables and figures are numbered sequentially by chapter beginning with one. Citations in text follow the style commonly used in scientific literature; full details are given in the References. These references are by no means comprehensive, but they include the material of particular pertinence to the book.

The fact is that symbolism is useful because it makes things difficult. What we wish to know is what can be proved from what. Now, in the beginnings, everything is self-evident; and it is very hard to see whether one self-evident proposition follows from another or not. Obviousness is always the enemy of correctness. Hence we invent some new and difficult symbolism, in which nothing seems obvious. Then we set up certain rules for operating on the symbols, and the whole thing becomes mechanical.

Bertrand Russell, *Mathematics and the Metaphysicians*

I

Applications of Core Theory to Market Exchange

1. Introduction

Let a group of traders own stocks of certain goods that they may exchange among themselves in any mutually agreeable way. Assume that every trader seeks maximum gain by exchange. The result is a redistribution of the initial stock among that group of traders who can agree on the terms of trade. It is as if such traders form a coalition that allocates or imputes the goods to its members. The freedom to trade with anyone is the same as the freedom to join any coalition. Therefore, a trader will join that coalition offering him the best terms. The forming and the dissolving of coalitions is equivalent to contracting and recontracting. The process continues until no one can make himself better-off by trade. The resulting set of coalitions and the allocations they prescribe constitute the core of the market. Thus, the core of the market contains all possible competitive equilibria including the particular version familiar as the intersection of supply and demand schedules, the classical competitive equilibrium. The core is the outcome of a competitive process which does not always result in the classical competitive equilibrium.

The classical competitive equilibrium is described as impersonal, and it is usually identified with the existence of a unique set of prices at which individual traders can buy or sell any amount without affecting these prices. This is assumed to result from the presence of a large number of traders. If there are few traders, then prices depend on individual actions, and no theory of market exchange to explain this phenomenon is widely accepted. Core theory fills this gap because it applies for any number of traders, and a market can have a nonempty core even if there are only two traders. The theory of the core determines a set of outcomes compatible with the largest amount of competition resulting from whatever number of traders is present. The traders do not passively accept prices but instead actively pursue their interest. Since buyers want low and sellers high prices and since all are in

3

competition, the outcome is a market equilibrium that generalizes the classical theory. Under some conditions the equilibrium determined by the core coincides with the intersection of the classical supply and demand schedules, but this is not always true. When it is not true, we need a new theory of competition to describe how markets work.

The classical analysis of a competitive equilibrium imposes conditions on trader preferences and production possibilities sufficient for the existence of prices giving a Pareto optimum. These assumptions rule out nonconvex indifference curves for consumers and nonconvex transformation curves for producers. Core theory provides more powerful tools revealing the nature of the competitive equilibrium while imposing fewer restrictions on the nature of preferences and production possibilities.

One can go a long way in core theory with simple mathematics if one is willing to measure the gains from trade in terms of a numeraire good from which consumers derive utility. This is equivalent to choosing a particular measure of consumer surplus consistent with the Marshallian partial equilibrium analysis. This approach has the considerable advantage of providing a certain function known in game theory as the characteristic function that is closely allied to Marshall's measure of consumer surplus. For these reasons I assume there is a numeraire commodity from which traders derive utility and which is used in exchange. It is also worth noting that this approach has for its basic elements the individual excess demand functions instead of the individual utility functions. The formation of coalitions and the construction of characteristic functions of coalitions amount to the construction of excess demand functions for groups of traders. Consequently, the approach herein is related to some other work in core theory as demand functions are related to utility functions.

The first part of this chapter studies the nature of the utility function and analyzes the gain from trade with a particular measure of consumer surplus. Next, we study a variety of market structures with the core. This leads to an analysis of the role of supply and demand schedules in core theory. Then there follows a treatment of transaction cost and some implications of core theory for market organization. Next, there is a study of multiunit trade which reveals the limitations of the classical analysis of a competitive equilibrium in terms of the intersection of supply and demand schedules. Finally, core theory is applied to some situations in which there are increasing returns to scale and public goods.

2. Consumer Surplus and Transferable Utility

For this study of market exchange it is necessary to postulate that every trader has a utility function which describes the utility he derives from his holdings of goods. For many purposes the precise form of the utility function is irrelevant. Thus, it is well known that many propositions of demand theory are invariant with respect to monotonic transformations of the utility function. Such propositions are independent of the form of the utility function, meaning that they do not require the measurability of utility. Similarly, certain important propositions of core theory are also independent of the measurability of utility. However, there are some propositions of core theory which rely on transferable utility, and it is thought that these propositions require measurable utility. I shall show that transferable utility does not imply measurability. It does imply that the gains from trade can be measured in terms of a numeraire commodity which itself yields utility. Hence, if the numeraire commodity does enter the utility function, many of the propositions dependent on transferable utility are in fact independent of the precise form of the utility function and, therefore, become more acceptable.[1] In this section I analyze the concept of transferable utility and show how it is related to one measure of consumer surplus.

The classical theory of demand starts by assuming a utility function

$$(1) \qquad u = f(q_1, q_2, \ldots, q_n)$$

that describes utility, u, as a function of the consumer holdings of the n goods, $q_j, j = 1, \ldots, n$. The consumer can buy any amount of these goods at fixed prices p_j subject to a budget constraint given by

$$(2) \qquad w = \Sigma p_j q_j$$

where w is interpreted as the consumer's wealth. Alternatively, one can define the q_j's as the *rates* of purchase so that w would be the consumer's *income*. In either case the demand functions which relate the p's and q's are derived by assuming the consumer chooses the q's to maximize u while satisfying the budget constraint. From now on I shall interpret the q's as stocks and w as wealth.

Assuming w is wealth, it has two interpretations. The first assumes that w is the value of the initial stocks of the various goods so that

$$(3) \qquad w = \Sigma p_j q_j^0,$$

where q_j^0 denotes the initial stock of good j. According to this interpretation, w is not itself a commodity; it is a symbol to represent the consumer's real wealth. Hence an increase in w means an increase in the stock of some good, assuming given prices. One may also interpret w as the initial stock of real

---

1. For instance, Shapley and Shubik (1966) prove a number of interesting propositions about trade when the participants have nonconvex preferences assuming a special kind of transferable utility which apparently assumes a constant marginal utility for the "money" used in exchange. However, their results do not actually depend on their special assumptions about transferable utility. It is enough to assume that the traders derive utility from the commodity used for exchange.

money balances. Thus, w represents a separate good that does not happen to enter the utility function (1). The demand functions are the same for both interpretations of w. However, if w does represent the stock of real money balances which do not enter the utility function, then the desired real money balance would be indeterminate. This raises a serious objection against the hypothesis that real balances do not enter the consumer's utility function.

In order to measure the gains from trade uniquely in terms of the numeraire, one must postulate a more general utility function than (1). Let us agree to call the numeraire commodity money so that the nominal price of this commodity is always 1 by hypothesis. In the more general utility function the consumer derives utility from the real stock of money as well as from the stocks of other goods. Thus the more general utility function, written as follows:

(4) $u = \phi(q_1, q_2, ..., q_n, w)$

includes (1) as a special case.

Given the consumer's initial wealth, I define the gains from trade to be the difference between the maximum amount of w that a consumer would be willing to pay for a given bundle of goods and the amount that he actually pays. In the case of two goods, W and Q, this measure of the gains from trade is readily shown in a diagram. Consider figure 1.1. The quantity of W is measured on the vertical axis, and the quantity of Q on the horizontal axis. Let the consumer begin with an amount OW_0 of W and nothing of Q.

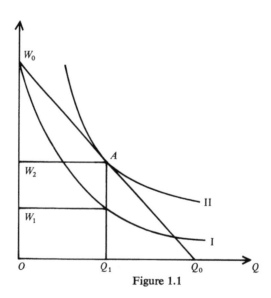

Figure 1.1

Hence, his initial position is represented by the point W_0 on the indifference curve I. Assume that the consumer can exchange W for Q at a constant price given by the slope of $W_0 Q_0$ so that the budget constraint is the straight line $W_0 Q_0$. Hence the consumer prefers the point A. The maximum amount that the consumer is willing to pay for OQ_1 is $W_0 W_1$ while the amount that he actually pays is $W_0 W_2$. Therefore, the consumer surplus measured in terms of the numeraire good is $W_2 W_1$. The maximum amount that he would be willing to pay, $W_0 W_1$, is determined by the condition that the consumer is just as well-off with trade as he was initially. It is plain that this amount $W_0 W_1$ for a given quantity of Q, OQ_1 is uniquely determined by the utility function, which represents the consumer's taste, and his initial position, which represents his initial wealth. It is by no means necessary to suppose, as some writers urge, that the marginal utility of the numeraire be a positive constant in order to have a unique measure of consumer surplus. On the contrary, the assumption of a constant marginal utility for the numeraire leads to an absurdity as we shall now see.

Suppose a special form of the utility function (4) in which the marginal utility of w is a positive constant. Let

(5)
$$u = f(q_1, \ldots, q_n) + cw.$$

For this utility function, a rise in real wealth will not increase the quantity demanded of any commodity except w! To prove this, let us derive the demand functions. The budget constraint is

(6)
$$\Sigma \, p_j q_j + w = w^0$$

where w^0 is the consumer's initial stock of w. Form the Lagrangian function as follows:

(7)
$$cw + f(q_1, \ldots, q_n) + \lambda[(w^0 - w) - \Sigma p_j q_j].$$

If there is a bundle of goods that maximizes the utility given by (5) and satisfies the budget constraint (6), then the partial derivatives of the Lagrangian (7) with respect to the q's and w must be zero. Hence, for a maximum it is necessary that

(8)
$$c - \lambda = 0 \Rightarrow \lambda = c,$$

and

(9)
$$\frac{\partial f}{\partial q_j} - \lambda p_j = 0, j = 1, \ldots, n \Rightarrow \frac{\partial f}{\partial q_j} = c p_j.$$

Equations (8) and (9) determine implicit functions for the quantities in terms of the p's and the initial wealth. We want the partial derivatives of the q's and of w with respect to w^0. From (6),

(10)
$$\Sigma p_j \frac{\partial q_j}{\partial w^0} + \frac{\partial w}{\partial w^0} = 1,$$

and from (9),

(11)
$$\Sigma \frac{\partial^2 f}{\partial q_i \partial q_j} \frac{\partial q_i}{\partial w^0} = 0.$$

Let

$$f_1 = \frac{\partial f}{\partial q_i} \quad \text{and} \quad f_{ij} = \frac{\partial^2 f}{\partial q_i \partial q_j}.$$

Therefore, in matrix form (10) and (11) imply that

(12)
$$\begin{bmatrix} f_{11} & f_{12} & \cdots & f_{1n} & 0 \\ f_{21} & f_{22} & \cdots & f_{2n} & 0 \\ & & \cdots & & \\ f_{n1} & f_{n2} & \cdots & f_{nn} & 0 \\ f_1 & f_2 & \cdots & f_n & c \end{bmatrix} \begin{bmatrix} \dfrac{\partial q_1}{\partial w^0} \\ \\ \cdots \\ \dfrac{\partial q_n}{\partial w^0} \\ \dfrac{\partial w}{\partial w^0} \end{bmatrix} = \begin{bmatrix} 0 \\ 0 \\ \cdots \\ 0 \\ c \end{bmatrix}.$$

Equation (12) uses (8) and (9), which imply that $f_j = c p_j$. It readily follows from (12) that

$$\frac{\partial q_j}{\partial w^0} = 0 \qquad\qquad (j = 1, \ldots, n),$$

(13)

$$\frac{\partial w}{\partial w^0} = 1.$$

Hence, the utility function (5) implies that the demand for q_j is completely inelastic with respect to w^0 and that the demand for w has a slope of 1 with respect to w^0. By a similar argument it follows that the demand for q_j is perfectly inelastic with respect to the initial amount, q_i^0, and that changes in q_i^0 only cause changes in the quantity of w demanded. Thus the slope of the demand for q_j with respect to real wealth is zero, and the slope of the demand for real balances with respect to real wealth is 1. These implications are refuted by every empirical demand study. Hence, (5) is an unacceptable utility function.[2]

Fortunately, a unique measure of the gains from trade in terms of the numeraire does not require a special utility function such as (5). One need

---

2. These are not new results. Samuelson proved them and a number of related propositions long ago (Samuelson 1966, vol. 1, bk. 1, chap. 5).

only assume that the traders derive utility from their real money holdings. Although some will reject this assumption, it is at least consistent with empirical demand studies. Moreover, this measure of the gains from trade is in principle an observable quantity in the same sense that the demand function is observable. Hence this measure is independent of the form of the utility function. One can also argue that consumers derive utility from their real money balances because these provide the service among others of being a temporary abode of purchasing power (see the appendix to this chapter for more on consumer surplus).

Our study of exchange begins with the most elementary case. Assume that a certain commodity Q is to be exchanged for money W. Let there be two kinds of traders, one, the owners, who each possess one unit of the good Q and the other, the nonowners, who each possess different amounts of money and are only willing to buy at most one unit of the good. Hence an owner is a potential seller of Q, and a nonowner is a potential buyer. Denote owner i by A_i and nonowner j by B_j. Assume that A_i refuses any offer below a_i and that B_j will not pay more than b_j. Figures 1.2 and 1.3 illustrate this situation. A_i's initial position is the point Q_1 in figure 1.2, and his reservation price is a_i, the distance from the origin to W_0. B_j's initial position is W_0' in figure 1.3, and his maximum price for one unit of the good is b_j, the distance from W_0' to the origin. If the two parties can agree on terms of trade, then A_i's holdings will be a point on the W-axis above W_0 and B_j's holdings will be a point on the vertical line $Q_1'R_1$. Let x_i denote A_i's final imputation. Since A_i must be at least as well off after trade as before, $x_i \geq a_i$. Denote B_j's final imputation by y_j, which represents B_j's gain from trade, his consumer surplus. Hence, $y_j \geq 0$. Now there is a possible ambiguity if $y_j = 0$. This can either mean that the buyer paid so high a price for the good that he is literally no better-off after trade than before; or it can mean that he remains in his initial position so that he has none of the good and a zero consumer surplus. It is not possible to resolve this ambiguity in general, and in fact both possibilities are important in certain trading situations.

It is also instructive to analyze this situation for the special case in which the numeraire has a constant positive marginal utility to explain the appeal of this assumption despite its unfortunate implications. If the marginal utility of W is constant, then the indifference curve II is parallel to indifference curve I (see figure 1.3). Hence the distance from the origin to W_1', denoted c_j, satisfies

$$(14) \qquad\qquad c_j = b_j + y_j.$$

The term b_j can be interpreted as the utility which B_j derives from one unit of Q, and y_j be interpreted as the utility of the numeraire. Hence, c_j represents the utility of the final position to B_j. Since $c_j \geq b_j$, it follows that $y_j \geq 0$. A similar interpretation applies to A_i. However, if the marginal utility of

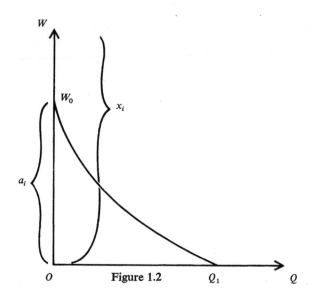

Figure 1.2

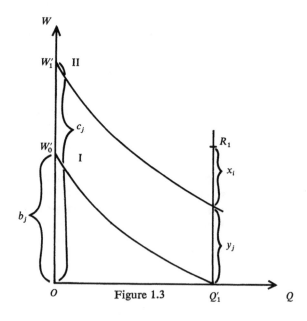

Figure 1.3

the numeraire is not a positive constant, then equation (14) would not hold, and it would not be possible to interpret $b_j + y_j$ as the utility of B_j's final holdings. Nevertheless, we may still interpret y_j as B_j's consumer surplus.

Figures 1.2 and 1.3 illustrate another point. Suppose the owner, whose utility is shown in figure 1.2, contemplates trade with the nonowner, whose utility is shown in figure 1.3. If there is trade because $a_i < b_j$ implies that it will benefit both parties, then the final imputation of the initial nonowner, the buyer, must lie on the vertical line $Q'_1 R_1$, where $OW'_0 = Q'_1 R_1$. Then x_i is the amount received by the initial owner; $b_j - y_j$ is the amount paid by the initial nonowner; and if there is trade, these two must be equal, that is, $x_i = b_j - y_j$. However, there is no presumption that trade must occur even if it benefits both parties. We are free to assume whether or not the two parties behave "rationally" in this regard. If trade occurs whenever it benefits both parties, then there is group rationality in game theory or, equivalently, Pareto optimality in economic theory. It seems obvious to have group rationality when there are only two parties but less so as the number of traders increases. In complicated situations involving many people the assumption of group rationality is often unwarranted. These issues are considered more fully in section 8 below.

The preceding discussion of consumer surplus is adequate for analyzing trade when the participants seek or offer at most one unit of the good. For multiunit trade where the parties seek or offer one or more units there is a larger variety of possible preferences. Their nature is more appropriately discussed in section 7 of this chapter and again in chapter II.

3. Some Simple Trading Situations

Assume there is one owner and two nonowners.[3] The owner refuses any offer below a_1. B_1, nonowner 1, will not bid above b_1, and B_2, nonowner 2, will not bid above b_2. For any trade to occur it is necessary that

(1) $$a_1 \leqslant \max(b_1, b_2).$$

The owner's reservation price is a_1, and the b's are the nonowners' limit prices.

Just as in studying the behavior of individual traders it is convenient to use reservation and limit prices, so too it is convenient to study the behavior of a group of traders with a suitable generalization of these prices. The

---

3. One may think of two buyers for a house. This example is thoroughly analyzed as a three person game by von Neumann and Morgenstern (1947). Based on their work Luce and Raiffa (1957) give a more elementary treatment. Both references assume group rationality, a concept further discussed in sec. 7. Von Neumann and Morgenstern call the imputations of the core, the "common sense" solutions.

appropriate generalization uses the characteristic function of game theory. The characteristic function, denoted by $v(S)$, gives the best collective imputation of a coalition S, a group of traders, under the most adverse conditions. The most adverse conditions arise when the members of the coalition S are confined to trade among themselves. If coalition S forms and if all of the remaining traders combine to form the complementary coalition denoted by $-S$, then $v(S)$ describes the collective imputation of S under these conditions. S and $-S$ are disjoint coalitions which together include all of the traders. Characteristic functions are mappings from sets of traders to real numbers having the following two properties:

  i. $v(\emptyset) = 0$, $\emptyset$ denotes the empty set;
  ii. $v(R \cup S) \geq v(R) + v(S)$, if $R \cap S = \emptyset$, that is, R and S are disjoint.

The first condition means that an empty coalition obtains nothing. The second condition means that an alliance between two nonoverlapping groups must do at least as well as each group separately. The technical name for ii is superadditivity.

We shall apply characteristic functions to a trading situation. First, consider the coalition consisting of the owner alone. Since he can refuse to sell, he will not accept an imputation below a_1, and acting alone, that is without trading, he cannot obtain more than a_1. Therefore, the value of the characteristic function for the coalition consisting of the lone trader A_1 is

(2) $$v(A_1) = a_1.$$

A coalition consisting of a single nonowner B_j cannot obtain any of the good by himself. Observe that the characteristic function describes the maximum valuation of the goods owned by the coalition. Since the B's have none of the goods initially by hypothesis and can only obtain some by forming a coalition with an initial owner, an A, it follows that

(3) $$v(B_1) = v(B_2) = 0.$$

Next consider the coalition consisting of B_1 and A_1. This coalition collectively owns one unit of the good, which it cannot be forced to sell. Moreover, it would not accept a price below the highest reservation price of either of its members, where a_1 is A_1's reservation price, and b_1 is B_1's limit or reservation price. (From now on I shall use the terms limit and reservation price interchangeably; no ambiguity will result.) To prove this, consider the argument as follows. Let the coalition receive an offer of w for its one unit so that $a_1 < w < b_1$. But then B_1 would be willing to pay A_1 an amount up to b_1 in order to keep the good for the coalition. A_1 would be better-off accepting B_1's side payment, and B_1, having the good, would be no worse-off. A similar argument applies if $b_1 < w < a_1$, except that A_1 would pay B_1 to keep the good for the coalition. This proves that $\max(a_1, b_1)$

gives the value of the characteristic function for the coalition (A_1, B_1) since

$$v(A_1, B_1) = \max{(a_1, b_1)} \geqslant v(A_1) + v(B_1),$$

or, more generally,

(4) $$v(A_1, B_j) = \max{(a_1, b_j)} \qquad (j = 1, 2,),$$

and

$$\max{(a_1, b_j)} \geqslant v(A_1) + v(B_j).$$

Similarly,

(5) $$v(B_1, B_2) = \max{(0, 0)} = 0 \geqslant v(B_1) + v(B_2).$$

Finally, there is the coalition of all three traders. It is artificial to describe what this coalition can obtain under the most adverse conditions because it includes all traders, so we need a different argument. Clearly,

(6) $$v(A_1, B_1, B_2) \leqslant \max{(a_1, b_1, b_2)}.$$

Next we use property ii of characteristic functions which implies that

$$v(A_1, B_1, B_2) \geqslant v(A_1, B_1) + v(B_2) = \max{(a_1, b_1)} + 0,$$

and

$$v(A_1, B_1, B_2) \geqslant v(A_1, B_2) + v(B_1) = \max{(a_1, b_2)} + 0.$$

Therefore,

(7) $$v(A_1, B_1, B_2) \geqslant \max{[\max{(a_1, b_1)} + \max{(a_1, b_2)}]}.$$

But

$$\max{[\max{(a_1, b_1)} + \max{(a_1, b_2)}]} = \max{(a_1, b_1, b_2)}.$$

Together with (6) this implies that

(8) $$v(A_1, B_1, B_2) = \max{(a_1, b_1, b_2)}.$$

Let x_1 denote the final imputation of A_1 and y_j the final imputation of B_j. The assumption that every trader attempts to make himself as well-off as possible implies that the imputations in the core satisfy the inequalities as follows:

(9)
$$\begin{cases} x_1 \geqslant v(A_1) = a_1, \\ y_1, y_2 \geqslant 0 = v(B_1) = v(B_2), \\ x_1 + y_1 \geqslant v(A_1, B_1), \\ x_1 + y_2 \geqslant v(A_1, B_2), \\ y_1 + y_2 \geqslant v(B_1, B_2). \end{cases}$$

In addition the imputations cannot exceed the maximum valuation of the group so that

$$(10) \qquad\qquad x_1 + y_1 + y_2 \leqslant v(A_1, B_1, B_2).$$

If there is equality in (10) so that the sum of the imputations equals the maximum valuation, then we say there is *group rationality* or, equivalently, *Pareto optimality*. The inequality (10) is the *feasibility constraint*, and it must be satisfied by any set of imputations. In this example it is easy to verify that the core constraints, (9), together with feasibility, (10), imply group rationality. This is a consequence of the nature of the characteristic function expressed in (4) and (8). Thus,

$$v(A_1, B_1, B_2) = \max \{v(A_1, B_1), v(A_1, B_2)\}.$$

Without loss of generality we can assume that

$$v(A_1, B_2) \geqslant v(A_1, B_1).$$

Therefore,

$$(11) \qquad\quad x_1 + y_1 + y_2 \geqslant v(A_1, B_2) + v(B_1) = v(A_1, B_1, B_2) + 0.$$

Equations (10) and (11) imply that

$$(12) \qquad\qquad x_1 + y_1 + y_2 = v(A_1, B_1, B_2),$$

so that there must be group rationality.

To illustrate a particular solution for this example assume that

$$(13) \qquad\qquad\qquad a_1 < b_1 < b_2.$$

Hence, $v(A_1, B_1) = b_1$, $v(A_1, B_2) = b_2$, and $v(A_1, B_1, B_2) = b_2$. Since $x_1 + y_2 \geqslant b_2$, (12) and (13) imply that $y_1 \leqslant 0$. But according to (9), $y_1 \geqslant 0$. Therefore,

$$(14) \qquad\qquad\qquad\qquad y_1 = 0.$$

The core inequalities reduce to the system as follows:

$$(15) \qquad \left. \begin{array}{l} x_1 > b_1 > a_1, \\[4pt] x_1 + y_2 = b_2, \end{array} \right\} \Rightarrow \left\{ \begin{array}{l} b_1 < x_1 < b_2, \\[4pt] y_2 = b_2 - x_1. \end{array} \right.$$

These results have a familiar interpretation in terms of supply and demand schedules. Consider figure 1.4. The demand schedule is $B_2 b_2 B_1 b_1 D$. The most eager buyer B_2 is willing to pay up to OB_2. At any price below OP_1, there is a demand for two units. The supply schedule is the step function $OAa_1 S$. Nothing is offered for sale at any price below OA, and one unit is offered for sale at any price above OA. The supply and demand schedules intersect in the interval $B_1 b_2$. This interval coincides with the range of core imputations in (15). Notice that B_1 fails to obtain the good and that his presence serves only to narrow the price range.

It is instructive to consider collusion between the two buyers. Let B_2, the more eager buyer, agree to refrain from bidding so that the less eager buyer B_1 can obtain the good for a lower price. For successful collusion B_2 must pay B_1 a price below b_1 since B_2 can obtain the good without collusion at a price above b_1. Hence, B_1 must pay A_1 less than b_1. Let B_1 pay A_1 a price p such that $a_1 < p < b_1$ and let B_2 pay B_1 a price p^* such that $p < p^* < b_1$. This scheme cannot succeed because the current owner A_1 can sell the good to B_2 at a price below p^* but above b_1, and both would be made better-off. Hence collusion between B_1 and B_2 is defeated, and the imputations are forced into the core.

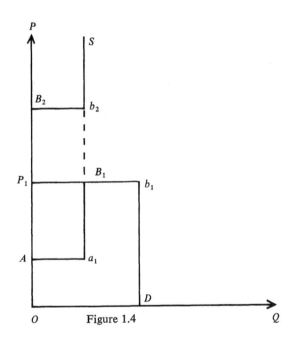

Figure 1.4

The next example is slightly more complicated because it includes two sellers, A_1, A_2, and two nonowners, B_1, B_2, so that at most two units can be exchanged. The values of the characteristic function for all possible coalitions are as follows:

$$v(A_1) = a_1, \; v(A_2) = a_2;$$
$$v(B_1) = v(B_2) = 0;$$
$$v(A_1, A_2) = a_1 + a_2;$$

$$v(A_i, B_j) = \max(a_i, b_j) \qquad\qquad (i, j = 1, 2);$$

$$v(B_1, B_2) = \max(0, 0) = 0;$$

$$v(A_1, A_2, B_1) = \max(a_1 + a_2, a_1 + b_1, a_2 + b_1);$$

$$v(A_1, A_2, B_2) = \max(a_1 + a_2, a_1 + b_2, a_2 + b_2);$$

$$v(A_1, B_1, B_2) = \max(a_1, b_1, b_2);$$

$$v(A_2, B_1, B_2) = \max(a_2, b_1, b_2).$$

It is tedious to write out the value of the characteristic function for the coalition of all traders; and, moreover, it is illuminating to adopt a more general formulation. Thus, let

$$A = \{A_i | i = 1, 2\} \quad \text{and} \quad a = \{a_i | i = 1, 2\};$$

$$B = \{B_j | j = 1, 2\} \quad \text{and} \quad b = \{b_j | j = 1, 2\}.$$

Let I denote the coalition of all traders so that $I = A \cup B$. Let $z = z_1 + z_2$, where z_h, $h = 1, 2$, denotes the imputation of a final owner; $v(I) = \max z$ subject to z_1, z_2 in $a \cup b$.

In this example there are $2^{2+2} - 1 = 15$ possible coalitions, excluding the empty coalition, so that 15 inequalities determine the imputations in the core. Let x_i denote the imputation of A_i and y_j the imputation of B_j. The imputations are in the core if they satisfy $\Sigma_{i, j \text{ in } S}(x_i + y_j) \geqslant v(S)$ for all coalitions $S \subset I$ and if $\Sigma_{i, j \text{ in } I}(x_i + y_j) \leqslant v(I)$.

To illustrate the nature of the core in this case, assume that $a_1 < a_2 < b_1 < b_2$. Thus, both nonowners value the good more than the initial owners. Our first objective is to prove group rationality. The basic core constraints are as follows:

(16)
$$\begin{cases} x_1 \geqslant a_1, & x_1 + y_1 \geqslant b_1, \\ x_2 \geqslant a_2, & x_1 + y_2 \geqslant b_2, \\ y_1 \geqslant 0, & x_2 + y_1 \geqslant b_1, \\ y_2 \geqslant 0, & x_2 + y_2 \geqslant b_2. \end{cases}$$

Feasibility gives

(17)
$$x_1 + x_2 + y_1 + y_2 \leqslant b_1 + b_2.$$

Since $x_1 + y_1 \geqslant b_1$ and $x_2 + y_2 \geqslant b_2$, summing we obtain

(18)
$$x_1 + x_2 + y_1 + y_2 \geqslant b_1 + b_2.$$

Therefore, (17) and (18) together imply that

(19)
$$x_1 + x_2 + y_1 + y_2 = b_1 + b_2.$$

Hence there is group rationality in this case. Now from (16) and (19) we obtain

(20)
$$\begin{cases} x_2+y_2 = b_2, \\ x_2+y_1 = b_1, \\ x_1+y_2 = b_2, \\ x_1+y_1 = b_1, \end{cases} \Rightarrow x_1-x_2 = 0 \Rightarrow x_1 = x_2 = x > a_2.$$

It is readily verified that the imputations which satisfy (20) also satisfy *all* of the core constraints. Conversely, any set of imputations in the core satisfies (16) and (17). Hence the imputations in this case are in the core if and only if they satisfy (16) and (17).

We can illustrate the core in terms of the intersection of supply and demand schedules as shown in figure 1.5. The demand schedule is the step function $B_2b_2B_1b_1D$, and the supply schedule is the step function $A_1a_1A_2a_2S$. At the equilibrium two units are exchanged for a price in the interval a_2b_1. It is also instructive to change the data of the problem and compare the core imputations with the equilibrium determined by the intersections of the appropriate supply and demand schedules.

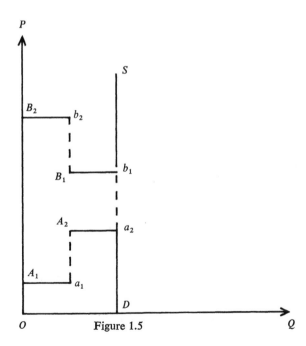

Figure 1.5

These examples suggest a general result that relates core theory to the intersection of supply and demand schedules. Assume there are m initial owners and n initial nonowners such that the former have exactly one unit each, and the latter desire at most one unit each. Thus initial owners are potential sellers and initial nonowners are potential buyers. All have (possibly) different reservation prices. These reservation prices do not necessarily form a continuum nor need there be an infinite number of traders to determine a unique equilibrium price and quantity. The supply and demand functions are step functions illustrated in figure 1.6. In fact this is the same kind of

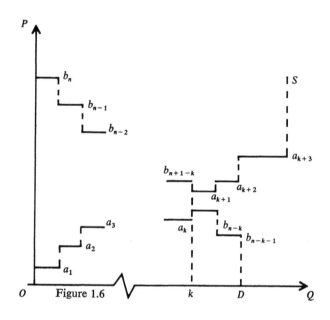

Figure 1.6

market described by Böhm-Bawerk in his celebrated example of horse trading.[4] The price is determined by Böhm-Bawerk's marginal pairs. Let k units be traded. The price must lie above b_{n-k} to exclude the most eager nonowner B_{n-k}, and it must lie below a_{k+1} to exclude the most eager of the extra marginal owners A_{k+1}. If either of the two owners, A_k or A_{k+1}, had the same reservation price or if either of the two nonowners, B_{n+1-k} or B_{n-k}, had the same reservation price, then the equilibrium price would be uniquely determined because at least one of the two schedules, the supply schedule in the former case or the demand schedule in the latter case, would be continuous at the point of intersection. But then the identity of one of the final owners, and one of the final nonowners would be indeterminate.

4. See Böhm-Bawerk (1930, bk. 4, chaps. 1–6).

The geometry can also illustrate other possibilities. Consider figure 1.7. Although there is an infinite number of traders in the case shown in figure 1.7, there is no unique equilibrium price because both the supply and the demand schedules are discontinuous at the equilibrium quantity. These examples illustrate the proposition that an infinite number of traders is neither necessary nor sufficient for a unique equilibrium price.

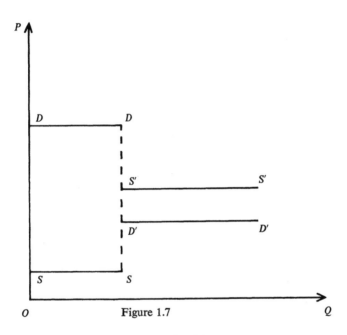

Figure 1.7

4. *m* Owners and *n* Nonowners

We now analyze the case in which there are m owners, A_i, $i = 1, ..., m$, and n nonowners, B_j, $j = 1, ..., n$, such that each owner has at most one unit to sell and each nonowner desires to buy at most one unit. This special case is important for the subsequent analysis of multiple trade.

Now there are $2^{n+m} - 1$ coalitions, excluding the empty coalition. The core consists of the set of imputations $\{x_i, y_j\}$, where x_i is the imputation of A_i and y_j is the imputation of B_j such that

(1)
$$\sum_{i,j \text{ in } S} (x_i + y_j) \geqslant v(S), \quad S \subset I,$$

$$\sum_{i,j \text{ in } I} (x_i + y_j) \leqslant v(I) \qquad \text{(feasibility)}.$$

In the first set of inequalities, S ranges over all possible coalitions except I.

The main result of this section is that the imputations of the core are equivalent to the equilibrium determined by the intersection of the supply and demand schedules. Hence in effect the market "solves" the $2^{n+m}-1$ inequalities that determine the core.

It is convenient to begin with a general proposition of game theory applicable to market exchange. Consider an n-person game and denote the imputation of player i by x_i. Such a game is said to be constant sum with sum c if

$$\text{(2)} \qquad\qquad v(S)+v(-S) = c$$

for *all* possible coalitions S, where $S \cup (-S) = I$. Therefore, in a constant sum game, one coalition gains at the expense of the other. We also need the concept of an *inessential* game. An n-person game is said to be inessential if for all coalitions S with $S = \cup \, S_i, S_i \cap S_j = \varnothing, i \neq j$,

$$\text{(3)} \qquad\qquad v(S) = \sum_i v(S_i).$$

Denote $v(\{i\})$ by v_i. Hence in an inessential game $v(I) = \Sigma_I v_i$. This means that it never pays for coalitions to form in an inessential game.

It is not hard to show that the only constant sum games that can have imputations in the core must be inessential.

LEMMA 1. *If a constant sum game has a nonempty core then it is inessential.*
Proof. For all coalitions S in a constant sum game $v(S)+v(-S) = c = v(I)$. By hypothesis the imputations are in the core and are feasible so that

$$\sum_S x_i \geqslant v(S) \quad \text{for all} \quad S \subset I,$$

$$\sum_I x_i \leqslant v(I) \qquad\qquad\qquad\qquad \text{(feasibility)}.$$

Therefore,

$$\sum_S x_i + \sum_{-S} x_i \geqslant v(S)+v(-S) = v(I).$$

Together with feasibility this implies group rationality so that

$$\sum_I x_i = v(I).$$

The next step is to prove that the game must be inessential. Since there is group rationality, together with

$$\sum_S x_i \geqslant v(S) \quad \text{and} \quad \sum_{-S} x_i \geqslant v(-S),$$

it follows that

$$\sum_S x_i = v(S) \quad \text{and} \quad \sum_{-S} x_i = v(-S)$$

for all coalitions S. Now choose S to be the coalition consisting of a single player and it readily follows that $x_i = v_i$. But then the game is inessential. QED.[5]

This result has important implications for market exchange. If the participants in a market can agree on trades, then this means that it pays for certain coalitions to form so that the game representing the market is essential. If all possible coalitions can form so that the market is competitive, then the imputations must be in the core. Hence the game cannot be constant sum. Therefore, in a competitive market in which there is trade the corresponding game is not constant sum. However, if no one would gain from trade, then coalitions would not form and the game would be inessential. Consequently, the participants are content with their initial endowments. In this case the corresponding game is constant sum. Therefore, trade in a competitive market converts an initial allocation of goods into a final allocation such that the market game representing the final allocation is a constant sum game.

We now apply these results to study exchange in a competitive market. Assume the owners' reservation prices are ordered as follows:

$$a_1 \leqslant a_2 \leqslant \ldots \leqslant a_m$$

and that the nonowners' limit prices are ordered as follows:

$$b_1 \leqslant b_2 \leqslant \ldots \leqslant b_n.$$

Since there is no trade in a competitive inessential market, in such a case we would find

$$\max \{b_j\} < \min \{a_i\}.$$

Exchange can only occur in an essential market. Necessarily, in an essential market

(4) $$\max \{b_j\} \geqslant \min \{a_i\}.$$

Conversely, if (4) is satisfied, there is a $k \geqslant 1$ such that

(5) $$a_k \leqslant b_{n-(k-1)} < a_{k+1}.$$

It follows that

$$v(A_k, B_{n-(k-1)}) = \max (a_k, b_{n-(k-1)}) = b_{n-(k-1)}.$$

Therefore, owner A_k and nonowner $B_{n-(k-1)}$ mutually benefit from exchange, which means that a coalition of these two would form. Hence the market is essential. This proves the following lemma.

LEMMA 2. *In an inessential market only the initial allocation of goods yields*

---

5. This result is well known. It is almost explicit in Luce and Raiffa (1957, p. 194) and is implied by von Neumann and Morgenstern (1947, secs. 61.5 and 64.2).

imputations satisfying the core constraints. A necessary and a sufficient condition for an essential market — that is, for trade to occur — is that inequality (4) *is satisfied.*

The next result establishes the identity of the final owners in the core and shows that the final allocation of the goods is such as to maximize the aggregate valuation, and it shows that all trade occurs at a common price. We shall see that as a result there is group rationality.

THEOREM 1. *Let the market be essential. Then there is a $k \geq 1$ such that the final owners in the core are $B_{n-(i-1)}$, $i = 1, ..., k$ and A_i with $i = k+1$, ..., m, $\Sigma_{i, j \, in \, I} (x_i + y_j) = v(I)$ (group rationality), and*

$$x_i = \begin{cases} x, & \text{if } i = 1, ..., k, \\ a_i, & \text{if } i > k. \end{cases}$$

The imputations of the initial nonowners, B_j, satisfy the equations as follows:

$$y_{n-(i-1)} + x = b_{n-(i-1)}, \qquad \text{for } i = 1, ..., k,$$

$$y_{n-(i-1)} = 0, \qquad \text{for } i = k+1, ..., n.$$

Proof. By lemma 2, the hypothesis implies that (4) is satisfied, which implies there is a $k \geq 1$ such that

(6) $$b_{n-(i-1)} - a_i \begin{cases} \geq \\ < \end{cases} 0 \qquad \begin{cases} \text{for } i = 1, ..., k, \\ \text{for } i = k+1, ..., \min(m, n). \end{cases}$$

To establish the identity of the final owners, it is sufficient to calculate $v(I)$ and to prove group rationality. Let

$$A = \{A_i / i = 1, ..., m\} \quad \text{and} \quad a = \{a_i / i = 1, ..., m\};$$

$$B = \{B_j / j = 1, ..., n\} \quad \text{and} \quad b = \{b_j / j = 1, ..., n\};$$

$$I = A \cup B.$$

Let $z = \Sigma_1^m z_i$, where z_i denotes the imputation of the ith final owner. Then

(7) $$v(I) = \max z \quad \text{subject to} \quad z_i \text{ in } a \cup b.$$

It follows from (6) that

$$z_i = \begin{cases} b_{n-(i-1)}, & \text{if } i = 1, ..., k, \\ a_i, & \text{if } i = k+1, ..., m. \end{cases}$$

Hence

(8) $$v(I) = \sum_1^k b_{n-(i-1)} + \sum_{k+1}^m a_i.$$

Since the imputations must be feasible, we must also have

$$\text{(9)} \qquad\qquad \sum_{i,\,j \text{ in } I} (x_i + y_j) \leqslant v(I).$$

Assume that $m \leqslant n$ and consider the coalition of pairs $(A_i, B_{n-(i-1)})$ for $i = 1, \ldots, m$, and the coalitions of the single nonowners $B_{n-(i-1)}$ for $i = m+1, \ldots, n$. Since the imputations are in the core by hypothesis, for each of the coalitions $(A_i, B_{n-(i-1)})$ with $i = 1, \ldots, m$,

$$\text{(10)} \qquad x_i + y_{n-(i-1)} \geqslant v(A_i, B_{n-(i-1)}) = \begin{cases} b_{n-(i-1)}, & \text{if } i = 1, \ldots, k, \\[2mm] a_i, & \text{if } i = k+1, \ldots, m. \end{cases}$$

For the coalitions involving the single nonowners, $B_{n-(i-1)}$, $i = m+1, \ldots, n$, we have

$$\text{(11)} \qquad\qquad y_{n-(i-1)} \geqslant v(B_{n-(i-1)}) = 0.$$

Summing the inequalities in (10) and (11), we obtain

$$\text{(12)} \qquad \sum_1^m x_i + \sum_1^n y_{n-(i-1)} \geqslant \sum_1^k b_{n-(i-1)} + \sum_{k+1}^m a_i.$$

By (8) the right-hand side of (12) equals $v(I)$. Hence (9) and (12) imply

$$\text{(13)} \qquad\qquad \sum_{i,\,j \text{ in } I} (x_i + y_j) = v(I),$$

which means there is group rationality.

If we had $m > n$, then instead of (11) we would use the inequalities implied by the single coalitions of the initial owners A_i, $i = n+1, \ldots, m$, as follows: $x_i \geqslant v(A_i)$. A similar argument would then imply (13). This proves group rationality.

The next step in the proof is to show that there must be equality in (10) for $i = 1, \ldots, m$. This would establish the identity of the final owners as determined by the imputations being in the core. First, assume $m \leqslant n$. The idea of the proof is to sum (10) and (11) over all but one i. Thus summing (10) and (11) over $i = 2, \ldots, m$, gives

$$\sum_2^m x_i + \sum_2^m y_{n-(i-1)} \geqslant \sum_2^k b_{n-(i-1)} + \sum_{k+1}^m a_i,$$

and then adding the inequalities for (11) gives

$$\text{(14)} \qquad \sum_2^m x_i + \sum_2^n y_{n-(i-1)} \geqslant \sum_2^k b_{n-(i-1)} + \sum_{k+1}^m a_i.$$

Together (13) and (14) imply $x_1 + y_n \leqslant b_n$. But (10) implies that $x_1 + y_n \geqslant b_n$, so for $i = 1$ there is equality in (10). The same procedure gives equality in

(10) for $i = 1, \ldots, m$. Taken in conjunction with (13), it follows that

(15) $$y_{n-(i-1)} = 0, \qquad \text{for } i = m, \ldots, n-1.$$

Finally, since

(16) $$x_i + y_{n-(i-1)} = a_i, \qquad \text{for } i = k+1, \ldots, m,$$

and $x_i \geqslant v(A_i) = a_i$, we conclude that

(17) $$y_{n-(i-1)} = 0, \qquad \text{for } i = k+1, \ldots, m-1.$$

If we adopt the convention that a zero consumer surplus $y_j = 0$ means that the nonowner fails to obtain any of the good, then (17) shows that the nonowners, $B_{n-(i-1)}$, $i = k+1, \ldots, n-1$, obtain none of the good, and (16) shows that initial owners, A_i, $i = k+1, \ldots, m$, retain possession of the good.

To complete the proof of the first assertion, we need no longer distinguish between the cases in which max $(m, n) = m$ and max $(m, n) = n$. Consider the initial owners, A_i, $i = 1, \ldots, k$, and the initial nonowners, $B_{n-(i-1)}$, $i = 1, \ldots, k$. Any of these nonowners values the good more than any one of the initial owners. We must prove that exchange occurs so that these nonowners each obtain one unit from the initial owners if the imputations satisfy the core constraints. The proof is by contradiction. Suppose one of the nonowners, $B_{n-(i-1)}$, $i = 1, \ldots, k$, fails to obtain one unit. Then at least one of the initial owners, A_i, $i = 1, \ldots, k$, must retain possession so that his final imputation would be $x_i = a_i$. But equality in (10) gives

$$y_{n-(i-1)} = b_{n-(i-1)} - a_i > 0,$$

so that initial nonowner $B_{n-(i-1)}$ must have obtained one unit of the good. Hence there must be some other nonowner, say $B_{n-(i'-1)}$, who fails to obtain any of the good. Therefore, his final imputation would be $y_{n-(i'-1)} = 0$. Consider a coalition between the initial owner A_i who retains possession and the initial nonowner $B_{n-(i'-1)}$ who fails to obtain one unit. For their imputations to be in the core, it must be true that

$$x_i + y_{n-(i'-1)} \geqslant v(A_i, B_{n-(i'-1)}) = b_{n-(i'-1)}.$$

Substituting the values for their imputations as given above, it must be true that $a_i + 0 \geqslant b_{n-(i'-1)}$, which contradicts (6) since both i and i' are not more than k. This contradiction proves that all $B_{n-(i-1)}$ such that $i = 1, \ldots, k$ must obtain possession of the good in order to satisfy the imputations of the core. This completes the proof of the first assertion of the theorem.

The third assertion of the theorem is that $x_i = x$ for all $i = 1, \ldots, k$. Since it has been shown that there is equality in (10), that is,

(18) $$x_i + y_{n-(i-1)} = \begin{cases} b_{n-(i-1)}, & \text{for } i = 1, \ldots, k, \\ a_i, & \text{for } i = k+1, \ldots, m, \end{cases}$$

it follows that if $m \geqslant 1$ and $k \geqslant 1$, then

(19) $$x_i + y_{n-(i-1)} = b_{n-(i-1)}$$

and

(20) $$x_{i+1} + y_{n-i} = b_{n-i}$$

for all $i \leqslant k - 1$. In addition, if these imputations satisfy the core constraints, then

(21) $$x_i + y_{n-i} \geqslant v(A_i, B_{n-i}) = b_{n-i}$$

and

(22) $$x_{i+1} + y_{n-(i-1)} \geqslant v(A_{i+1}, B_{n-(i-1)}) = b_{n-(i-1)}.$$

Substitute the value of $y_{n-(i-1)}$ given by (19) into (22) and obtain

(23) $$x_{i+1} - x_i \geqslant 0.$$

Next substitute the value of y_{n-i} given by (20) into (21) and obtain

(24) $$x_i - x_{i+1} \geqslant 0.$$

Therefore,

(25) $$x_i = x_{i+1}, \qquad \text{for all} \quad i = 1, \ldots, k-1.$$

Now if m or $k = 1$ there is nothing to prove. Hence this completes the proof of the third assertion of the theorem.

To prove the fourth assertion of the theorem substitute the common value of x_i for $i = 1, \ldots, k$ in the first equation of (18) and obtain

(26) $$y_{n-(i-1)} + x = b_{n-(i-1)}, \qquad \text{for} \quad i = 1, \ldots, k.$$

Equation (17) completes the proof of the fourth assertion of the theorem. This concludes the proof of theorem 1. QED.

After so long a proof, an economic interpretation of the results is desirable. Since x_i is the receipt per unit of those initial owners A_i who sell one unit each, it is the price of the commodity. The second assertion of the theorem therefore means that all of the sellers obtain a common price. However, it is, of course, false that all buyers obtain the same imputation y_j. This is clear from (26), which says that every buyer can obtain a different consumer surplus. It is true that market exchange results in a redistribution of the ownership of the given stock of m units of the commodity in such a way as to maximize the aggregate valuation of the good. Thus, a_i denotes an initial owner's limit price and b_j an initial nonowner's limit price. The result of market exchange is to place the m units of the good into the hands of those m traders for whom the limit price is highest. This is what (7) means. The theorem does not quite establish that the imputations in the core are the same as those given by the intersection of the appropriate supply and demand schedules. To prove this, it is first necessary to determine the range of prices implied by the imputations of the core. This is accomplished in the following corollary.

COROLLARY.

(27) $$a_k \leqslant x < b_{n-(k-1)}$$

and

(28) $$v(A_k, B_{n-k}) \leqslant x \leqslant a_{k+1} + b_{n-(k-1)} - v(A_{k+1}, B_{n-(k-1)}),$$

where x denotes the common imputation of the sellers, A_i, $i = 1, ..., k$.
Proof. To prove (27), rewrite (26) so that

$$y_{n-(i-1)} = b_{n-(i-1)} - x_i > 0, \qquad \text{for all } i = 1, ..., k.$$

In addition, $x_i \geqslant a_i$, for $i = 1, ..., k$. Therefore,

$$\min_{i=1,...,k} b_{n-(i-1)} \geqslant x \geqslant \max_{i=1,...,k} a_i, \qquad \text{for } i = 1, ..., k.$$

Since

$$\min_{i=1,...,k} b_{n-(i-1)} = b_{n-(k-1)} \quad \text{and} \quad \max_{i=1,...,k} a_i = a_k,$$

this implies (27).

To prove (28), consider the two core constraints as follows:

(29) $$x_k + y_{n-k} \geqslant v(A_k, B_{n-k})$$

and

(30) $$x_{k+1} + y_{n-(k-1)} \geqslant v(A_{k+1}, B_{n-(k-1)}).$$

By theorem 1, $y_{n-k} = 0$ so that (29) implies that

(31) $$x_k \geqslant v(A_k, B_{n-k}).$$

Equation (26) for $i = k$ gives

(32) $$y_{n-(k-1)} + x_k = b_{n-(k-1)}.$$

Substitute the value of $y_{n-(k-1)}$ given in (32) into (30) and obtain

(33) $$x_{k+1} + b_{n-(k-1)} - x_k \geqslant v(A_{k+1}, B_{n-(k-1)}).$$

By theorem 1, $x_{k+1} = a_{k+1}$. Therefore, (32) implies

$$a_{k+1} + b_{n-(k-1)} - v(A_{k+1}, B_{n-(k-1)}) > x_k = x,$$

which together with (31) gives (28). QED.

The inequalities of (27) and (28) coincide with those determined by Böhm-Bawerk's "marginal pairs" in his celebrated analysis of a market.[6] Figure

---

6. Böhm-Bawerk (1930).

1.6 illustrates the narrowest possible limits on the price that can be imposed by (27) and (28). The widest possible limits, $a_k < x < b_{n-(k-1)}$ occur if $v(A_{k+1}, B_{n-(k-1)}) = a_{k+1}$.

The quantity transacted and the price range determined by theorem 1 and its corollary are readily calculated from the intersection of the pertinent supply and demand schedules. The demand schedule gives the price bids as a nonincreasing step function of the quantity demanded. Thus the demand for one unit is that of the nonowner B_n, who is willing to pay the most and will bid up to b_n. The demand for two units is determined by B_n together with the next most eager nonowner B_{n-1}. Thus, B_{n-1} is willing to bid up to $b_{n-1} < b_n$. Hence at any price between b_{n-1} and b_{n-2}, there would be a demand for two units. In general the demand for j units corresponds to a price below $b_{n-(j-1)}$ and above b_{n-j}. The nondecreasing property of the demand function is a consequence of the fact that any nonowner willing to obtain one unit at a price below his limit price remains an effective bidder at all such prices. If there are n nonowners, then the demand schedule intersects the quantity axis at $q = n$. The demand schedule is continuous for any range of quantities such that the buyers have a common limit price.

The supply schedule gives the price offered as a nondecreasing step function of the quantity offered. This is a consequence of the fact that any owner A_i who offers to sell one unit at a price above a_i remains an effective supplier at all prices above a_i. Hence the supply schedule cumulates the suppliers as the price rises.

According to these constructions, there cannot be backward bending supply schedules nor positively sloped demand schedules. The equilibrium determined by the intersection of the two schedules is necessarily stable. Upon reflection this is not surprising. Imputations in the core have the property that no one can improve his situation by recontracting, which amounts to asserting that equilibria in the core are stable.

It must be kept in mind that this interpretation refers to single-unit trade. In multiunit trade, initial owners have more than one unit to offer and initial nonowners are each willing to buy more than one unit. As we shall see in section 7, the imputations in the core with multiunit trade do not necessarily coincide with those implied by the intersection of supply and demand schedules of the classical treatment.

The construction of the supply and the demand schedules with the equilibrium determined by their intersection should be regarded as a mathematical algorithm that conveniently derives some of the properties of the core. We must subsequently show that all of the core constraints are satisfied by the imputations determined by the intersection. Both theorem 1 and its corollary use only those core constraints involving coalitions between pairs (A_i, B_j), feasibility (9), and individual rationality

$$x_i \geqslant v(A_i) = a_i, \, y_j \geqslant v(B_j) = 0.$$

It remains to be shown that the imputations satisfying these conditions also satisfy the remaining inequalities prescribed by the core. This is the subject of the next section.

5. The Basic Core Constraints

A primary coalition is defined to be either a coalition of a single trader or a coalition of a pair (A_i, B_j). Imputations in the core satisfy the primary constraints as follows:

$$
\begin{aligned}
&(1) & x_i &\geqslant v(A_i) = a_i, \\
&(2) & y_j &\geqslant v(B_j) = 0, \\
&(3) & x_i + y_j &\geqslant v(A_i, B_j),
\end{aligned}
\qquad
\begin{aligned}
&(i = 1, \ldots, m), \\
&(j = 1, \ldots, n).
\end{aligned}
$$

In addition, any imputations are feasible if they satisfy

$$
(4) \qquad \sum_{i,\,j\,\mathrm{in}\,I} (x_i + y_j) \leqslant v(I).
$$

The first result is as follows:

THEOREM 2. *Let the market be essential. The imputations* $\{x_i\}$ *and* $\{y_j\}$ *are in the core if and only if they satisfy the primary constraints and are feasible.*

Proof. Sufficiency follows by definition of the core. To prove necessity one must show that if the imputations are feasible and satisfy the primary constraints, then they satisfy

$$
(5) \qquad \sum_{i,\,j\,\mathrm{in}\,S} (x_i + y_j) \geqslant v(S)
$$

for all coalitions $S \subset I$. Suppose S contains m_i owners and n_j nonowners, where $m_i = 1, 2, \ldots, m$ and $n_j = 1, 2, \ldots, n$. Then S can be represented as follows: $S = A_{(i)} \cup B_{(j)}$ such that $A_{(i)}$ contains the m_i owners in S, and $B_{(j)}$ contains the n_j nonowners in S. Let $a_{(i)}$ be the set of a_i's corresponding to the A_i's who are in S and let $b_{(j)}$ be the set of b_j's corresponding to the B_j's who are in $B_{(j)}$. Let $z = \sum_1^{m_i} z_h$. Then

$$
v(S) = \max z \quad \text{subject to} \quad z_h \ \text{in} \ a_{(i)} \cup b_{(j)}.
$$

Order the members of $a_{(i)}$ as follows:

$$
a_{i,1} \leqslant a_{i,2} \leqslant \ldots \leqslant a_{i,m_i}.
$$

Similarly, order the members of $b_{(j)}$ as follows:

$$
b_{j,1} \leqslant b_{j,2} \leqslant \ldots \leqslant b_{j,n_j}.
$$

Next form coalitions between pairs of owners and nonowners in S as follows:

$$
(A_{i,h}, B_{j,n_j-h}), \quad h = 1, \ldots, \min(m_i, n_j).
$$

Sum over the corresponding primary constraints to obtain

(6) $$\sum_{h=1}^{\min(m_i, n_j)} [x_{i,h} + y_{j,n_j-h}] \geq \sum_{h=1}^{\min(m_i, n_j)} v(A_{i,h}, B_{j,n_j-h}).$$

If $m_i \leq n_j$, then sum over the primary constraints (2) as follows:

(7) $$\sum_{h=m_i+1}^{n_j} y_{j,n_j-h} \geq \sum_{h=m_i+1}^{n_j} v(B_{j,n_j-h}).$$

Adding (6) and (7) gives

(8) $$\sum_{i,j \text{ in } S} (x_i + y_j) \geq \sum_{h=1}^{\min(m_i, n_j)} v(A_{i,h}, B_{j,n_j-h}) + \sum_{m_i+1}^{n_j} v(B_{j,n_j-h}).$$

It is readily verified that the right-hand side of (8) is precisely $v(S)$. If $m_i > n_j$, then sum over the primary constraints in (1) as follows:

(9) $$\sum_{h=n_j+1}^{m_i} x_{i,h} \geq \sum_{h=n_j+1}^{m_i} v(A_{i,h}).$$

Adding (6) and (9) gives

(10) $$\sum_{i,j \text{ in } S} (x_i + y_j) \geq \sum_{h=1}^{\min(m_i, n_j)} v(A_{i,h}, B_{j,n_j-h}) + \sum_{h=n_j+1}^{m_i} v(A_{i,h}).$$

In this case also the right-hand side of (10) is $v(S)$. Hence (8) and (9) imply (5). QED.

The idea behind this formal proof is quite simple. To prove necessity, choose a coalition S. Then construct a market for only those traders who are members of S. In this way one can apply the pertinent parts of the proof of theorem 1, which applies to any given market.

Theorem 2 implies that the conclusions of theorem 1 and its corollary remain valid if the imputations satisfy the primary core constraints and are feasible. The second important conclusion from theorem 2 is that the equilibrium determined by the intersection of the supply and demand schedules does satisfy all of the core constraints. Hence in this case in which the owners each have one unit of the good and the nonowners each want at most one unit of the good, the analysis based on the core gives precisely the same results as the classical theory. This is not always true. Moreover, even when the conclusions are the same, it is for different reasons. The equilibrium determined by the core results from bargaining among traders in pursuit of self-interest. Coalition formation or, equivalently, contracting is the mechanism that narrows the range of outcomes and forces imputations into the core. The result is a uniquely determined quantity traded and a price constrained to lie within certain limits. There is a unique equilibrium if the supply and demand schedules intersect at a point of continuity of at least one of these schedules.

If m or n exceeds one, there are $(m+1)(n+1)-1$ primary constraints. Together with the feasibility requirement, this gives $(m+1)(n+1)$ inequalities to determine the imputations in the core. However, the total number of possible constraints is $2^{m+n}-1$ so that we see that a market can find its equilibrium more efficiently than would appear at first. Moreover, some of the primary core constraints are redundant. This fact is implicit in the proofs of theorems 1 and 2. The basic primary constraints and the feasibility requirement give the minimum number of inequalities forcing imputations into the core. Our next task is to discover the basic primary core constraints.

Before proving the next theorem about the basic primary constraints, let us adopt a device that avoids the awkward necessity of distinguishing the cases $m \leqslant n$ and $m > n$ in the course of proofs. This device adds dummy traders who cannot affect either the quantity traded or the price range, but which enables us to assume $m = n$. Assume in the original market that $m < n$. If so, then add $n-m$ dummy owners, A_i, $i = m+1, ..., n$, such that $a_i = a_m + c$, $c > 0$. Hence in this case $v(A_i) = v(A_m) + c$, and c can always be chosen large enough to ensure that no dummy trader can profitably enter a coalition with an initial nonowner. If $m > n$, then it is necessary to add dummy traders who are nonowners, B_j, with $j = n+1, ..., m$ such that the corresponding b_j satisfies $b_j = 0$ with $j = n+1, ..., m$. The latter condition guarantees that a dummy nonowner can never buy anything. Henceforth, unless it is stated otherwise, it is assumed that the number of initial owners m equals the number of initial nonowners n. The following theorem gives a set of basic primary constraints and the number of basic primary constraints.

THEOREM 3. *Let the market be essential. Hence k units are exchanged with $k \geqslant 1$. Let $m = n$. Let the imputations be feasible so that they satisfy (4). In addition to (1) and (2) a set of basic primary constraints is as follows:*

$$(11) \qquad x_i + y_{m-(i-1)} \geqslant v(A_i, B_{m-(i-1)}) \qquad (i = 1, ..., m),$$

$$(12) \qquad x_i + y_{m-i} \geqslant v(A_i, B_{m-i}) \qquad (i = 1, ..., k),$$

$$(13) \qquad x_i + y_{m-(i-2)} \geqslant v(A_i, B_{m-(i-2)}) \qquad (i = 1, ..., k+1).$$

The number of basic primary constraints is $3m+2k+1$ if $m > k$ and is $3m+2k-1$ if $m = k$.

Proof. The hard work for the proof of this theorem has already been done in proving theorem 1. We need only determine which core constraints were necessary for the results stated in theorem 1. The first result in theorem 1 asserts that the imputations of the core satisfy 4(13) [equation (13) in sec. 4, this chap.] This result depends on feasibility and the inequalities (1), (2), and (11). Next, theorem 1 asserts that there must be equality in (11). This fact together with 4(13) establishes the identity of the final owners in the core. The proof of equality in (11) depends on 4(13) and on the inequalities given in (11). The second assertion of theorem 1 is that all sellers receive the same price. This fact, proven after 4(18) relies on the primary constraints 4(21) and 4(22) for $i = 1, ..., k$. These constraints are precisely the same as the ones given in (12) and (13) above. Hence all of the conclusions of theorem 1 follow from the hypotheses of theorem 3. Since the corollary of theorem 1 uses precisely the same basic primary constraints as the theorem itself, this corollary is also a consequence of the hypotheses of theorem 3. Finally, it is necessary to show that the results of theorem 2 follow from the hypotheses of theorem 3. This readily follows from the fact that the proof of (5) uses a subset of the basic primary constraints given in (11). With all these results, it is easy to show that if the imputations are feasible and satisfy the basic primary constraints, then they satisfy all of the primary constraints. Hence, if the imputations are feasible and satisfy the basic primary constraints, then they are in the core. QED.

Next, we prove the following corollary.

COROLLARY. *In an essential market the imputations are in the core if and only if they satisfy the basic primary constraints.*

Proof. Sufficiency is obvious since if the imputations are in the core they satisfy the basic primary constraints. Necessity is theorem 3. QED.

By an appropriate permutation of the basic primary constraints given in (11) to (13), one can generate other sets of such constraints which, together with (1) and (2), can determine the imputations of the core. Thus the set of primary basic constraints is not unique though the number of basic primary constraints is. The set of basic primary constraints given in theorem 3 is the most convenient to use. These results serve two purposes. First, they facilitate the analysis of the efficiency of a market, and, second, they aid the extension of the results on exchange to multiunit trade, the topics of the next two sections.

6. Market Efficiency and Honest Brokers

Theorem 3 shows that the number of basic primary contacts between owners and nonowners sufficient to force the imputations into the core is $m+2k+1$ if $m > k$. This number should be compared with $2^{2m}-2-2m$, which is the number of possible coalitions excluding the empty coalition, the coalition among all traders, and the coalitions of single traders. Confining contacts to the basic primary pairs would considerably increase the efficiency of the market because fewer contacts among traders would suffice to force imputations in the core. Nevertheless, it is true that the number of basic primary contacts is approximately twice the number of traders, possibly a large number. If contacts and contracting were costless, no one would care about the number of contacts sufficient for a competitive equilibrium. Since trading does absorb resources from other pursuits, traders consider competing uses for the resources that they allocate to market activities. As a result there is less negotiating than in a world with costless exchange. The purpose of this section is to analyze the consequences of these costs and to consider their implications for the efficient organization of markets.

One way to represent these costs is to assume that each market participant limits himself to a given number of contacts and acts on the basis of these. This implies that the market breaks into submarkets, possibly overlapping, such that the number of participants in every submarket depends on the number of contacts desired by the traders. An obvious implication is that prices in the submarkets are not necessarily the same.[7]

The partitioning of the total market into submarkets due to the existence of transaction costs is formally equivalent to removing some constraints that determine the imputations in the core for the whole market. The existence of transaction costs can alter the market equilibrium in two ways. First, it can change the identity of the final owners, and, second, it can change the final imputations with respect to those in the core for the total market. Therefore, the aggregate value of the commodity to the final owners is less than the maximum aggregate value that would be realized with zero transaction costs. This does not imply a social loss if the difference between the aggregate valuation and the maximum valuation is less than the transaction costs that would be necessary to attain the maximum. Moreover, although there are positive transaction costs, the final owners may be the same as in a total market with zero transaction costs. When this happens the aggregate valuation is maximized though the final imputations are not in the core for the whole market. A maximum aggregate valuation implies a socially optimal ownership, although the distribution of wealth is not the same as in the core, and not all sellers obtain the same price. These possibilities are now illustrated for two owners and two nonowners.

In the first case assume there are only two primary contacts, (A_1, B_2) and (A_2, B_1). The imputations satisfy the constraints as follows:

(1) $$x_i \geq v(A_i) = a_i \quad \text{and} \quad y_j \geq v(B_j) = 0,$$

(2) $$x_1 + y_2 \geq v(A_1, B_2) = b_2,$$

(3) $$x_2 + y_1 \geq v(A_2, B_1) = b_1,$$

---

7. Shapley and Shubik (1966) provide a different approach to the problem of transaction cost. They define two kinds of cores, a strong ε-core, which has a fixed cost of coalition formation independent of the number of persons in the coalition, and a weak ε-core, which has a cost of coalition formation proportional to the number of members.

(4) $$\sum_{T} (x_i + y_j) \leqslant v(I) = b_1 + b_2.$$

Thus, two of the basic primary constraints are missing, (A_1, B_1) and (A_2, B_2). Nevertheless, the final owners will be the same as those in the core. Inequalities (2), (3), and (4) imply that there must be equality in (4). Given (2) and equality in (4), it follows that $x_2 + y_1 \leqslant v(A_2, B_1)$, so that there must be equality in (3). Given equality in both (3) and (4), it follows that there must be equality in (2). Consequently,

(5) $$x_1 + y_2 = v(A_1, B_2),$$
and
(6) $$x_2 + y_1 = v(A_2, B_1).$$

The latter two equations and (1) imply that both owners, who value the good less than the nonowners, will sell it to the nonowners. However, both sellers may not receive the same price nor will the price range coincide with the one determined by the core. This example shows that the final owners are the same as those in the core though some of the primary constraints are missing.

In the second example assume there are only two primary contacts, (A_1, B_1) and (A_2, B_1). Thus, in addition to (1) and (4) the constraints are as follows:

(7) $$x_1 + y_1 \geqslant v(A_1, B_1) = b_1,$$
and
(8) $$x_2 + y_1 \geqslant v(A_2, B_1) = b_1.$$

However, from (7) and (8) one cannot infer equality in (4). Moreover, there is only one exchange, A_1 sells to B_1 at a price p such that $a_1 < p < a_2$. Neither the final owners nor the final imputations are the same as in the core. This example shows that with some missing constraints the final owners are such that the aggregate valuation of goods is not maximized because there is inequality in (4).

These two examples illustrate the two consequences of positive transaction costs. The first shows that positive transaction costs need not prevent the final owners from being the same as those in the core. The second shows that transaction costs can change the identity of the final owners as well as the final imputations. The effect on the identity of the final owners is more serious because trade should cause a transfer of ownership of goods to those who most value them.

Continuing our study of market efficiency, let us analyze the outcome of random contacts between pairs of traders. We shall see that such contacts are very inefficient, which explains why in markets of any size there are intermediaries who act as agents or brokers for the owners and nonowners. Intermediaries can increase the efficiency of the market and facilitate the attainment of a competitive equilibrium provided they represent their

clients appropriately. To complete the analysis of the effects of inter-
mediation, we shall study exchange when there is demand for and supply
of more than one unit of the good by the owners and nonowners. It turns
out that in this case the competitive equilibrium is not generally determined
by the intersection of the classical supply and demand schedules.

Assume there are random contacts between pairs of traders (A_i, B_j).
We maintain the assumption that the number of owners m equals the
number of nonowners. In this analysis it is helpful to use an $m \times m$ matrix
whose entries are either zero or one. A contact between an owner A_i and
a nonowner B_j is represented by a one in row i and column j. Lack of
contact is represented by a zero in the cell. Let owners and nonowners be
distinguishable in advance, which overstates the efficiency of random
contact. If owners and nonowners could not be distinguished in advance,
some contacts would be wasted, which lowers market efficiency. Let there
be recontracting. Hence the final equilibrium is independent of the chrono-
logical sequence of the contacts between traders. This assumption enables
us to avoid a difficult analysis of the dynamics.

The easiest problem to solve is the calculation of the probability that m
chance contacts result in a transfer of ownership to those individuals who
would be the final owners in the imputations determined by the core. To
simplify the problem further, assume that every trader is a member of only
one pair of contacts. Hence the matrix has m ones and $m^2 - m$ zeros. The
number of possible matrixes equals the number of combinations of m^2
objects m at a time, which is denoted by

$$\binom{m^2}{m} = m^2! / (m^2 - m)! \, m!$$

It is necessary to count the number of ways that m contacts can result in a
transfer of ownership to the correct parties. Suppose the quantity transacted
in the core is $k \geqslant 1$. An intramarginal initial owner is an A_i with $i = 1, \ldots, k$
and an extramarginal initial owner is an A_i with $i = k+1, \ldots, m$. Thus an
intramarginal initial owner is a trader who sells the unit he owns and ends
up in the core as a nonowner. An extramarginal initial owner begins and
ends with one unit so that he makes no exchange in the core. An intramarginal initial nonowner is a B_j such that $j = m-(k-1), \ldots, m$, while
an extramarginal initial nonowner is a B_j with $j = 1, \ldots, m-k$. In the core
an intramarginal initial nonowner buys one unit, while an extramarginal
initial nonowner remains a nonowner. Partition the matrix representing
the pattern of contacts so that the northeast $k \times k$ submatrix corresponds
to the intramarginal traders while the southwest $(m-k) \times (m-k)$ submatrix
corresponds to the extramarginal traders. Any set of contacts among the
extramarginal traders cannot result in trade because any A_i in this set
values the good more than any B_j in the set. However, any pattern of
contacts among the intramarginal traders does result in exchange such that

the final owners are the same as those in the core. This is because any B_j in this set values the good more than any A_i in the set. The number of ways that contacts can occur within these two groups of traders and result in the final core owners is given by

$$\binom{k^2}{k} \binom{(m-k)^2}{(m-k)}.$$

Hence the probability that m pairs of contacts between an initial owner and an initial nonowner results in the final core owners is given by

$$\binom{k^2}{k} \binom{(m-k)^2}{(m-k)} \div \binom{m^2}{m}$$

This probability is a minimum if $k = m/2$, and m is even or if $k = (m+1)/2$ and m is odd. Given a constant ratio of k to m, as m increases the probability that m equally likely random contacts between A_i and B_j results in final owners who are in the core rapidly approaches zero. Hence such random contacts rapidly become less efficient with rising m. For the probability of obtaining the final core owners to remain constant as m rises, the contacts per trader must be an increasing function of m. Consequently, random contacts are inefficient for obtaining the final core owners.

The probability that random contacts secure the same imputations as the core is even less than the probability of obtaining the final core owners by chance because it is the product of the probability of obtaining the final core owners by chance and the probability of obtaining the additional primary contacts necessary to force the imputations into the core with a common price. General results are hard to obtain. The calculation requires a count of the number of permutations of basic primary contacts, one set of which is shown in the hypothesis of theorem 3, a theorem which illuminates the problem. It shows that the intramarginal traders must average nearly three contacts each for the imputations to be in the core. Moreover, any contact between intramarginal and extramarginal traders must be accompanied by appropriate contacts within the group of intramarginal traders to forestall blocking the core imputations. Consideration of a few numerical examples persuades one that chance contacts rarely force the imputations into the core.

These facts explain several aspects of real markets including the very existence of organized exchanges. In addition, we shall see how brokers can increase the efficiency of real markets if they faithfully obey certain instructions of their clients. Brokers can facilitate attainment of the core imputations with very few contacts. Let us see how.

Appropriate instructions to brokers exploit a property of competitive equilibria as yet undiscussed. We have seen that if every owner is willing to sell at most one unit and if every nonowner is willing to buy at most one unit, then the imputations in the core can be derived from the intersection of the supply and demand curves. Therefore, we can discover additional properties of the core from a study of the geometry of the situation. Figure 1.6 shows that the competitive equilibrium has the property that the maximum net quantity of goods is exchanged at a common price compatible with every trader's limit price. This is the relevant property for determining the instructions to the brokers. Every broker should be told to trade as much as he can at a common price without violating any of his clients' limit prices.

How would the brokers operate? Let a broker represent m owners of the commodity, A_i, $i = 1, ..., m$. Since every owner has a reservation price a_i, his imputation must satisfy $x_i \geqslant a_i$. His broker must not violate this constraint. In addition the broker must sell as many units as possible at a common price x. Similarly, let an honest broker represent the nonowners B_j subject to the instructions that no nonowner pays more than b_j, his reservation price, that all buyers pay the same price, and that the broker buys as many units as possible without violating the first two instructions. Suppose the quantity transacted in the core would be k units. Let $f(x)$ denote the quantity offered by the owners' broker at the common price x and let $g(x)$ denote the quantity demanded by the nonowners' broker. The term $f(x)$ is a nondecreasing step function of x and $g(x)$ is a nonincreasing step function of x. The quantity transacted at any given price x is

$$\min_{f,\,g} \{f(x), g(x)\}.$$

The broker's instructions are that a maximum quantity is to be exchanged at a common price x. Hence the quantity exchanged satisfies

(9) $\max_{x} \min_{f,\,g} \{f(x), g(x)\}$,

and it is easy to verify that this determines the quantity transacted to be $f(x) = g(x) = k$. This proves theorem 4.

THEOREM 4. *Let an honest broker represent the m owners, A_i, $i = 1, ..., m$ and be given instructions as follows:*
     i. $x_i \geqslant a_i$;
     ii. all sellers are to receive the same price x;
     iii. subject to i and ii, maximize the quantity sold.
Let another honest broker represent the m nonowners, B_j, $j = 1, ..., m$, and be given the instructions as follows:
     i'. for any buyer B_j the price x must satisfy $x \leqslant b_j$;
     ii'. all buyers are to pay the same price x;
     iii'. subject to i' and ii', buy as many units as possible.
The set of imputations that satisfies these conditions is precisely the core.

It follows from this result that $2m + 1$ contacts are sufficient to secure the core imputations, m contacts between a broker and each of his clients, and one contact between the two brokers. Although this requires more contacts than the number of basic primary constraints of the pairs (A_i, B_j) as described in theorem 3, this procedure is much more efficient than random contacts. Chiefly responsible for this result is the fact that the identity of the brokers is assumed to be known in advance to the prospective buyers and sellers. A clearing house, which is informed of the traders' limit prices, can calculate the supply and demand schedules $f(x)$ and $g(x)$ and then solve (9) to determine the equilibrium. However, both the brokers and a clearinghouse require a method of settling upon a mutually agreeable price within the range implied by the core. If the traders deal directly with each other, such a rule is unnecessary.

Although the brokers raise market efficiency by reducing the number of contacts necessary to attain the competitive equilibrium, their services are not free. Nor is this all. The broker's clients must compensate him in such a way that he is induced to serve them honestly. Not only should the broker's fee depend on the quantity transacted according to theorem 4, but also it should depend on the client's gain by exchange, which is the difference between the actual price and the client's reservation price. Moreover, some of the efficiency from brokerage would be lost if there must be individual negotiation between the broker and his prospective client to determine the brokerage fee since the prospective client may have to contact several brokers in order to obtain the best terms. In addition, we should deduct the cost of the resources allocated to the task of ensuring that the brokers carry out their instruction in calculating the net gain from brokerage. Competition among brokers and the accumulation of experience instills the confidence necessary to render the device of brokerage a workable method of approximating a competitive equilibrium.

According to this analysis, at most two brokers are needed. However, there undoubtedly are limits to the number of transactions that an individual broker can handle, and, presumably, brokerage costs per unit traded is an increasing function of the number of transactions. Hence for large markets it is more efficient to have several brokers who are all subject to the instructions given in theorem 4 and who each have a limited clientele. Then contacts among the brokers result in a set of imputations in the core. However, a new factor appears because each broker is equivalent to a trader willing to buy or to sell more than one unit of the commodity. We now study exchange under these conditions.

7. Multiunit Trade

Imputations in the core have new properties when the individuals own or wish to own more than one unit of the good, multiunit trade, distinguishing it from the classical approach based on the intersection of supply and demand schedules. Some of these were pointed out by von Neumann and Morgenstern who show that while the quantity traded may be the same as in Böhm-Bawerk's theory, all of the traders do not necessarily face a common price per unit, and the average transaction price is not confined to as narrow a band as is implied by Böhm-Bawerk's marginal pairs.[8] Moreover, as we shall see, with multiunit trade there may not be Pareto optimality (equals group rationality in game theory).

An even more important problem with multiunit trade is the possibility of an empty core. The core is said to be empty if no feasible set of trades can give rise to imputations capable of satisfying the core constraints. The core cannot be empty with single-unit trade as shown above. With multiunit trade it will be shown that the market always has a nonempty core if all of the traders have convex preferences, meaning that their excess demand schedules are negatively sloped or, equivalently, that their indifference curves are convex to the origin. Therefore, the presence of some traders with non-negatively sloping excess demand schedules is necessary for a market with an empty core. In this section it is assumed that all traders have negatively sloped excess demand schedules; chapter II contains a discussion of the general case.

With multiunit trade one must also show that the core constraints are implementable by means of a set of feasible exchanges. We shall postpone analysis of this question to the next chapter, where more powerful mathematical tools are available to to treat it.

We begin with a formal statement of multiunit trade. Assume there are m initial owners, A_i, $i = 1, ..., m$, and n initial nonowners, B_j, $j = 1, ..., n$. The latter are the potential buyers; the former are the potential sellers. This is a convenient albeit inessential distinction. Let A_i own m_i units of the good whose limit prices are ordered as follows:

(1) $$a_{i1} \leqslant a_{i2} \leqslant ... \leqslant a_{im_i}$$

Therefore, A_i is willing to sell s_i units, $s_i \leqslant m_i$, for an aggregate sum that must be at least equal to

$$\sum_{s=1}^{s_i} a_{is}.$$

The assumption about convex preferences means that the larger is the owner's initial stock, the lower is the price acceptable for selling one more unit. Concave preferences imply an increasing marginal rate of substitution so that the system of inequalities in (1) would be reversed, that is, $a_{i,s} \geqslant a_{i,s+1}$. An arbitrary set of preferences would be represented by a sequence of a's that is not monotonically ordered. Before giving some examples, let us describe the preferences of the initial nonowners.

---

8. Von Neumann and Morgenstern (1947, pp. 555–86).

A nonowner B_j is a potential buyer who is willing to buy up to n_j units at limit prices ordered as follows:

$$(2) \qquad\qquad b_{j1} \leqslant b_{j2} \leqslant \ldots \leqslant b_{jn_j}.$$

Hence, B_j is willing to buy t_j units. $t_j \leqslant n_j$, for a sum not to exceed

$$\sum_{t=1}^{t_j} b_{j, n_j - (t-1)}.$$

When the limit prices of the buyer satisfy (2), the buyer is said to exhibit decreasing marginal rates of substitution, convex preferences, or, equivalently, a negatively sloped excess demand schedule.

Let us now consider a few examples to illustrate these ideas. Suppose an owner starts with two units of a good. Assume he is willing to sell the first unit at a lower price than the second. For example, if he would accept 5 for the first unit and 7 for the second, then his reservation price is 5 for one unit, and the lowest acceptable sum for 2 units is 12 so that his preferences are convex. With concave preferences the owner would value the first unit at 5 and the second at 7. In either case the smallest acceptable amount for both units is 12. For a buyer to have concave preferences means his willingness to pay a higher price for the second than for the first unit. To illustrate how such preferences affect trade assume there are two sellers and one buyer with the following limit prices:

	A_1	A_2	B
first unit	5	7	6
second unit	7	5	0

Let each A_i have two units of the good. The buyer will pay up to 6 for one unit. A_1 would sell one unit for at least 7, but A_2 would accept any price above 5 for one unit. Trade is possible between A_2 and B but not between A_1 and B. The values of the characteristic function are given by

$$v(A_1, B) = \max \{5+7, 5+6\} = 12,$$
$$v(A_2, B) = \max \{7+5, 7+6\} = 13.$$

These are derived as follows. Let A_1 and B form a coalition. Since this coalition has two units of the good, it can allocate these in only two ways; give both units to A_1 or give one unit to A_1 and one to B. In the first case the coalition would not accept less than 12 for the sale of both units while in the second case it would not accept less than 11. Therefore, 12 is the value of the characteristic function for the coalition (A_1, B). Similarly, the coalition (A_2, B) can ensure itself at least 13.

In general, if not all traders have negatively sloped excess demand functions, then there may not exist a set of feasible trades to implement the core constraints. One of the principal results of this section is to show that a market has a nonempty core if all traders have convex preferences (equals negatively sloped excess demand functions). In chapter II, using more powerful methods, we shall see nonempty cores for a somewhat broader range of cases—first, if all of the traders have concave preferences and, second, if all of the initial owners have convex preferences while all of the initial non-owners have concave preferences (thm. II.3). For the present it is assumed that the limit prices are in the order shown in (1) and (2) so that all traders have negatively sloping excess demand. Therefore, no buyer would accept an additional unit for a price above the average limit price of the preceding units, and, similarly, no seller would offer an additional unit at a price below the average reservation price of the preceding units.

Let x_i and y_j denote the final imputations of A_i and B_j, respectively:

$$(3) \qquad\qquad x_i \geqslant v(A_i) = \sum_{s=1}^{m_i} a_{is},$$

and

$$(4) \qquad\qquad y_j \geqslant v(B_j) = 0$$

are two sets of basic constraints representing individual rationality that must be satisfied by imputations in the core. In addition, the other core constraints are given by

$$(5) \qquad\qquad \sum_{i,\,j \text{ in } S} (x_i + y_j) \geqslant v(S)$$

for all coalitions S composed of subsets of A's and B's. Define

$$A = \cup_{i=1}^{m} A_i \quad \text{and} \quad B = \cup_{j=1}^{n} B_j.$$

Corresponding to these two sets of traders are the sets of a's and b's given by

$$a = \{a_{is}/i = 1, ..., m; s = 1, ..., m_i\},$$

and

$$b = \{b_{jt}/j = 1, ..., n; \; t = 1, ..., n_j\}.$$

The total amount of goods available for exchange is M, where M satisfies

$$(6) \qquad\qquad M = \sum_{1}^{m} m_i.$$

Let z_h denote the imputation of final owner h and define z as follows:

$$(7) \qquad\qquad z = \sum_{1}^{M_I} z_h,$$

$$(8) \qquad\qquad v(I) = \max z \quad \text{subject to} \quad z_h \; \text{in} \; a \cup b \qquad \text{(cf. sec. 4).}$$

If the imputations are feasible, then

(9)
$$\sum_{i,\,j \text{ in } I} (x_i + y_j) \leqslant v(I),$$

where $I = A \cup B$, the set of all traders.

Next, consider an arbitrary coalition S consisting of the union of A_S and B_S, where A_S and B_S are subsets of A and B, respectively. Let M_S denote the number of owners in A_S and define

$$z(S) = \sum_{h=1}^{M_S} z_{hS},$$

where $z_{h,S}$ denotes the imputation of a final owner in S.

(10)
$$v(S) = \max z(S) \quad \text{subject to} \quad z_{h,S} \quad \text{in} \quad a_S \cup b_S,$$

where a_S is the subset of a corresponding to A_S and b_S is the subset of b corresponding to B_S. Observe that if A_S is empty, then $M_S = 0$, and $v(S) = v(B_S) = 0$ as required. To develop the new aspects of multiunit trade, it is helpful to work through some examples.

In the first example there are two nonowners who are each willing to buy at most one unit, and there is one owner willing to sell at most two units. The imputations in the core satisfy the following inequalities:

(11)
$$x_1 \geqslant v(A_1) = a_1 + a_2, \quad y_j \geqslant v(B_j) = 0 \qquad (j = 1, 2),$$

(12)
$$x_1 + y_1 \geqslant v(A_1, B_1) = a_2 + b_1,$$

(13)
$$x_1 + y_2 \geqslant v(A_1, B_2) = a_2 + b_2,$$

(14)
$$x_1 + y_1 + y_2 \leqslant v(A_1, B_1, B_2).$$

In the case of single-unit trade the imputations necessarily sum to the upper bound shown in (14) giving equality there. When there is equality in (14), the market is said to exhibit "group rationality" in game theory and "Pareto optimality" in economic theory. However, the core imputations with multiunit trade do not always result in equality in (14). To prove this, we need only exhibit a set of imputations satisfying (11) to (13) such that the maximum $v(I)$ is not attained, which is easily done. Let

(15)
$$x_1 = a_1 + a_2 + c$$

with $c \geqslant 0$ as required by (11). With this value for x_1, (12) implies that

(16)
$$a_1 + c + y_1 \geqslant b_1,$$

and (13) implies that

(17)
$$a_1 + c + y_2 \geqslant b_2.$$

Summing (16) and (17) gives

(18)
$$2(a_1 + c) + y_1 + y_2 \geqslant b_1 + b_2,$$

which implies that

(19) $$y_1 + y_2 \geqslant b_1 + b_2 - 2(a_1 + c).$$

It follows from (15) that

(20) $$x_1 + y_1 + y_2 \geqslant b_1 + b_2 - 2(a_1 + c) + a_1 + a_2 + c,$$

so that

(21) $$x_1 + y_1 + y_2 \geqslant b_1 + b_2 + a_2 - a_1 - c.$$

If $a_2 - a_1 - c \geqslant 0$, then (21) implies that

(22) $$x_1 + y_1 + y_2 \geqslant b_1 + b_2,$$

so that there would have to be equality in (14). But if $a_2 - a_1 - c < 0$, then, depending on the size of c, there can be inequality in (14). The only possible general conclusion about the core imputations is that

(23) $$a_2 \leqslant x_1 - (b_1 - y_1) \leqslant b_2,$$

(24) $$a_2 \leqslant x_1 - (b_2 - y_2) \leqslant b_1.$$

Since $b_j - y_j$ is the price paid by B_j, (23) means that A receives a price from B_2 that is greater than a_2 and less than b_2 while he receives a price from B_1 that is also above a_2 and below b_1. The seller does not necessarily receive the same price from both.

Compare the present case with one in which there would be two owners, A_1 and A_2, who would each be willing to sell at most one unit at prices not below a_1 and a_2, respectively. In this case there would be a common price p such that $a_2 < p < b_1$. The buyer who values the good most would obtain the most consumer surplus. However, when there is only one seller with two units to sell, this does not happen. It is as if the seller prevents competition between the two units of the good that he owns. As a result the seller obtains a larger imputation than the sum of the imputations of two separate owners who have the same limit prices as the single owner.

The next example reveals even more striking differences between the core imputations and the conclusions derived from the intersection of the supply and demand schedules. Let there be two owners and one nonowner. The first owner A_1 has two units that he values in the order $a_{11} < a_{12}$. The second owner has three units that he values in the order $a_{21} < a_{22} < a_{23}$. The nonowner is willing to buy up to three units at limit prices as follows: $b_1 < b_2 < b_3$. Let the relation between the a's and b's be given by

$$a_{12} > b_3 > b_2 > a_{23} > a_{22} > b_1 > a_{11} > a_{21}.$$

The equilibrium determined by the intersection of the supply and demand schedules implies that two units would be sold at a price p such that

(25) $$b_1 < p < a_{22}.$$

(See fig. 1.8.) Core theory gives a different result.

For the imputations to be in the core they must satisfy

(26) $$y \geqslant v(B_1) = 0,$$

(27) $$x_1 \geqslant v(A_1) = a_{11} + a_{12},$$

(28) $$x_2 \geqslant v(A_2) = a_{21} + a_{22} + a_{23},$$

(29) $$x_1 + y \geqslant v(A_1, B_1) = a_{12} + b_3,$$

(30) $$x_2 + y \geqslant v(A_2, B_1) = a_{23} + b_3 + b_2.$$

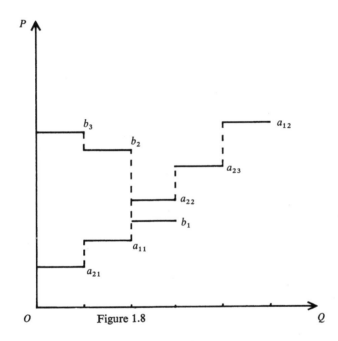

Figure 1.8

Finally, for the imputations to be feasible, they must satisfy

(31) $$x_1 + x_2 + y \leqslant v(I) = a_{12} + a_{23} + a_{22} + b_3 + b_2.$$

As in the preceding example, the core constraints do not necessarily imply that there is group rationality, that is, equality in (31). In other words a set of imputations satisfying (26) to (30) does not necessarily imply equality in (31).

With single-unit trade where all traders either bought or sold at most one unit each, one could prove that if the imputations are in the core so that

$$\sum_{i,\, j \text{ in } S} (x_i + y_j) \geqslant v(S) \quad \text{for all} \quad S \subset I \quad \text{with} \quad S \neq I,$$

it would follow that

$$\sum_{i,\,j \text{ in } I} (x_i + y_j) \geq v(I).$$

Together with feasibility this yields the result that $\Sigma_I (x_i + y_j) = v(I)$, so that there would be group rationality. However, with multiunit trade there need not be group rationality even if all of the core constraints are satisfied, demonstrating that the kind of competition represented by the core does not inevitably result in group rationality.

Before continuing this analysis of group rationality, let us derive further implications about the imputations in the second example. Let

$$(32) \qquad\qquad x_1 = a_{11} + a_{12} + c_1 \qquad\qquad (c_1 \geq 0),$$

$$(33) \qquad\qquad x_2 = a_{21} + a_{22} + a_{23} + c_2 \qquad\qquad (c_2 \geq 0).$$

These two conditions merely restate (27) and (28). Equation (29) implies that

$$(34) \qquad\qquad y \geq b_3 - a_{11} - c_1 \geq 0,$$

and (30) implies that

$$(35) \qquad\qquad y \geq b_3 + b_2 - a_{21} - a_{22} - c_2 \geq 0.$$

The latter two inequalities impose an upper bound on the c's, which sets an upper limit on the gains from trade for the two initial owners. If

$$(36) \qquad\qquad b_3 - a_{11} > c_1$$

and

$$(37) \qquad\qquad b_3 + b_2 - a_{21} - a_{22} > c_2,$$

then $y > 0$. Hence trade does actually occur. However, we cannot determine who gets what or how much. Figure 1.8 helps understand the difference between this analysis and the conventional supply and demand approach. In the conventional approach one assumes that every unit is separately owned or separately sought. Such reasoning ignores the fact that there is only one nonowner and two owners. Core theory requires explicit consideration of every trader's resources and of the number of traders. Thus the inequalities in (36) and (37) represent the limits imposed by a pair of bilateral negotiations, one pair between A_1 and B_1 and the other between A_2 and B_1. All that we know from these limits is that A_1 will sell at most one unit.

It is not difficult to see what changes in the market would force the core imputations into the narrow band determined by the intersection of the supply and demand schedules. Let there be additional traders who are willing to buy or sell one unit each. Thus let there be three buyers, $B_j^*, j = 1, 2, 3$, whose limit prices are b_j and 5 sellers A_{is}^* with limit prices a_{is} such that each of these 8 traders is willing to trade at most one unit at prices constrained by their limits. These traders force the imputations into the core, and there is a common price within the limits shown in (25). Each A_{is}^* competes directly with A_i, and each B_j^* competes directly with B_j. The competition widens the range of choice open to every participant and thereby forces all of the sellers to obtain a common price within a narrower band than prevails in the market originally. Initially the participants had the power to make all or none offers. The new traders destroy this power. Finally, the presence of these new traders implies that the final imputations satisfy the criterion of group rationality confirming that those persons will obtain the good who most value it.

Before stating and proving the next theorem about multiunit trade using these ideas, it is necessary to discuss group rationality. The feasibility requirement, (9), $\Sigma_I (x_i+y_j) \leqslant v(I)$ explicitly states that the imputations cannot exceed the maximum valuation attainable by allocating the existing stock to those market participants who most value the good. Group rationality means that the imputations yield equality in (9). Formally, there is group rationality if and only if $\{x_i, y_j\}$ satisfy

$$(38) \qquad \sum_I (x_i+y_j) = v(I).$$

Theorem 1 refers to a situation in which every trader owns or wishes to own at most one unit of the good, and in this case there must be group rationality. Therefore, one need not assume group rationality because it is an inevitable consequence of the core constraints. In contrast, the core imputations with multiunit trade are not always group rational so that individual self-interest as represented by the core constraints

$$(39) \qquad \sum_S (x_i+y_j) \geqslant v(S) \quad \text{with} \quad S \subset I \quad \text{and} \quad S \neq I$$

does not generally imply (38).

Von Neumann and Morgenstern in their original work attempt to deduce group rationality by introducing a fictitious player who enters no coalition with the real players but who obtains the difference between the maximum $v(I)$ and the sum of the imputations of the real participants. If it is also assumed that the real players behave so as to minimize the amount going to the fictitious player, then group rationality results. Such an argument extends the rationale of the core to an indefensible position. Thus consider the coalition S of real traders where $S \subset I$ and $S \neq I$. We assert that

$$\sum_S (x_i+y_j) \geqslant v(S),$$

because through their own efforts S can obtain at least $v(S)$ and at worst if the strongest anticoalition forms, namely $-S$, then the coalition S can obtain no more than $v(S)$. But the maximal anticoalition of the set of all traders is the empty set which is literally powerless. Thus von Neumann and Morgenstern invent the fictitious extra trader who gets what the empty set would were it not empty. This argument is almost sheer theology and unacceptable.

It is far better not to assume group rationality and instead to discover when it is a consequence of the core constraints.[9]

In view of these remarks a different approach to group rationality is needed.

DEFINITION. *A collection* $T = \{S_i/i = 1, \ldots, u\}$ *decomposes the market if* $I = \cup_1^u S_i, S_i \cap S_{i'} = \varnothing, \text{for } i \neq i' \text{ and}$

$$(40) \qquad\qquad v(I) = \sum_1^u v(S_i).$$

In other words the collection T decomposes the market if every trader belongs to exactly one of the coalitions S_i and if the sum of the characteristic values for all of the coalitions $v(S_i)$ equals $v(I)$. In addition, if a market has a decomposing collection T, then for a nonempty core it is necessary that

$$(41) \qquad\qquad v(\cup_i S_i) = \sum v(S_i)$$

for all possible $\cup_i S_i$. If condition (41) were not satisfied, then because of the superadditivity of the characteristic function we would have

$$v(\cup_i S_i) > \sum v(S_i)$$

so that the core would be empty. Hence (41) is necessary for a nonempty core if the collection T decomposes or splits the market.

THEOREM 5. *If a market with a nonempty core has a decomposing collection of coalitions, denoted by T, then there is group rationality.*

Proof. Since the imputations are in the core, it follows that for every S_k in T

$$\sum_{S_k} (x_i + y_j) \geqslant v(S_k).$$

Summing over all of the S_k in T gives

$$\sum_k \sum_{S_k} (x_i + y_j) \geqslant \sum_k v(S_k) = v(I).$$

By hypothesis the core is not empty so that the imputations satisfying the core constraints are feasible. Therefore, there must be group rationality. QED.

The theorem asserts that the existence of a decomposing collection for a market with a nonempty core is sufficient for group rationality. It is, however, by no means a necessary condition. Thus both examples in this section illustrate that there can be group rationality without satisfying (40). In a single-unit trade market, the basic primary coalitions described in theorem 3 constitute a decomposing collection that satisfies the conditions of theorem 5.

---

9. Von Neumann and Morgenstern deal with this problem when they extend the analysis from zero-sum to nonzero-sum games (1947, sec. 56).

The next theorem describes the outcome of multiunit trade when all of the traders have convex preferences.

THEOREM 6. *In market 1 let m owners, A_i, each have m_i units of a good, $i = 1, ..., m$, such that A_i values his holdings according to the order shown in (1) so that every A_i has a negatively sloped excess demand function (convex preferences). Let there be n initial nonowners, $B_j, j = 1, ..., n$, such that each B_j is willing to buy up to n_j units at limit prices shown in (2) so that every B_j has a negatively sloped excess demand function also. Assume the imputations are in the core so that they satisfy (3), (4), and (5).*

In market 2 there is single-unit trade such that for every trader A_i in market 1 there are m_i initial owners A_{is}^*, $i = 1, ..., m$ and $s = 1, ..., m_i$, each with one unit and a reservation price a_{is}. Also for each nonowner B_j in market 1 there are n_j nonowners B_{jt}^* in market 2, where $j = 1, ..., n;$ $t = 1, ..., n_j$, who are each willing to buy at most one unit at limit prices b_{jt}. The imputations in market 2 satisfy the core constraints.

It follows that

    i. *the core of market 1 includes the core of market 2; the two cores are identical if $m_i = n_j = 1$ for all i and j;*

    ii. *if there is a decomposing collection of coalitions for market 1, then the same quantity is traded in both markets, any A_i is a final owner in market 1 if some corresponding A_{is}^* is a final owner in market 2, and, similarly, B_j is a final owner if the corresponding B_{jt}^* is a final owner in market 2.*

Proof. To prove i, observe that the core of market 1 is determined by a set of inequalities that is a subset of those inequalities which determine the core of market 2. Consequently every imputation in the core of market 2 is also in the core of market 1 as asserted. If $m_i = n_j = 1$ for all i and j then every trader owns or wishes to own at most one unit thereby making both markets identical. This completes the proof of i.

To prove ii, observe that by hypothesis and by theorem 5 there is group rationality in market 1. Theorem 2 applies to market 2 so that there is group rationality in market 2. Moreover, the maximum valuation given by $v(I)$ is the same in both markets, and it is attained in both because of group rationality in both. Therefore, in both markets the same quantity is exchanged, and there is the correspondence asserted between the final owners in the two markets. QED.

The equilibrium in market 2 of the theorem is precisely the one determined by the intersection of the classical supply and demand analysis, while in market 1 because of less competition there is a wider range of possible outcomes. This results from the fact that in market 1

$$A_i = \cup_s A_{is}^* \quad \text{and} \quad B_j = \cup_t B_{jt}^*,$$

so that in market 1 the traders are permanent unions of the corresponding traders in market 2. In market 1 a trader does not seek or offer units one at a time unless forced to do so by competition. In market 2 there is a one-to-one correspondence between units of the good and traders, unlike market 1. In market 1 all trades are not necessarily at the same unit price, and even the common unit prices capable of satisfying the core constraints of market 1 are confined to a wider range than is true of the common unit price of market 2. In market 2 the traders seek or offer one unit of the good, and the core implies that trades occur at a common unit price. To apply supply and demand analysis to market 1 where there is multiunit trade would be inconsistent with the assumption that traders pursue their self-interest and offer better terms only when they have to because of competition. Finally, there is the important fact that what makes it possible to establish a correspondence between the given multiunit market 2 and the single-unit market 1 is the assumption that all traders in market 2 have negatively sloped excess demand schedules. Therefore, convexity of the preferences is necessary for this correspondence. It is easy to verify that convex preferences are also sufficient for this correspondence.

These results also advance our understanding of brokerage. Assume a market with multiunit traders having negatively sloped excess demand schedules. Direct trade among themselves gives imputations in the core of market 1. Let brokers represent several traders such that every broker is instructed to obey his clients' reservation prices. Since there are fewer brokers than clients, there are fewer constraints determining the core, which widens the price limits making possible discrepancies between the actual and the optimal quantity exchanged. Consequently, aggregation of traders into broker-representatives can reduce competition in the market. However, competition for the services of the brokers can restore the beneficial effects of core imputations in the original market. Moreover, brokerage improves the functioning of the market because the brokers are the focal points for the traders.

8. Increasing Returns and Public Goods from the Viewpoint of the Core

We now apply core theory to a famous problem involving both production and exchange. The problem is to determine when a certain indivisible good should be produced. The usual example is a bridge.[10] It is assumed that the bridge must exceed a given size and that no cost is incurred when anyone uses the bridge. The latter assumption is unimportant and mainly dramatizes

10. This problem has an extensive literature. The bridge example and its analysis is in Hotelling (1938). See also Telser (1969a) for additional discussion and references.

the point that even if bridge users paid a toll this would not solve the problem. Indeed, a toll would be positively harmful because it would lead to a waste of resources by discouraging the use of the bridge and encouraging other costly means of crossing. Most analysts agree that the bridge should be built if it makes some people better off and none worse off; but how can this be known? We are first inclined to confine measurement of the benefits from the bridge to its actual users, but this may underestimate the benefit by neglecting gains to third parties. For instance, consumers might obtain cheaper lettuce if it is now less costly to ship lettuce by truck across the bridge. Should the lettuce growers be forced to contribute toward the cost of building the bridge or should all consumers be forced to contribute regardless of their taste for lettuce? Can a market mechanism reveal whether the bridge should be built? The bridge example is difficult because the possible benefits are so widespread. To clarify the pertinent issues, we shall pose the problem with another mode of transportation, an airplane.

Assume an airplane can carry two passengers at a cost per trip between two given cities of $500. Anyone willing to pay this amount can charter the airplane by himself. If there are such persons, then, occasionally, the plane makes the trip half empty. Flights will occur if there are pairs who are willing to pay a total of $500 although neither is willing to pay $500 alone. It is also possible that one passenger would travel free if the other is willing to bear the total cost of the trip, a case covered by the subsequent analysis.

This situation is amenable to analysis using core theory. A seller A_1 will not accept less than a for the sale of either one or two units so that his reservation price does not depend on how many units that he sells. B_j, a potential buyer, is willing to pay at most b_j per trip. Since the airplane can carry at most two passengers, the number of possible coalitions between buyers and sellers is limited to two; one seller and one buyer, and one seller with two buyers. The imputations are in the core if they satisfy the following inequalities:

$$y_j \geqslant \quad v(B_j) \quad = 0,$$
$$x_1 \geqslant \quad v(A_1) \quad = a,$$
$$x_1 + y_j \geqslant \quad v(A_1, B_j) \quad = \max\{a, b_j\} \qquad (j = 1, 2),$$
$$x_1 + y_1 + y_2 \leqslant v(A_1, B_1, B_2) = \max\{a, b_1 + b_2\}.$$

If either $a < b_2$ or $a < b_1 + b_2$, then flights will occur at a price between the seller's lower limit and the buyer's upper limit.

It is by no means true that in this class of situations the core is always nonempty. By slightly changing the conditions we can readily have an empty core as shown in section II.3, chapter II, for two identical planes and three identical potential passengers no one of whom is willing to defray the cost of a trip by himself. In the remainder of this section we shall assume the core is not empty and take up this question in the next chapter.

We now consider sufficient conditions for the existence of an air travel service between the two cities. This aspect of the problem gives a formal analysis for determining when a bridge should be built. Assume an entrepreneur contemplates whether he should offer air passenger service between the two cities. The airplane costs $10,000, lasts for ten years when it will collapse beyond repair, and can make 200 flights per year carrying, at most, two passengers per flight. Assume the entrepreneur has an alternative use for his capital that yields 10 percent per year. Hence he will not enter the air travel business unless the sum of gross receipts over the next ten years exceeds the out of pocket expenses by at least $20,000, neglecting discounting.

The problem consists of three stages. First, suppose the plane owner can lease the plane to someone who will operate the air travel service. If the plane owner can expect total rental receipts above $20,000 over the next ten years, then he is willing to buy the plane. Next consider the situation from the rentor's viewpoint. For simplicity, assume the rental is $2,000 per year. Actually it can be less because the excess above $1,000 can be amortized at 10 percent per year. For the rentor to be willing to lease the airplane for a year, he must gross more than $2,000 above his running expenses in, at most, 200 flights. Hence he must expect to make up to 200 contracts with one or two passengers per contract at sufficiently favorable terms to yield the desired amount. Finally, suppose the passengers are willing to book flights for up to one year in advance. The passengers and the potential rentor of the plane can then easily determine whether there are mutually satisfactory terms. Assuming the rentor finds it profitable to lease the plane and assuming this can be arranged in advance for every year of the plane's life, the entrepreneur would not hesitate to buy the plane for $10,000. Thus there is a sequence of agreements, one between the plane owner and the rentors who will operate the air travel service and another between the operators of the air travel service and the passengers, that determine whether the ultimate users of the air travel service value it enough to induce its provision. This analysis confines itself to the direct users of the air travel service. However, one can also include the indirect benefits by supposing another sequence of contracts between the passengers and third parties who benefit from their travel. Thus the third parties may have business with the passengers which helps determine the passengers' limit prices for plane tickets. Possibly, travellers negotiate with each other to cover the expenses of their trip, and some people may actually travel free. Contracts between the operators of the travel service and the passengers provide a mechanism for measuring the benefit. Similarly, indirect benefits are most easily registered in the terms of contractual agreements between the passengers and third parties. Such benefits help determine the passengers' limit prices. In these ways the market is a device for measuring the benefits to see whether the service is worth providing.

In the above situation the passengers must bear the ultimate uncertainty because they must book their flights long in advance, but not all passengers are willing or able to do this because they cannot form reasonable estimates either of their demand for the air travel service or of how much to value it. Individual estimates are more difficult than are estimates for a large group of passengers. Both airplane owners and entrepreneurs operating air travel services learn how to predict demand accurately so that they are willing to bear the risk of making long-term investments. Therefore, one may justifiably continue the analysis as if there actually were forward contracts with the passengers.

We seek formal conditions such that the entreprenuer would be willing to buy the airplane and offer the air travel service. Assume for convenience, that both operations are undertaken by the same firm. For this purpose we require the constraints defining the *ex ante* core. These *ex ante* constraints correspond to the terms that would be satisfied by all possible forward contracts between owner-operators of airplanes and potential passengers. If in fact there are no forward contracts, then the constraints of the *ex ante* core are assumed to represent the terms *expected* by the owner-operator.

Assume the airplane can make up to T trips and can carry up to two passengers per trip. In the preceding numerical example T is 2,000. The owner's expected or *ex ante* imputation on trip i is x_i and satisfies $x_i \geq a, i = 1, ..., T$, where a is the running cost per trip, \$500 in the example above. On trip i there are n potential buyers, $B_{ij}, j = 1, ..., n$, with limit prices in the following order: $b_{i,j} < b_{i,j+1}$. The *ex ante* imputation of buyer j on trip i is y_{ij} and satisfies

$$y_{ij} \geq v(B_{ij}) = 0, i = 1, ..., T, \quad \text{and} \quad j = 1, ..., n.$$

The *ex ante* constraints which determine the imputations of the core are given by the inequalities as follows:

(1) $\qquad\qquad x_i + y_{ij} \geq v(A_1, B_{ij}) = \max \{a, b_{ij} \quad \text{given} \quad i\},$

(2)
$$x_i + y_{ij} + y_{ij'} \geq v(A_1, B_{ij}, B_{ij'}) = \max \{a, b_{ij} + b_{ij'}, j, j' = 1, ... \, n, \quad \text{given} \quad i\},$$

(3) $\qquad \sum_{j=1}^{n} y_{ij} \leq v(A_1, \cup_j B_{ij}) = \max \{a, b_{in} + b_{i,n-1} \quad \text{given} \quad i\},$

(4) $\qquad\qquad \sum_{i=1}^{t} x_i \geq ta + C \quad \text{for some} \quad t \quad \text{with} \quad 1 \leq t \leq T.$

The imputations that satisfy these constraints are in the *ex ante* core. The constant C in (4) represents the fixed cost of owning the airplane including the opportunity cost of an alternative investment. Constraint (1) refers to a coalition between the airplane owner-operator and a potential passenger on a given trip i. Constraint (2) refers to a coalition with two potential passengers on a given trip i. Constraint (3) represents feasibility because the imputations on a given trip i cannot exceed $v(A_1, \cup_j B_{ij})$.

In this case there is group rationality because of (2) for $j = n-1$ and $j' = n$, provided $n \geqslant 3$. Hence there must be equality in (3) if $n \geqslant 3$. After some reductions it follows that

$$y_{ij} = 0 \quad \text{for every} \quad i \quad \text{and} \quad j = 1, \dots, n-2,$$

(5) $$x_i + y_{i,n-1} + y_{i,n} = v(A_1, B_{in}, B_{i,n-1}),$$

(6) $$y_{i,n-(j-1)} \leqslant v(A_1, B_{i,n}, B_{i,n-1}) - v(A_1, B_{n-(j-1)}) \quad (j = 1, 2).$$

For the plane service to be offered in at least one period, it is necessary and sufficient that there exist a set of t pairs, $b_{in} + b_{i,n-1}$, such that

(7) $$\sum_{i \text{ in } t} (b_{i,n} + b_{i,n-1}) \geqslant ta + C.$$

The constraints that determine the *ex post* core which assumes the existence of the air travel service are even simpler. *Ex post*, inequality (4) does not apply because the owner-operator of the plane service will make a flight if the receipts from that flight merely defray the running expenses. Moreover, the theory implies that not all passengers necessarily pay the same price. In this case a buyer's imputation satisfies (6), and there is still group rationality, (5), if $n \geqslant 3$. Since the plane has a limited life, the air travel service will be provided indefinitely only if the imputations in the core also satisfy (4).

The situation with more than one potential operator of the service is considerably more complicated because, depending on the demand conditions, the core can be empty. The complication arises because the cost conditions imply increasing returns, falling average cost, analogous to non-convex preferences. The problem of when the core is not empty is better deferred to the next chapter. For the present we may proceed formally by deriving some results pertaining to the case in which the core is not empty. If the core is not empty and there are two planes, then there must be a common price on each trip but not necessarily the same price on different trips. Roughly speaking, for a nonempty core it is sufficient that there is enough demand to prevent too much idle capacity on the planes. The constraints defining the core are given as follows:

(8) $$x_{ik} \geqslant v(A_{ik}) = a, i = 1, \dots, T \quad \text{and} \quad k = 1, 2,$$

where A_{ik} represents seller k on trip i. Instead of the single inequality (4), there are now the two inequalities

(9) $$\sum_{i \text{ in } t_1} x_{i1} \geqslant t_1 a + C, \quad \text{for some} \quad t_1 \quad \text{with} \quad 1 \leqslant t_1 \leqslant T,$$

and

(10) $$\sum_{i \text{ in } t_2} x_{i2} \geqslant t_2 a + C \quad \text{for some} \quad t_2 \quad \text{with} \quad 1 \leqslant t_2 \leqslant T,$$

and t_1, t_2 are not necessarily equal. Given that there are t_k's capable of satisfying (9) and (10), the two planes would operate in every period thereby preventing price discrimination on any trip i, as we shall see. The remaining constraints that define the core are as follows:

(11) $$x_{ik} + y_{ij} \geqslant v(A_{ik}, B_{ij}) = \max\{a, b_{ij}\},$$

(12) $$x_{ik} + y_{ij_1} + y_{ij_2} \geqslant v(A_{ik}, B_{ij_1}, B_{ij_2}) = \max\{a, b_{ij_1} + b_{ij_2};$$
$$j_1 \neq j_2, j_1, j_2 = 1, \ldots, n\},$$

(13) $$\sum_k x_{ik} + y_{ij_1} + y_{ij_2} + y_{ij_3} \geqslant v(\cup_k A_{ik}, B_{ij_1}, B_{ij_2}, B_{ij_3}),$$

(14) $$\sum_k x_{ik} + y_{ij_1} + y_{ij_2} + y_{ij_3} + y_{ij_4} \geqslant v(\cup_k A_{ik}, B_{ij_1}, B_{ij_2}, B_{ij_3}, B_{ij_4}),$$

(15) $$\sum_{k=1}^{2} x_{ik} + \sum_{j=1}^{n} y_{ij} \leqslant v(\cup_k A_{ik}, \cup_j B_{ij}) = \max\{2a, a + b_{in} + b_{i,n-1};$$
$$\sum_{j=1}^{3} b_{i,n-(j-1)}; \sum_{j=1}^{4} b_{i,n-(j-1)}\}.$$

The last inequality represents the feasibility constraint. Group rationality is readily demonstrated from (14) by setting $j_1 = n$, $j_2 = n-1$, $j_3 = n-2$ and $j_4 = n-3$. It then follows that the right-hand term of (14) is the same as the right-hand term of (15) so that there must be equality in (15) giving group rationality, namely,

(16) $$\sum_{k=1}^{2} x_{ik} + \sum_{j=1}^{n} y_{ij} = v(\cup_k A_{ik}, \cup_j B_{ij}).$$

From this one concludes that

(17) $$y_{ij} = 0 \quad \text{for all} \quad i \quad \text{and for} \quad j = 1, 2, \ldots, n-4.$$

To prove that $x_{i1} = x_{i2}$ for all i, it is simplest to use (12). Reconsider (12) and the combination of subscripts j_1, j_2 for the values n, $n-1$, $n-2$, $n-3$. Taken in conjunction with (16), an implication of equality exists in (12) for these combinations. It readily follows that both sellers must obtain a common imputation on every flight, that is

(18) $$x_{i1} = x_{i2} \quad \text{for all} \quad i,$$

provided, of course, that both (9) and (10) are satisfied.

If both planes operated on any trip i, then the common price x_i satisfies

(19) $$x_i \leqslant b_{i,n-2} + b_{i,n-3}.$$

There are also conditions involving the extramarginal buyers, but these are of no interest for present purposes. If only one plane operates on trip i, then the price x_i satisfies

(20) $$x_i < b_{in} + b_{i,n-1} \quad \text{and} \quad b_{in} + b_{i,n-1} \geqslant b_{i,n-2} + b_{i,n-3}.$$

Therefore, the upper price limit with two planes operating never exceeds the upper price limit with one plane operating for any given flight i. For two planes to operate in every period, it is necessary and sufficient that

(21) $$2(Ta+C) < 2\sum_{i=1}^{T}(b_{i,\,n-3}+b_{i,\,n-2}).$$

For only one plane to operate on every possible flight, it is necessary and sufficient that both

(22) $$Ta+C < \sum_{i=1}^{T}(b_{i,\,n}+b_{i,\,n-1})$$

and

(23) $$Ta+C \geqslant \sum_{1}^{T}(b_{i,\,n-3}+b_{i,\,n-2})$$

are satisfied.

Figure 1.9 illustrates the solution. The lower demand schedule, $b_n \ldots b_{n-3}$ shows that single passengers alone would be unwilling to pay for a flight. The upper demand schedule gives the demand by pairs of most eager passengers. According to this figure, the market would support only one plane because while b_n+b_{n-1} lies above the cost schedule aa, $b_{n-2}+b_{n-3}$

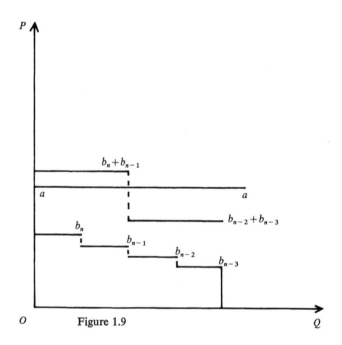

O Figure 1.9 Q

lies below it. With a larger demand the market could support two planes. The present situation differs from the preceding ones only insofar as the effective demand entails pairs instead of single buyers.

Once the air travel service exists, constraints (9) and (10) representing long-run costs do not apply and flights will occur on any day in which receipts would be enough to cover running costs. However, for the indefinite provision of the service the passengers must be willing to pay enough to defray all costs including the cost of replacing the airplane. Although passengers have temporary bargaining power to drive prices down to short-run cost, the continual exercise of this power would eventually destroy the air travel service.

Perhaps the most important result of this analysis is to show the effects of competition among the operators of the air travel service. With only one plane in service there can be price discrimination among customers but not with two or more planes. The more planes there are, the lower the receipts per plane per trip. Total receipts per plane cannot fall below long-run cost or the plane will not be replaced. With enough competing air travel services for the two cities, receipts per firm will fall to the lowest level compatible with the firm's survival and the state of demand. As figure 1.9 shows, profits are not driven to zero because of discontinuities. These conclusions depend on the hypothesis that the core is not empty. If the core is empty, then no competitive equilibrium exists and the market mechanism would not provide the service.

If passengers can arrange long-term contracts with the air travel service, then they can solicit bids from a large number of firms capable of operating the service although there is only enough demand for one firm to run the service. Competition among potential operators would force the price down to the lowest level compatible with demand and long-run cost represented by the right-hand side of (9). Even with only one firm in actual operation there can be competitive prices. However, forward contracting or its equivalent is necessary for this conclusion. Otherwise, with a level of demand only large enough to support few planes and because of indivisibilities, entry of new firms would be unprofitable although existing firms secure revenue above their long-run cost.

A similar analysis applies for bridge construction. A private firm would build and maintain a bridge under the same kind of conditions as would induce an entrepreneur to run an air travel service between cities. In either case the entrepreneur must expect enough revenue to defray total expenses. For example, the bridge could operate like a private club, a bridge club. Those who would benefit from the bridge could see to its construction. They could raise the necessary funds by private subscription and could refuse the use of the bridge to anyone failing to contribute. Since the contributions are voluntary, their size provides a measure of the worth of the bridge to the subscribers. The group can hire experts to advise them on the kind of bridge to be built, and they can solicit bids from bridge construction companies, awarding a contract to the lowest bidder. These technicalities raise no obstacle. The affair corresponds to the making of forward contracts between the owner-operator of the air travel service and his customers.

The main problem arises from the necessity of measuring the benefit of the bridge before it is built. This is why subscribers may wish to prevent nonsubscribers from using the bridge. Unless use is confined to subscribers, some persons who would benefit from having the bridge might not contribute and would rely on others' benevolence to pay for the bridge. Consequently subscribers may wish to restrict use of the bridge to themselves not so much out of a sense of fairness but because in this way they can secure a more accurate measure of how much the group values the bridge to see whether it is worth its cost. There are objections to this procedure arising from its neglecting the cost of preventing nonsubscribers from using the bridge. If this cost is very great, then it may be better for the subscribers to allow anyone the use of the bridge while recognizing that this results in fewer bridges being built than in a world without these costs. A more vivid example of this problem is fire protection. Let a group of households establish their own fire department, financed by their voluntary subscriptions and prevent nonsubscribers from obtaining the services of the fire department. A fire may start in a nonsubscriber's house next to a subscriber. If so, the subscriber may prefer to have the fire department serve his neighbor rather than risk the loss of his own house. These considerations explain why there are communities of like-minded persons requiring all of their members to pay for certain common services as a condition of belonging to the community. Such communities are like local government.

Even if nonsubscribers can be economically excluded from the use of the bridge, there remains the problem of how the fixed cost should be shared among the subscribers. For example, in the case of the bridge, should the more frequent users be compelled to contribute more than the less frequent users? Thus local inhabitants may use the bridge more than distant inhabitants. Should local residents pay more? In practice residents and tourists are often treated differently. How does ability to pay apply? Should the rich contribute more than the poor to the bridge? Similar problems arise in the air travel service example. The effective demand per trip depends on the total passenger demand, but the theory does not determine how much an individual passenger does or ought to pay. Any method of sharing the cost of the trip among the passengers is consistent with the theory if the total revenue is acceptable to the owner-operator. The upper bound on the total revenue of the air travel service depends on the existence of competition among actual or potential operators.

This analysis explains why direct users of services are often the ones who pay for it despite benefits to third parties. It is cheaper to collect fees from the direct users than from third parties. Moreover, it is often easier for direct users to determine the size of their benefit, and, sometimes, they do recoup some of their cost from indirect beneficiaries.

Another way to provide for bridge construction and maintenance is to turn it over to private enterprise. This would be formally equivalent to the operation of the air travel service by entrepreneurs when there are no explicit forward contracts. The expectation of a positive net return as determined by the constraints of the *ex ante* core may give the necessary incentive. However, in this case some form of price regulation may be required to use the resources efficiently and advance the public interest. For example, a two-part toll may be devised such that the first part is independent of the frequency of use and in the aggregate covers the fixed cost while the other part depends on use and covers the running expense. The first part corresponds to the contributions by subscribers. Therefore, it raises the question of whether every subscriber should pay the same fee and, if not, how the users should share the total cost.

A more detailed formal analysis of these problems with core theory may yield interesting new results. One of the important questions is to determine when the core representing this situation would not be empty. Such an analysis would have to incorporate more of the relevant constraints governing the production and consumption of public goods. Since this lies outside the scope of our work, we shall not pursue the matter here. We shall study one important aspect of this problem, the emptiness of the core when there is increasing returns to scale as in the bridge or airplane example. It is shown in section II.3 that with decreasing average cost and a capacity constraint the core can be empty so that no competitive equilibrium exists. Therefore, an institution such as a competitive market cannot resolve the conflicting interest of the parties and some other kind of social institution would be necessary.

9. A Brief Historical Note

Surprisingly, a formal analysis of market exchange does not appear in economic literature until 1881 when Edgeworth published his *Mathematical Psychics*. Böhm-Bawerk's *Positive Theory of Capital* (first German edition, 1889) analyzes a market where each trader exchanges at most one unit of a good subject to a constraint on his limit price. His theory implies a range of equilibrium prices determined by his celebrated marginal pairs and clearly states the nature of the equilibrium determined by the intersection of supply and demand schedules.

There was no important advance over Böhm-Bawerk's analysis until 1944 when von Neumann and Morgenstern published *The Theory of Games and Economic Behavior* which devotes the final two chapters to a game theoretic analysis of Böhm-Bawerk's problem. This analysis studies exchange in terms of the core constraints without naming them explicitly and derives wider price limits than Böhm-Bawerk. By assuming group rationality, von Neumann and Morgenstern show that their theory and Böhm-Bawerk's imply that the same quantity will be exchanged. Despite its power and originality, this work had little impact on economics for fifteen years. Revival of interest had to await appreciation of Edgeworth's remarkable work of eighty years earlier.

Edgeworth studies exchange in terms of contracting and recontracting, which corresponds to coalition formation in game theory. Each of Edgeworth's traders seeks the best possible terms of trade. Assuming that the traders' preferences are described by a continuous, ordinal utility function, Edgeworth gives limits for the terms of trade. These limits shrink to the competitive price as the number of traders increases.

Shubik (1959a) showed the relevance of game theory for Edgeworth's model. He gave rigorous proofs for some of Edgeworth's results using the concept of the core explicitly. This analysis was extended by Debreu and Scarf (1963) and by Aumann (1964 and 1966).

The conclusions of Debreu and Scarf are restrictive in two respects. First, the indifference curves of the traders are assumed to be convex so that given the choice between two bundles of goods with equal utility, the consumer would prefer the arithmetic mean of the two. Although convexity is a traditional assumption of the neoclassical analysis, it does not always yield implications in conformity with reality. Second, they assume a finite number of trader types such that the number of traders of each type increases proportionately. This rules out dominance by one type as the number of traders increases. Aumann uses a mathematical device which permits him to treat an infinite number of traders at the outset and allows his traders to have indifference curves of almost any shape. Since Aumann assumes a continuum of traders, so that there are as many traders as there are real numbers and assumes a finite number of goods with a finite amount of each good, almost all of his traders have virtually no wealth. Nevertheless, these assumptions imply the existence of a competitive equilibrium coinciding with the core.

One may interpret Aumann's results to mean that individual preferences are irrelevant for the existence of a competitive equilibrium if those with peculiar tastes are not too numerous. Thus the market supply and demand schedules do not necessarily reflect any individual's supply or demand schedules. It is interesting to note that Michael Farrell stressed this point in his important 1959 article. Using familiar geometrical diagrams he shows how there can be a competitive market equilibrium although the traders have nonconvex indifference curves thereby demonstrating that convexity is not necessary for a competitive equilibrium. Rothenberg (1960) provides a detailed analysis of some of these issues and see also Farrell (1961).

My model partakes of some of the properties of Farrell's and Aumann's work, but it does not assume continuous individual utility functions and divisible goods. It requires trade to be conducted with money valued in itself. The uniqueness of an equilibrium price is a consequence of continuous market supply and demand schedules and not of continuous individual preferences. This approach emphasizes that a market equilibrium cannot be determined in general by imagining trade between representative buyers and sellers. Moreover, an infinite number of traders is unnecessary for a unique competitive equilibrium.

One of the chief merits of core theory is its capacity to solve new problems involving increasing returns, indivisibilities, public goods, and many others. Conventional analyses conducted with supply and demand schedules are not always reliable guides for the study of a competitive equilibrium nor do they reveal how competition or markets work. The set of all possible contracts resulting from the traders' pursuit of their interest is fundamental to a competitive market. The collision of self-interests forces the imputations into the core, and thus the core does represent competition.

One of the important results in this chapter refers to multiunit trade. If traders have negatively sloped excess demand schedules, then there exists an equivalence between a multiunit trade market and a certain single-unit trade market such that prescribed coalitions of single-unit traders in the latter correspond to multiunit traders in the original market. As a result, we can study several important properties of the core in a multiunit trade market by dealing with the more elementary single-unit trade market to which it corresponds. The salient aspect of single-unit trade is the one-to-one relation between units of the good and traders. Consequently, we can show that the multiunit trade market has a nonempty core and the existence of a common unit price, which need not be unique, capable of satisfying all of the core constraints. Nevertheless, in multiunit trade though all the traders have negatively sloped excess demand schedules and the imputations are in the core, it does not follow that there must be group rationality and a common price per unit although these properties characterize the core for the corresponding single-unit trade market.

The one-to-one correspondence between traders and goods arising from convex preferences (negatively sloped excess demand functions) is only sufficient but is by no means necessary for a nonempty core. In addition convexity has an implication that may perhaps not pertain to other markets with a nonempty core, namely, a common unit price capable of satisfying all of the core constraints.

Shapley and Shubik (1969) in a highly stimulating article apply core theory to the study of externalities. One of their examples is of considerable interest not only in revealing an aspect of externalities but also in understanding market exchange in economic theory. Following Edgeworth, we postulate a market with n individuals each of whom has an initial bundle of goods denoted by σ^i, $i = 1, \ldots, n$, where each σ^i is an m-vector.

$$(1) \qquad U_i(x^i) = u$$

expresses the utility derived by person i from a bundle of goods x^i. Observe that i's utility depends only on his own bundle and not on the goods held by others; direct externalities exist if instead of (1) the utility of i would be expressed by $U_i(x^1, ..., x^n) = u$. A utility maximizing person faced with a given m-vector of prices p chooses a preferred bundle of goods x^{*i} such that for all x^i satisfying the budget constraint

$$(2) \qquad\qquad (p, x^i) = (p, \sigma^i)$$

$[(p, x^i)$ denotes the scalar product $\Sigma_j p_j x^i_j]$, we have

$$(3) \qquad\qquad U_i(x^i) \leqslant U_i(x^{*i}),$$

and for any bundle x^i such that

$$(4) \qquad\qquad U_i(x^i) > U_i(x^{*i}),$$

$$(5) \qquad\qquad (p, x^i) > (p, x^{*i}).$$

Hence no one can afford a bundle better than x^{*i}. Let us call a set of positive prices capable of satisfying (2) to (5) and the condition as follows:

$$(6) \qquad\qquad \sum_{i=1}^{n} (\sigma^i - x^{*i}) = 0,$$

a classical competitive price equilibrium. Shapley has proved the following:

SHAPLEY'S THEOREM. *A set of preferred allocations of goods $\{x^{*i}/i = 1, ... n\}$ corresponding to a classical competitive price equilibrium must be in the core.*[11]

Proof. We must show that no coalition can block x^{*i}. Suppose on the contrary that a coalition S can block. Then there would be an allocation of goods x^i for i in S such that

$$(7) \qquad\qquad \sum_{i \text{ in } S} (\sigma^i - x^i) = 0$$

with

$$U_i(x^i) \geqslant U_i(x^{*i})$$

and strict inequality for at least one i in S. Therefore, $(p, x^i) \geqslant (p, \sigma^i)$ and there is strict inequality for at least one i so that

$$\sum_i (p, x^i) > \sum_i (p, \sigma^i),$$

---

11. Shapley's theorem is stated with a proof due to Shapley, in fact the one followed in the text, in Scarf (1962, p. 130). See also Debreu and Scarf (1963, thm. 1).

It should be added that the utility functions are assumed to have the conventional properties attributed to them by the neoclassical analysis—insatiability, concavity, and continuity.

which would give a contradiction of (7) since the prices in a classical competitive price equilibrium are positive. QED.

This important result means that in the absence of externalities a classical competitive price equilibrium generates a set of allocations in the core so that a nonempty core is *necessary* for a classical competitive price equilibrium. Observe that in the proof of Shapley's theorem a coalition attempts to block an allocation by using only its own resources, which is the meaning of condition (7).

Now let us consider an example of an externality due to Shapley and Shubik (1969) where the core is empty, and yet there is a classical competitive price equilibrium. In the example there are *n* people who each have one bag of garbage. Assume they all have the same utility function as follows:

$$(8) \qquad U_i(x^i) = -x^i,$$

where now x^i denotes the number of bags of garbage person i has in a proposed allocation. Define the characteristic function to be $v(I) = -n$ and

$$(9) \qquad v(S) = -(n - |S|), \qquad\qquad \text{if } |S| < n,$$

where $|S|$ denotes the number of members of the coalition S. Thus the best that a coalition can do under the most adverse conditions is to receive the bags of garbage from all of the nonmembers of S. It is easy to verify that the core is empty in this case by considering the core constraints for all coalitions composed of $n-1$ members. Each of these coalitions can dump their garbage on the one nonmember, and it would be impossible to satisfy simultaneously all of the core constraints for these n possible coalitions. Nevertheless, it is also true that there is a positive price p such that if every person is required to pay a recipient of garbage p for each bag of garbage delivered, then everyone will keep his own garbage and thereby maximize his utility subject to a budget constraint. At first blush this example seems to contradict Shapley's theorem above, but the rub lies in the meaning of a blocking coalition in this case. It is no longer true that (7) holds. Instead a blocking coalition is only subject to the constraints

$$0 \leqslant \sum_{i \text{ in } S} x^i \leqslant n \quad \text{and} \quad \sum_{i \text{ in } I} x^i = n,$$

although $\sum_S \sigma^i = |S|$. That is, we can no longer calculate the optimal allocation of a blocking coalition on the basis of its own resources giving rise to an externality. We shall encounter an extended discussion of related phenomena in section V.5, the sharing of the cartel return.[12]

---

12. The relevance of the Shapley-Shubik example for the classical notion of an equilibrium and its relation to the core emerged in the course of a conversation I had with David Starret.

One of the main results due to Shapley and Shubik (1969) is to show that with external economies the core of a market is nonempty if and only if the core is nonempty in a corresponding market in which each trader receives his "share" of the external economies. However, the situation with external diseconomies is entirely different as shown by the garbage example.

Appendix: Consumer Surplus

The controversial subject of consumer surplus deserves more attention than has been devoted to it in the text.[13] In this appendix we shall consider some of the sources of disagreement about this concept and some of the reasons for using it despite the need for caution in doing so.

We begin with a consumer's utility function and assume there are two goods, Q and W.

$$(1) \qquad\qquad u = \phi(q, w)$$

is the consumer's utility function. Let $(0, w_0)$ represent the consumer's initial position which corresponds to the point R in figure 1.10. The equation of the indifference curve passing through the points R and N satisfies

$$(2) \qquad\qquad \phi(q, w) = \phi(0, w_0) = u_0.$$

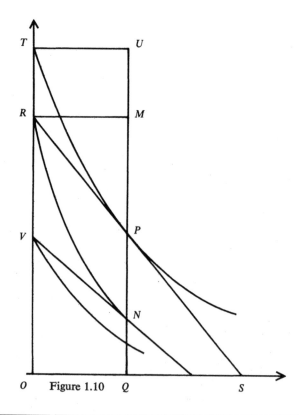

Figure 1.10

13. For good discussions of consumer surplus see Hicks (1946, pp. 38–41, and 1956, chaps. 10 and 11), and Friedman's essay on the Marshallian demand curve (1953).

We can derive from the indifference curve the maximum price that a consumer is willing to pay for an increment of Q at a point (q, w) that satisfies (1). This is given by

(3)
$$\frac{\Delta w}{\Delta q} = - \left(\frac{\partial \phi}{\partial q}\right) \bigg/ \left(\frac{\partial \phi}{\partial w}\right) \bigg|_{u_0}.$$

If we solve (2) for w as a function of q and u_0, we can write

(4)
$$w = F(q, u_0)$$

so that

(5)
$$\frac{\partial F}{\partial q}\bigg|_{u_0} = \lim_{\Delta q \to 0} \frac{F(q + \Delta q, u_0) - F(q, u_0)}{\Delta q}$$

gives the demand function for q holding u constant. Thus the slope of the indifference curve at a given point (q, w) can be interpreted as the highest price a consumer would be willing to pay for a small increment of the good in terms of the numeraire W. It follows that

(6)
$$\int_0^q \frac{\partial F}{\partial r} dr = F(q, u_0) - F(0, u_0),$$

which corresponds to the length of MN, gives the largest amount the consumer would be willing to pay for OQ of the good rather than do without. This is equivalent to the well-known result that the area under the demand curve given by $OCEQ$ in figure 1.11 measures the same magnitude as the length of MN. The demand schedule CE maintains real income constant in the sense that $u = u_0$ for all points on CE.

Much of the controversy about consumer surplus raises the same issues as in the debate about the most useful concept of a demand schedule. (See Friedman's essay on the Marshallian demand curve, 1953.) Let us study the response of a consumer whose initial position is the point R in figure 1.10 and who has trading opportunities represented by all of the points inside or on the triangle $ORPS$. The slope of RS gives the constant price per unit facing the consumer. The consumer will prefer the point P because it yields the most utility consistent with the constraints. He pays MP for OQ. The consumer's surplus with respect to the initial point R is PN, but this does not necessarily maximize consumer surplus. The maximum consumer surplus with respect to the initial point R is that quantity of the good which gives the largest value to the vertical distance from RS to the indifference curve RN. This occurs at the point where the slope of the indifference curve equals the slope of RS, which is not necessarily at the point P. Hence for the initial point R the maximum consumer surplus can require a different quantity than the one which maximizes utility and satisfies the trading constraints.

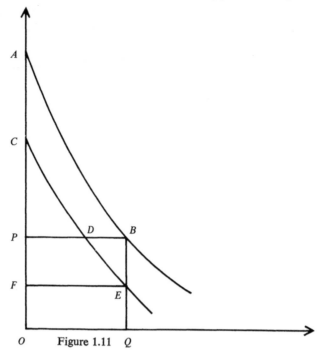

Figure 1.11

The difficulty is due to the fact that by permitting the consumer trading opportunities along *RS* it is as if the consumer starts from *T* instead of from *R*. Hence his real wealth is higher by having the freedom to determine how much to buy at a constant price. With respect to the indifference curve *TP* there is a sense in which *P* maximizes both utility and consumer surplus. Thus there is an appropriate indifference curve having this property just as there is an appropriate demand curve for predicting the consequences of shifts in supply schedules. We should use the demand schedule derived from the slope of the indifference curve passing through the point *TP* if we wish to preserve an equivalence between maximizing consumer surplus and maximizing utility subject to the trading constraints. The postulated trading opportunity allows the consumer to behave as if his initial point is at *T* instead of at *R* and there is a two-part price of which the first part is a lump sum amount *RT* independent of the quantity and the second part is a constant price per unit equal to the slope of *RS*. Let the consumer maximize his surplus subject to these trading conditions. He will then choose the point *P* because it yields the maximum consumer surplus of zero while all other attainable points would yield a negative consumer surplus. In fact the consumer does not pay the lump sum amount *RT* although it does measure the consumer surplus for the postulated trading opportunities and it is also the maximum consumer surplus. *RT* need not equal *NP*. This is acceptable because the two different measures of consumer surplus refer to two different kinds of trading opportunities. The surplus measured by *NP* assumes the consumer has an all-or-none choice of the quantity *OQ*. The surplus measured by *RT* = *UM* assumes that the consumer can trade along *RS*. Therefore, the measure of consumer surplus should refer to the appropriate circumstances. There is no all-purpose measure.

In figure 1.11 the area *APB* equals the consumer surplus *RT*. The consumer surplus for a comparable trading set with respect to the initial point *R* in figure 1.10 requires a lower price *QE* for which the consumer surplus would be the area *CEF*. For this to be true the consumer must be able to buy any quantity at the constant price *QE*. Hence he must be able to trade along the line *VN* in figure 1.10 where the slope of *VN* equals the price *QE* in figure 1.11. Just as *T* is the virtual initial point corresponding to *R* so *R* is the virtual initial point corresponding to *V*. The demand schedule *AB* in figure 1.11 is derived from the slope of the indifference curve *TP* in figure 1.10, and the demand schedule *CE* in figure 1.11 corresponds to the slope of the indifference curve *RN* in figure 1.10.

The situation confronting an individual must be consistent with equilibrium for all individuals. Among the advantages of the concept of surplus is that we can derive the market equilibrium as the solution of a maximum problem. This is most easily illustrated for a case in which there are two kinds of traders. Let

$$F(q) = \int_0^q f(r)\, dr,$$

(7)

$$G(s) = \int_0^s g(r) dr$$

define the surplus for each group of traders assuming an all-or-none choice of *q* for the first and an all-or-none choice of *s* for the second. Let the total available stock be *K* so that

(8) $$s + q = K \qquad (s, q \geqslant 0).$$

The problem is to find *s* and *q* that maximizes *F(q) + G(s)* subject to the constraint (8). Form the Lagrangian function as follows:

(9) $$L(q, s; \alpha, \beta, \lambda) = F(q) + G(s) + \lambda(K - s - q) + \alpha s + \beta q,$$

where λ, α, and β are the Lagrangian multipliers with $\alpha, \beta \geqslant 0$. If *f* and *g* are continuous so that *F(q)* and *G(s)* are continuously differentiable then the Kuhn-Tucker theorem for the necessary conditions of a maximum subject to inequality constraints applies with the results as follows:

(10) $$f(q) - \lambda + \beta = 0, \quad q\beta = 0, q[f(q) - \lambda + \beta] = 0,$$
$$g(s) - \lambda + \alpha = 0, \quad s\alpha = 0, s[g(s) - \lambda + \alpha] = 0.$$

There are only two possible cases. In the first the entire stock is given to one group of traders, and in the second it is divided between the two groups.

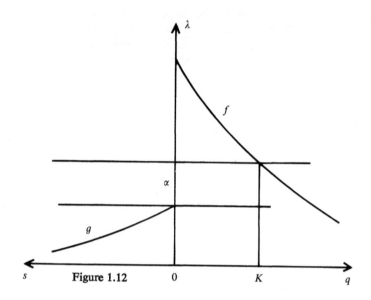

Figure 1.12

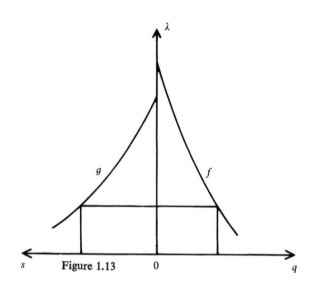

Figure 1.13

Figure 1.12 illustrates the first and figure 1.13 the second. Thus suppose for the first case that the optimal solution requires that $q = K$ and $s = 0$. Then $\beta = 0$, $f(K) = \lambda$ and $g(0) + \alpha \leqslant \lambda$. In the second case $f(q) = g(s) = \lambda$, and $\alpha = \beta = 0$. Suppose that the demand functions can be inverted and written as follows: $q = f^{-1}(\lambda)$ and $s = g^{-1}(\lambda)$.

(11)
$$K = f^{-1}(\lambda) + g^{-1}(\lambda)$$

gives the equilibrium conditions in a familiar form. The left-hand side is the total available supply while the right-hand side represents the aggregate demand. Hence the solution of the maximum problem gives the classical result. The definition of the characteristic function and the core constraints extend these ideas to a larger market.

II

Further Applications of Core Theory to Market Exchange

1. Introduction

The set of imputations in the core satisfy a number of inequalities of the type

$$(1) \qquad \sum_{i \text{ in } S} x_i \geqslant v(S),$$

where x_i is the imputation of individual i, and $v(S)$ denotes the value of the characteristic function for the coalition S. The term $v(S)$ gives the largest amount that the members of S can guarantee themselves under the most adverse conditions. Therefore, $v(S)$ gives the largest amount that the members of S can obtain by their own resources. If the individuals may join any coalition, then their imputations must satisfy the system of inequalities given by (1). In addition it must be true that

$$(2) \qquad \sum_{I} x_i \leqslant v(I),$$

where I is the set of all individuals. It is natural, therefore, to study the linear programming problem

$$(3) \qquad \min \ \sum x_i \ \text{subject to (1),} \qquad \text{for all} \quad S \subset I \quad \text{and} \quad S \neq I.$$

The solution of this problem gives the smallest amount that the individuals can obtain while satisfying all of the core constraints given by (1). Hence for the core to be nonempty it is necessary that

$$(4) \qquad \min \ \sum x_i \leqslant v(I).$$

In addition to this primal linear programming problem there is a dual problem of great importance in studying market exchange especially for multiunit trade. The analysis of single-unit trade given in the preceding chapter could be thorough using simple mathematical tools. However, more powerful tools are required to study multiunit trade adequately. To this end

68

1. *Introduction*

we shall develop the analysis of the dual linear programming problem, the source of the important concept known as a balanced game, which always has a nonempty core.

Several problems in multiunit trade do not arise in single-unit trade markets. First, given the assumptions about the nature of the traders' preferences, under what conditions will there exist a feasible set of imputations to satisfy the core constraints? That is, when will there be a nonempty core? To answer this question, it is not only necessary to study the conditions leading to feasible solutions of the core inequalities (1), but it is also necessary to determine whether there is a set of feasible trades consistent with these inequalities. That is, it must be possible to implement these inequalities with a set of allocations of the goods such that the sum of the stocks in the possession of all traders is equal to the initial stock. A second important question concerns group rationality. Assuming that the market has a nonempty core and that there are feasible trades to implement the core, when will there be group rationality? Third, under what conditions will there be a common price per unit as in the classical concept of a competitive equilibrium assuming a nonempty core. Fourth, how do changes in the number of traders affect the emptiness of the core? More precisely, suppose that the core of a given market is empty. If the number of traders of every type in this market increases so as to maintain a constant proportional relation, will it be true that the sequence of markets thereby generated all have empty cores? In the last section of this chapter it is shown that if there are n types of traders in the market defined in terms of their preferences and initial stocks, and if there are m traders of each type, then, in general, for a countable sequence of markets, the cores will be nonempty.

There are several other questions related to the effect of numbers on the equilibrium. Will the entry of more traders in a market have a smaller effect, the greater the number of traders already present? Finally, does a trader always receive the value of his marginal contribution to the market? The last question deserves some explanation since it is not often discussed in economic theory. Let $v(I_n)$ denote the value of the characteristic function of the coalition of all traders in an n-person market. Let another trader enter the market so that the characteristic function for the coalition of all traders becomes $v(I_{n+1})$. The marginal contribution of the $(n+1)$th trader is defined as follows: $v(I_{n+1}) - v(I_n)$. Since x_{n+1} is this trader's imputation, the question is one of determining when $x_{n+1} = v(I_{n+1}) - v(I_n)$, if the core is not empty and if the imputations are in the core.

The next section develops the relations between core theory and linear programming. It explains the concept of a balanced collection of coalitions and shows its relation to the duality theorem of linear programming. In this section these concepts are applied to the study of single-unit trade. Section 3 studies the necessary conditions for a nonempty core. Section 4 considers the problem of feasible trades, which is analogous to determining the conditions under which the intersection of supply and demand schedules gives the competitive equilibrium. Section 5 continues the analysis of multi-unit trade assuming the traders have convex preferences so that the core is not empty. Section 6 studies how the number and diversity of traders affects the equilibrium. Finally, in section 7 we resume the study of the emptiness of the core and show how this is related to the number of traders.

In concluding this introduction we shall agree that the relation $S \subset I$ for all coalitions S means *proper* inclusion so that $S \neq I$, where, as always, I denotes the set of all traders. Sometimes, for emphasis we shall say $S \subset I$ and $S \neq I$.

2. Balanced Collections of Coalitions

In applying linear programming methods to the theory of the core it is convenient to begin with a general n-person game not necessarily representing a market. Let x_i denote the imputation of player i and let $v(S)$ denote the characteristic function of the coalition S. The set of imputations $\{x_i\}$ is said to be in the core if for all coalitions $S \subset I$ and $S \neq I$

$$(1) \qquad \sum_S x_i \geq v(S),$$

and

$$(2) \qquad \sum_I x_i \leq v(I),$$

where I denotes the coalition of all n players.

Now consider the primal linear programming problem as follows:

$$(3) \qquad \min_{x_i} \sum x_i \quad \text{subject to} \quad (1).$$

The core is not empty if and only if the solution of the primal problem satisfies

$$(4) \qquad \min \sum x_i \leq v(I).$$

According to this definition of the core the sum of the imputations of all players need not attain the upper bound $v(I)$. When the upper bound is attained so that

$$(5) \qquad \sum_I x_i = v(I),$$

then there is said to be group rationality. In economic applications this is equivalent to Pareto optimality. Therefore, in this approach the core can be nonempty without there being group rationality. Moreover, not every game has a nonempty core, as we shall see. Games with an empty core arise, roughly speaking, if $v(I)$ is too small to permit the imputations to satisfy the inequalities (1).

In addition to the *primal* linear programming problem there is the *dual* problem that leads to the concept of a balanced collection of coalitions. For every inequality in (1) define a nonnegative number δ_S, the dual variable, so that for every coalition S there is a δ_S. The dual problem is defined as follows:

(6) $$\max_{\delta_s} \sum \delta_S v(S) \quad \text{subject to} \quad \delta_S \geqslant 0$$

and

(7) $$\sum_{i \text{ in } S} \delta_S = 1, \quad \text{for all} \quad S \subset I \quad \text{with} \quad S \neq I.$$

The set of constraints (7) deserves some explanation. Every member i of the coalition S is assigned the same dual variable δ_S. Constraint (7) requires that the sum of the δ_S's assigned to a given player i summed over all of the coalitions to which he belongs must equal one. As we shall see below, this condition is important in ensuring the feasibility of the trades necessary to implement the core constraints. If the corresponding core constraint in (1) is an equality then $\delta_S \geqslant 0$, while if it is a strict inequality then $\delta_S = 0$. That is,

$$\sum_S \delta_S \sum_i [x_i - v(S)] = 0.$$

Hence in the latter case δ_S must equal zero, while in the former case it may equal zero.

The duality theorem of linear programming asserts that the solutions of the primal and dual problems give the same values to their respective objective functions. That is,

$$\max_{\delta_s} \Sigma \, \delta_S v(S) = \min_{x_i} \Sigma \, x_i.$$

Hence the core is not empty if and only if

(8) $$\min \Sigma \, x_i = \max \Sigma \, \delta_S v(S) \leqslant v(I).[1]$$

With this preparation we are ready for the concept of a balanced collection of coalitions.

DEFINITION. *Let T be a collection of coalitions* $\{S\}$. *T is called a balanced collection if there is a set of nonnegative numbers* $\{\delta_S\}$ *with a* δ_S *for every S in T such that (7) is satisfied for all i in S and for all S in T.*

---

1. For a statement and proof of the fundamental duality theorem of linear programming, see Gale (1960, thm. 3.1, p. 78).

DEFINITION. *Let a game with n players have a characteristic function $v(S)$ defined for all possible coalitions S. The game is called a balanced game if for every balanced collection*

$$\Sigma \; \delta_S v(S) \leqslant v(I).^2$$

Although it would now be desirable to provide an intuitive interpretation of these concepts of balance, it is better to defer this until we discuss the feasibility of trade in section 4. For the present we can acquire some understanding by using these formal concepts and especially by deriving some implications of balance. The first implication is immediate.

THEOREM 1. *A balanced n-person game with a characteristic function has a nonempty core.*

Proof. The definition of a balanced game together with the duality theorem of linear programming gives the desired conclusion. QED.

We should observe that in a balanced game if the combined imputations of the players is a minimum then some coalitions will in fact receive the smallest amount that they are willing to accept. If it also happens to be true that

$$\min \Sigma \; x_i = v(I),$$

so that there is necessarily group rationality, then the smallest acceptable amount is also the maximum attainable. This property will be seen to characterize the classical concept of a competitive equilibrium.

At this point it is helpful to consider some examples. The first one is a game with an empty core. Consider a three-person game for which the characteristic function is defined as follows:

$$v(1) = v(2) = v(3) = 0, \quad v(1, 2) = 2.01,$$

(9)

$$v(1, 3) = v(2, 3) = 2, \quad v(I) = 2.01 + c,$$

where c is an arbitrary nonnegative parameter. Whether or not this game has a nonempty core depends on the value of c. The inequalities that define

---

2. Scarf (1967) exploits the properties of a balanced collection of coalitions to prove that every balanced *n*-person market has a nonempty core assuming barter. There is neither money nor a characteristic function in his approach and traders value bundles of goods according to their ordinal utility function. Scarf goes on to show that convexity of trader preferences imply market balance and, therefore, a nonempty core. Telser (1971) proves that if in an *n*-person market every trader's ordinal utility function is a strictly increasing function of continuously divisible commodity bundles then the market has a nonempty core if and only if it is balanced. Shapley (1967) gives a formal analysis of balanced sets and contains valuable references to the literature. Shapley and Shubik (1968) give additional material on balanced games and markets and call money in the sense used herein "transferable utility."

2. *Balanced Collections of Coalitions*

the imputations in the core are as follows:

(10)
$$\begin{cases} x_1 & \geqslant 0, \\ \quad x_2 & \geqslant 0, \\ \qquad x_3 \geqslant 0, \\ x_1 + x_2 & \geqslant 2.01, \\ x_1 \quad + x_3 \geqslant 2, \\ \quad x_2 + x_3 \geqslant 2. \end{cases}$$

The primal problem is

(11) $$\min(x_1 + x_2 + x_3) \quad \text{subject to (10).}$$

In describing the dual problem it is sufficient to consider only the last three inequalities in (10). Corresponding to these three inequalities are the three dual variables, δ_1, δ_2, and δ_3. Thus, δ_1 corresponds to the coalition between players 1 and 2, δ_2 corresponds to the coalition between players 1 and 3, and δ_3 corresponds to the coalition between players 2 and 3. The dual problem is

(12) $$\max(2.01\delta_1 + 2\delta_2 + 2\delta_3),$$

subject to

(13)
$$\begin{cases} \delta_1, \ \delta_2, \ \delta_3 \geqslant 0, \\ \delta_1 + \delta_2 \qquad \leqslant 1, \\ \delta_1 \qquad + \delta_3 \leqslant 1, \\ \delta_2 + \delta_3 \leqslant 1. \end{cases}$$

For this simple problem one readily verifies that the solution of the primal problem is

$$x_1 = x_2 = 2.01/2, \quad x_3 = 2 - 2.01/2$$

and that the solution of the dual problem is

$$\delta_1 = \delta_2 = \delta_3 = 1/2.$$

Hence there is a nonempty core if and only if

(14) $$c \geqslant 2 - 2.01/2.$$

If c is too small, then the minimum value of the objective function would exceed the largest amount that is attainable by the coalition of all players which is impossible. In other words, if c is too small, the core is empty.

The second example is closer to our area of interest. Consider the multiple-trade example given in section I.7. In this example there is one owner A who can sell up to two units, and there are two potential customers B_1 and B_2 who each desire at most one unit. A values his two units at a_1 and a_2 so that he would be willing to sell one unit at a price not less than a_2 and two units at a price not less than $a_1 + a_2$. The limit prices for B_1 and B_2 are b_1 and b_2.

Assume that $a_1 < a_2 < b_1 < b_2$. The inequalities that define the imputations of the core are as follows:

(15)
$$\begin{aligned}
x &\geqslant v(A) &&= a_1 + a_2, \\
x + y_1 &\geqslant v(A, B_1) &&= a_2 + b_1, \\
x + y_2 &\geqslant v(A, B_2) &&= a_2 + b_2, \\
y_1, y_2 &\geqslant 0. &&
\end{aligned}$$

In addition the imputations must satisfy the constraint given by

(16)
$$x + y_1 + y_2 \leqslant v(I) = b_1 + b_2.$$

Assume that $a_1 > 0$ ensuring the existence of feasible solutions for the primal problem which is min $(x + y_1 + y_2)$ subject to (15). Hence there is an optimal solution.

The dual problem is quite simple. It is

$$\max \left[\delta_1(a_1 + a_2) + \delta_2(a_2 + b_1) + \delta_3(a_2 + b_2) \right]$$

subject to

(17) $\delta_1 + \delta_2 + \delta_3 = 1,$ and $\delta_j \geqslant 0.$

In this case the δ_j's refer, respectively, to the first three inequalities of (15). In fact the constraint (17) refers only to those coalitions to which A belongs because this gives the most binding constraint on the δ's. The dual problem is easily solved, and we find that $\delta_1 = \delta_2 = 0$, $\delta_3 = 1$ gives the maximum value of the objective function, which is $a_2 + b_2$. Corresponding to these values of the dual variables, we have two equations to solve for the imputations of the primal problem. These are

(18)
$$\begin{aligned}
x + y_1 + y_2 &= a_2 + b_2 \\
x \qquad + y_2 &= a_2 + b_2.
\end{aligned}$$

Hence, $y_1 = 0$, which means that B_1 fails to obtain any of the good in this solution. Moreover,

$$0 \leqslant y_2 = a_2 + b_2 - x_1 \leqslant a_2 + b_2 - (a_2 + b_1) = b_2 - b_1$$

because $\delta_2 = 0$ implies that

$$x + y_1 = x \geqslant a_2 + b_1.$$

This solution of the linear programming problem does not satisfy group rationality. The following set of equations yield a set of imputations in the core that attains group rationality.

(19)
$$\begin{aligned}
x \qquad\quad &= 2a_2, \\
x + y_1 &= a_2 + b_1, \\
x + y_2 &= a_2 + b_2.
\end{aligned}$$

The solution of this system implies that

$$x + y_1 + y_2 = v(I) = b_1 + b_2.$$

This results in an exchange of goods from A to the B's, which maximizes the aggregate valuation of the goods and thereby attains the best possible final allocation.

In the dual problem $y_2 \leqslant b_2 - b_1$, while the solution with group rationality gives $y_2 = b_2 - a_2 > b_2 - b_1$. Since y_2 is B_2's "consumer surplus", it follows that the dual linear programming problem gives both B's a smaller consumer surplus and the seller a larger imputation than is the case with group rationality.

Let us continue the analysis of this example by changing the supply conditions. Instead of having one seller with two units to offer, let us introduce two sellers, A_1 and A_2, who each have one unit to offer at the limit prices a_1 and a_2. In this case we shall see that the linear programming solution does satisfy group rationality and that the collection of two trader coalitions given by $\{(A_i, B_j)\}$ is balanced. The core constraints for the collection of coalitions $\{(A_i, B_j)\}$ are as follows:

$$x_1 + y_1 \geqslant b_1, \quad x_2 + y_1 \geqslant b_1,$$
$$x_1 + y_2 \geqslant b_2, \quad x_2 + y_2 \geqslant b_2.$$

Denote the dual variables by δ_{ij} so that δ_{ij} refers to the coalition (A_i, B_j). The dual problem is

$$\max [\delta_{11} b_1 + \delta_{12} b_2 + \delta_{21} b_1 + \delta_{22} b_2]$$

subject to

$$\sum_i \delta_{ij} = 1, \quad \text{and} \quad \sum_j \delta_{ij} = 1.$$

In this case the set of δ's that solves the dual problem is not unique. One solution is given by

(20) $$\delta_{ij} = 1/2 \qquad (i, j = 1, 2),$$

and another is given by

(21) $$\delta_{12} = \delta_{21} = 1, \quad \delta_{11} = \delta_{22} = 0.$$

For both sets of δ's the objective function attains the same maximum, namely, $b_1 + b_2$. However, the first solution implies that there must be equality for all four core constraints because all of the δ's are positive. This in turn gives the result that $x_1 = x_2$, which means that there must be a common price. The second solution, (21), yields only the result that there must be equality in the two core constraints as follows:

$$x_1 + y_2 = b_2, \quad x_2 + y_1 = b_1,$$

and from these one cannot conclude that there necessarily is a common price.

Finally, the example illustrates the important point that

$$\max \Sigma \; \delta_{ij} v(A_i, B_j) = \min \Sigma \; (x_i + y_j) = v(I).$$

Hence, for this balanced collection, there must be group rationality. Though the conclusion is not surprising, since the example is actually one of single-unit trade, for which the result is already known from the material in chapter I, it is gratifying to have another tool of analysis that can reach this conclusion by a different route. Moreover, this example illustrates the assertion of theorem I.6, which relates the core for multiunit trade to the corresponding maximally competitive core for single-unit trade. Before applying linear programming methods to multiunit trade, let us do so for single-unit trade with a new tool—balanced collections of coalitions.

Without loss of generality we may assume an equal number of potential buyers and sellers.[3] Denote the latter by A_i and the former by B_j with i, $j = 1, ..., m$. Let the A's and B's be indexed such that the sellers' reservation prices satisfy $a_i \leqslant a_{i+1}$, and the buyers' limit prices satisfy $b_j \leqslant b_{j+1}$. We shall see that the collection of pairs $\{(A_i, B_j)\}$, called the *primary pairs* in section I.5, is balanced, and that with single-unit trade the core imputations are group rational.

THEOREM 2. *Let* v_{ij} *denote the value of the characteristic function for the primary pairs* (A_i, B_j), *where*

$$(22) \qquad\qquad v_{ij} = max \; \{a_i, b_j\}.$$

If

$$(23) \qquad\qquad v(I) = \sum_1^k b_{m+1-i} + \sum_{k+1}^m a_i,$$

then

$$T = \{(A_i, B_j)/i, j = 1, ..., m\}$$

is a balanced collection and the core imputations are group rational.

Proof. To prove that T is a balanced collection, we must show that there is a set of nonnegative numbers $\{\delta_{ij}/i, j = 1, ..., m\}$ such that

$$(24) \qquad\qquad \sum_i \delta_{ij} = 1 \quad \text{and} \quad \sum_j \delta_{ij} = 1,$$

which yield the

$$\max \Sigma \; \delta_{ij} v_{ij} \leqslant v(I).$$

This is easily done because (23) implies that

$$v_{i, m+1-i} = \begin{cases} b_{m+1-i}, & \text{if } i = 1, ..., k, \\ \\ a_i, & \text{if } i = k+1, ..., m. \end{cases}$$

---

3. See section I.5.

Hence we may choose

$$\delta_{i,\,m+1-i} = \begin{cases} 1, & \text{if } i = 1, \ldots, m, \\ 0, & \text{otherwise.} \end{cases}$$

With this choice T is shown to be a balanced collection since the constraints are satisfied, and in addition

$$\sum \delta_{ij} v_{ij} = \sum_1^k b_{m+1-i} + \sum_{k+1}^m a_i = v(I),$$

which proves group rationality.

To complete the proof of the theorem one must show that the solution of the linear programming problem implies a set of imputations satisfying all of the core constraints. The duality theorem implies that for all pairwise coalitions

(25) $x_i + y_j \geqslant v_{ij}.$

In addition we must show that

(26) $\sum x_i + \sum y_j \geqslant v(S),$ for all $S \subset I, \quad S \neq I,$

given that in addition to (25) we have

(27) $x_i + y_{m+1-i} = v_{i,\,m+1-i}.$

To show that in fact (26) is satisfied, we may appeal to theorem I.2, which asserts that if the primary constraints are satisfied and feasible then (26) is also satisfied. QED.

As in the preceding example there is no unique set of δ's solving the dual linear programming problem. This is shown most easily with the aid of table 2.1. The table gives the values of v_{ij} for the intramarginal pairs of

TABLE 2.1
Values of the Characteristic Function for Single-Unit Trade and the
Primary Pairs (A_i, B_j)

	B_1	B_2	...	B_{m-k}	B_{m+1-k}	...	B_{m-1}	B_m
A_1					b_{m+1-k}	...	b_{m-1}	b_m
A_2					b_{m+1-k}	...	b_{m-1}	b_m
...								
A_k					b_{m+1-k}	...	b_{m-1}	b_m
A_{k+1}	a_{k+1}	a_{k+1}	...	a_{k+1}				
...								
A_m	a_m	a_m	...	a_m				

traders and for the extramarginal pairs. Recall that an intramarginal seller is an A_i such that $i = 1, \ldots, k$, and an intramarginal buyer is a B_j such that $j = m, m-1, \ldots, m+1-k$. All others are extramarginal. The number k denotes the number of units exchanged which is determined by the characteristic function for all traders, $v(I)$, shown in (23). Consequently,

$$v_{ij} \leq v_{i,\,j+1}, \quad \text{and} \quad v_{ij} \leq v_{i+1,\,j}.$$

Hence the largest numbers in the first k rows of table 2.1 appear in the columns that give the results for the intramarginal traders, and the largest v's in the remaining rows and columns appears for the coalitions among the extramarginal traders. We can find another set of δ's that maximizes the objective function by imposing equality in a subset of the basic primary coalitions given in section I.5 and repeated herein. A set of basic primary core constraints is as follows:

$$
\begin{array}{lll}
(28) & x_i + y_{m+1-i} \geq v_{i,\,m+1-i} & (i = 1, \ldots, m), \\
& x_i + y_{m-i} \ \ \geq v_{i,\,m-i} & (i = 1, \ldots, k), \\
& x_i + y_{m+2-i} \geq v_{i,\,m+2-i} & (i = 1, \ldots, k+1).
\end{array}
$$

A set of δ's that maximizes the objective function $\Sigma\, \delta_{ij}v_{ij}$ and also has the property of implying a common price per unit of the good is given by

$$
\begin{array}{lll}
(29) & \delta_{i,\,m-i} = 1/3 & (i = 1, \ldots, k-1), \\
& \delta_{i,\,m+1-i} = 1/3 & (i = 2, \ldots, k-1), \\
& \delta_{i,\,m+2-i} = 1/3 & (i = 2, \ldots, k). \\
& \delta_{1,\,m} = \delta_{k,\,m+1-k} = 2/3.
\end{array}
$$

The δ's appropriate for the extramarginal traders are

$$\delta_{i,\,m+1-i} = 1 \qquad\qquad (i = k+1, \ldots, m).$$

Finally, all other δ's are set equal to zero. It is readily verified that this choice of δ's satisfies the constraints given in (24) and maximizes the objective function. Two properties of this solution imply a common price per unit. First, for any strictly positive δ there must be equality in the corresponding core constraint. This implies that in a certain subset of the basic primary core constraints in (28) there must be equality. From these equations we may conclude that

$$
(30) \qquad\qquad\qquad x_i = x \qquad\qquad\qquad (i = 1, \ldots, k).
$$

Second, and more generally, a necessary condition for a common price is that the solution of the dual problem is not unique. Hence in the following material on multiunit trade it is important to determine when the solution of the dual problem is not unique.

3. Empty Cores

Multiunit trade raises the possibility of an empty core, which cannot happen with single-unit trade as shown by theorems I.2 and II.2. Moreover, even with multiunit trade the core is not empty if all traders have convex preferences, equivalent to negatively sloped excess demand functions, as is shown by theorem I.6. In general increasing returns are most conducive to an empty core, as we shall see. We begin with a detailed examination of an example of a market with an empty core.

Let each of two initial owners have one unit of a good and let each of two initial nonowners desire at most two units. The limit prices of the four traders are as follows:

$A_1^{(1)}$	$A_2^{(1)}$	B_1	B_2
2	6	$8+c$	6
6	2	6	$8+c$

The superscript for A_i denotes his initial holdings so that $A_1^{(1)}$ means that A_1 begins with one unit he is willing to sell for not less than 2, and he is willing to buy another unit for not more than 6. Had he two units, he would not accept less than 8 for both. Therefore, A_1 has a positively sloped excess demand function, or, equivalently, an increasing marginal rate of substitution between money and the given commodity, that is, concave preferences.[4]

---

4. We shall frequently use the terms concave and convex preferences. The reason for this usage can be seen from a diagram of an indifference map. Let the quantity of the good be plotted on the horizontal axis and the quantity of money on the vertical axis. If there is a decreasing marginal rate of substitution between the two, then the indifference curve is convex to the origin, and we say the preferences are convex. Increasing marginal rates of substitution correspond to concave preferences because the indifference curve would be concave to the origin. As an example of concave preferences consider a bibliophile who is indifferent between having none or all eight editions of Marshall's *Principles of Economics*. Also assume that this book collector would be willing to pay a higher price for one more edition the larger the number of editions he has already. This man has concave preferences.

There is also an analytic formulation familiar in economics drawing out the analogy with demand and marginal cost functions. Let

(i) $$F(q) = \int_0^q f(r)\, dr$$

define the maximum amount that a buyer would be willing to pay for a quantity q of the good rather than go without so that $f(q)$ corresponds to the buyer's demand function. If $f(q)$ is a continuous function of q, then

(ii) $$F'(q) = f(q),$$

and if $f(q)$ is a differentiable function of q, then

(iii) $$F''(q) = f'(q).$$

Therefore, $f'(q) < 0$, a negatively sloped demand, corresponds to convex preferences and $f'(q) > 0$, a positively sloped demand to concave preferences. One may also interpret $F(q)$ as the minimum acceptable amount to a seller offered the alternative of selling q or nothing. According to this interpretation, $F(q)$ represents the seller's total cost, and $f(q)$ the marginal cost. Decreasing average cost would correspond to concave preferences.

It is important to observe that for a given $f(q)$, $F(q)$ is determinate up to the choice of an arbitrary constant. See also the appendix, chapter I.

In terms of limit prices, convex preferences correspond to $b_{t+1} \leqslant b_t$, and concave preferences correspond to $b_{t+1} \geqslant b_t$.

Insatiability means that no finite bundle of goods is capable of yielding the consumer a maximum utility. Of course, a consumer can only demand a finite bundle regardless of the nature of his preferences because his total real wealth is finite.

A_2 also starts with one unit that he is willing to sell for at least 6, and he would be willing to buy one more unit for a price not above 2. Therefore, A_2 has a decreasing marginal rate of substitution between the good and money, convex preferences, or, equivalently, a negatively sloped excess demand.

The conditions of the example are summarized so far by the following values of the characteristic function:

(1) $$v(A_1) = 2, \quad \text{and} \quad v(A_2) = 6.$$

Let the two initial owners form a coalition so that they would jointly own two units. What is the least they would accept for the sale of both? At first the answer might seem to be 12, but this is false. The coalition can allocate its two units in only four different ways, and none of these achieves a valuation of 12. Thus A_1 cannot have a "marginal utility," so to speak, of 6 from the second unit without having the first. Therefore, the coalition of the two A's can obtain at most 8 by reallocation of its two units, and it would reject any offer below 8 for the sale of both. Consequently,

(2) $$v(A_1, A_2) = 8.$$

To compute $v(I)$, we must find an allocation of the two units maximizing the aggregate valuation derived from the possession of the good. Inspection of the table implies that

(3) $$v(I) = 14 + c, \qquad \qquad \text{if } c > -2,$$

which is attainable in several ways; either a B has both units, or A_2 and B_1 have one unit each.

Similarly,

(4)
$$v(A_1 A_2 B_1) = 14 + c, \quad v(A_1 B_1 B_2) = 8 + c,$$
$$v(A_1 A_2 B_2) = 14 + c, \quad v(A_2 B_1 B_2) = 8 + c.$$

The imputations are in the core only if they satisfy the following:

(5)
$$x_1 + x_2 + y_1 \geqslant 14 + c, \quad x_1 + y_1 + y_2 \geqslant 8 + c,$$
$$x_1 + x_2 + y_2 \geqslant 14 + c, \quad x_2 + y_1 + y_2 \geqslant 8 + c.$$

Summing these inequalities gives

$$3(x_1 + x_2 + y_1 + y_2) \geqslant 44 + 4c$$

so that

(6) $$x_1 + x_2 + y_1 + y_2 \geqslant (44 + 4c)/3 \; > \; 14 + c = v(I).$$

Therefore, the imputations are not feasible, and the core is empty. The term $v(I)$ is too small to satisfy the constraints resulting if traders can form all possible coalitions and the market lacks a competitive equilibrium.

One may speculate about the reasons for an empty core in this example. The fact that no unique allocation of the goods can achieve $v(I)$ does not explain the emptiness of the core as is readily shown by changing the 6 in row 1 column 2 and the 6 in row 2 column 3 both to 5.9. With these changes $v(I)$ is attainable in only one way by giving both units to B_2, but still the core is empty. Of course with a different initial allocation of the two units the core would not be empty, but this is a trivial fact. For instance, there are always initial allocations coinciding with those necessary to achieve $v(I)$, and in these cases not only is the core nonempty but also the market is inessential so that there would be no trade.

One can deny the possibility of an empty core by asserting the impossibility of nonnegatively sloped excess demand schedules, but this is unacceptable. For example, one may be indifferent between furnishing a room with antique or with modern furniture. Convex preferences, a negatively sloped excess demand, would imply that one would prefer a room furnished with a mixture of antiques and moderns while concave preferences imply that either extreme is better than a mixture. Hence concave preferences are not irrational and cannot be ruled out a priori. Moreover, in the example with at most two units per trader there could not be an empty core without the presence of some traders with concave preferences. However, there are some simple changes in the example that would preserve concave preferences and would imply a nonempty core.

Concave preferences for traders can be translated into increasing returns for producers. If the A's represent producers, then the a's represent the increments in the cost of producing additional units. Therefore, concave preferences become isomorphic to decreasing average cost, which one would not wish to exclude a priori. In fact, an empty core on this interpretation would mean that with increasing returns and a negatively sloped demand a competitive equilibrium does not exist. Therefore, it behooves us to accept the fact that equilibrium in the sense of the core does not always exist. Since the classical version of a competitive equilibrium is always included in the core, when the core is empty there is no classical competitive equilibrium either.

Returning to the problem of market exchange without production, we have seen that if all traders have convex preferences then the market has a nonempty core, but this gives a sufficient and not a necessary condition. The assumption of a nonempty core imposes no severe restrictions on the nature of trader preferences so that one may easily construct a wide variety of preferences for which the market has a nonempty core. Unfortunately, there are few general results in terms of the nature and diversity of trader preferences implied by the hypothesis of a market having a nonempty core. It is interesting, therefore, to consider some cases in which simple hypotheses about trader preferences imply a nonempty core.

In the first case all the traders have concave preferences. In terms of the initial owner reservation prices this means

(7)
$$a_{m+1} \geqslant (1/m) \sum_1^m a_s,$$

or, equivalently,

(8)
$$a_{s+1} \geqslant a_s,$$

with similar conditions for the b's. Moreover, one cannot guarantee a non-empty core if some traders are unwilling to absorb the whole stock. Thus consider

$A_1^{(2)}$	$A_2^{(1)}$	B_1	B_2
1	1	$4+c$	4
3	3	$5+c$	$5+c$
$9+2c$	$9+2c$	0	0

In this market

$$v(A_1 A_2 B_1) = 13+2c, \quad v(A_1 B_1 B_2) = 9+2c,$$
$$v(A_1 A_2 B_2) = 13+2c, \quad v(A_2 B_1 B_2) = 4+c.$$

Since $v(I)$ equals $13+2c$, it readily follows that the core is empty if $c > 0$. To ensure a nonempty core for a market where all of the traders have concave preferences, some traders must have a large enough demand to be willing to absorb the whole stock. Let us agree to describe such preferences as insatiable. If all traders have insatiable concave preferences, then there is no advantage in dividing the initial stock among several traders.

LEMMA 1. *If all traders have insatiable concave preferences and the a's and b's satisfy condition (8) for every trader, then the market has a nonempty core.*

Proof. Without loss of generality we may assume an equal number of initial owners and nonowners such that in the aggregate the owners have m units of the good. Let $a_{.i} = \Sigma_s a_{si}$ and $b_{.j} = \Sigma_t b_{tj}$. Hence,

$$v(I) = \max \{a_{.i}, b_{.j}\}.$$

Recall that a decomposing collection T is a collection of sets S_k such that $I = \cup_k S_k$ with $S_k \cap S_{k'} = \emptyset$ if $k \neq k'$, and

$$v(I) = \sum_T v(S_k)$$

(see sec. I.7). In this case there are only two possible decomposing collections as follows:

(i) $S_1 = (\cup_1^m A_i, B_1),$ $S_{j+1} = B_j$ $(j = 2, ..., m),$

(ii) $S_1 = \cup A_i,$ $S_{j+1} = B_j$ $(j = 1, ..., m).$

In both cases it is readily verified that

$$v(\cup S_k) = \Sigma\, v(S_k)$$

for all possible unions of the sets S_k in the collection T.

Consider first case (i). Let $X = \Sigma\, x_i$ and $Y = \Sigma\, y_j$. Let

$$X = \max\{v(\cup A_i),\, v(\cup A_i, \cup_2^m B_j)\},$$
$$X + y_1 = v(I),$$
$$y_j = 0, \qquad\qquad \text{for } j = 2, \ldots, m,$$
$$x = x_i, \qquad\qquad \text{for } i = 1, \ldots, m$$

[i.e., $x = (1/m)X$]. We must show that these choices of imputations satisfy all of the following core constraints:

(9)
$$\sum_R x_i + y_1 \geqslant v(\cup_R A_i, B_1),$$

and

(10)
$$\sum_R x_i \qquad \geqslant v(\cup_R A_i, \cup_j B_j).$$

It follows from the choice of the imputations that

$$\sum_R x_i = (|R|/m)\max\{v(\cup A_i),\, v(\cup A_i, \cup_2^m B_j)\},$$

where $|R|$ denotes the number of members of the coalition R. Therefore, we must show that

(11)
$$(|R|/m)\max\{\,\cdot\,\} + v(I) - \max\{\,\cdot\,\} \geqslant v(\cup_R A_i, B_1),$$

since

$$y_1 = v(I) - \max\{\,\cdot\,\}.$$

Now

$$v(\cup_R A_i, B_1) + (1 - |R|/m)\max\{\,\cdot\,\} \leqslant v(I) + (1 - |R|/m)\,v(I) = v(I),$$

which gives the desired inequality in (11) and yields (9). To establish (10) it is only necessary to prove that

$$\sum_R x_i = (|R|/m)\max\{\,\cdot\,\} \geqslant v(\cup_R A_i, \cup B_j),$$

which is obviously true, concluding the demonstration that all core constraints are satisfied. By the duality theorem of linear programming,

(12)
$$\min \Sigma\, (x_i + y_j) = v(I).$$

Next consider case (ii) and let

$$X = v(I),\, x_i = x, \quad\text{and}\quad y_j = 0 \qquad (j = 1, \ldots, m).$$

It follows that

$$\sum_R x_i = (|R|)/m\, v(I) \geqslant v(\cup_R A_i, \cup_j B_j).$$

Hence all of the core constraints are satisfied, and in this case (12) holds also. Therefore, in both cases the core is not empty. QED.

With the help of this lemma it is possible to prove a more interesting sufficient condition for a nonempty core.

THEOREM 3. *Let every initial owner have limit prices satisfying the order given by*

(13) $$a_{s+1} \leqslant a_s$$

and every initial nonowner have limit prices satisfying the order given by

(14) $$b_{s+1} \geqslant b_s.$$

Thus the initial nonowners have concave preferences while the initial owners have convex preferences. Then the core of this market is not empty.

Proof. In this situation there are only two possibilities. Either there is a decomposing collection T such that

(i) $$S_1 = (\cup_1^m A_i, B_1), \quad S_{j+1} = B_j \qquad (j = 2, \ldots, m),$$

or

(ii) $$S_1 = \cup_1^m A_i, \qquad S_{j+1} = B_j \qquad (j = 1, \ldots, m).$$

In the first case one of the initial nonowners absorbs the entire stock and in the second case the goods remain in the hands of the initial owners who may, however, exchange some units among themselves. Hence in the second case the core is obviously not empty. In the first case, which corresponds to the first case of the lemma, exactly the same imputations described in the lemma apply and give the result that all of the core constraints are satisfied. Hence in the first case, as well, the core is not empty. QED.

An example shows that the conjecture which asserts that the core is not empty if all B's have convex while all A's have concave preferences is false. Consider

$A_1^{(2)}$	$A_2^{(1)}$	B_1	B_2
4	3	7	6
6	5	5	3
8	11	0	0

In this case

$$v(I) = 19, \quad v(A_1 A_2 B_1) = 19, \quad v(A_1 B_1 B_2) = 13,$$
$$v(A_1 A_2 B_2) = 19, \quad v(A_2 B_1 B_2) = 7,$$

and it is readily verified that the core of this market is empty.

The fact that the core of a market is not empty if all traders have convex or concave preferences suggests the investigation of intermediate markets, where the preferences of the traders are given by arbitrary orderings of the limit prices and the markets can differ only with respect to the order of the limit prices of the participants, not with respect to the prices themselves. For example, let the three limit prices be 3, 6, and 9. There are four possible distinct orders as follows:

M_1	M_2	or	M_2	M_3
3	6		3	9
6	9		9	6
9	3		6	3

A trader whose prices are in the order given in M_1 has concave preferences while one whose prices are in the M_3 order has convex preferences. The two orders given in M_2 are neither concave nor convex. Given a market in which the traders have arbitrary orders of their limit prices, we can construct two extreme markets by a rearrangement of every trader's limit prices such that in M_1, the concave market, every trader has concave preferences while in the convex market M_3, every trader has convex preferences. We denote the given market by M_2. Thus M_2 differs from M_1 and M_3 only insofar as the limit prices are in a different order.

DEFINITION. *Let M_2 represent a given market with initial owners, the A's, and initial nonowners, the B's. The concave market M_1 is derived from the given market M_2 by a rearrangement of the limit prices of every trader such that*

$$(15) \qquad a_{s+1} \geqslant a_s \quad and \quad b_{t+1} \geqslant b_t.$$

The convex market M_3 is derived from the initial market M_2 by a rearrangement of the limit prices in M_2 such that

$$(16) \qquad a_{s+1} \leqslant a_s \quad and \quad b_{t+1} \leqslant b_t.^5$$

With the aid of the concave market M_1 and the convex market M_3 we shall gain some insight into how the shape of the traders' preferences affects the emptiness of the core. For brevity let $C(M_i) = \varnothing$ mean that the core of market M_i is empty with an appropriate interpretation of $C(M_i) \neq \varnothing$. It has been shown that $C(M_3) \neq \varnothing$. However, it is not true in general that $C(M_1) \neq \varnothing$ unless upon rearrangement it turns out that any of the traders with concave preferences is willing to absorb the whole stock available in the market. To ensure this insatiability, it is necessary to augment the sequence of limit prices for any trader by additional prices equal to the largest number in the rearranged sequence. Thus assume that in the concave market either any trader is insatiable after rearrangement so that he would absorb the whole stock or that the trader's sequence is augmented to make him insatiable. In either case this implies that $C(M_1) \neq \varnothing$.

Let $v_i(S)$ denote the value of the characteristic function for the coalition S in M_i. It is readily verified that

$$(17) \qquad v_1(S) \leqslant v_2(S) \leqslant v_3(S)$$

---

5. See n.4.

for all coalitions $S \subset I$. Thus a given market M_2 is always between the concave market M_1 and the convex market M_3 in the sense of the inequalities given in (17). An intuitive explanation for this order is as follows. For a given coalition S the traders differ in the three markets only with respect to the order of their limit prices. In the convex market M_3, it is always possible to allocate the goods in the possession of a coalition S that will maximize the sum of the aggregate valuation. However, in the concave market M_1, where everyone has concave preferences, this is not always possible, because, for a trader to attain his highest limit price, he would have to obtain all of the goods in the possession of the coalition. Since it is never advantageous to divide the given stock among several members of S with concave preferences, the maximum attainable aggregate valuation is always below the maximum upon rearrangement into a convex order. It follows from (17) that

(18) $\qquad v_2(S) = \alpha_S v_3(S) + (1 - \alpha_S) v_1(S), \qquad$ for $0 \leqslant \alpha_S \leqslant 1$.

In view of (17) one may conjecture that if $C(M_1) \neq \varnothing$ and $C(M_3) \neq \varnothing$ then since M_2 is between the concave and the convex markets, $C(M_2) \neq \varnothing$. However, counter-examples abound to disprove this conjecture. Consider a market M_2:

$A_1^{(1)}$	$A_2^{(1)}$	B_1	B_2	B_3
2	6	$8+c$	6	7
6	2	6	$8+c$	7

where $v_2(I) = 15 + c$. The corresponding concave market M_1 is given by

$A_1^{(1)}$	$A_2^{(1)}$	B_1	B_2	B_3
2	2	6	6	7
6	6	$8+c$	$8+c$	7

where $v_1(I) = 14 + c$.

For M_3 the preferences satisfy

$A_1^{(1)}$	$A_2^{(1)}$	B_1	B_2	B_3
6	6	$8+c$	$8+c$	7
2	2	6	6	7

and $v_3(I) = 16 + 2c$.

It is not hard to show that in these examples $C(M_1) \neq \varnothing$, and $C(M_3) \neq \varnothing$, while $C(M_2) = \varnothing$ if $c \geqslant 1$. However, it is easy to prove theorem 4.

THEOREM 4. *Let $C(M_1) \neq \varnothing$ and $C(M_3) \neq \varnothing$. Then, if*

(19) $\qquad max \; \Sigma \; \delta_S \alpha_S v_3(S) \leqslant \alpha_I v_3(I)$

and

(20) $\qquad max \; \Sigma \; \delta_S (1 - \alpha_S) v_1(S) \leqslant (1 - \alpha_I) v_1(I),$

then $C(M_2) \neq \varnothing$.

Proof.

$$\max \Sigma \, \delta_S v_2(S) = \max \Sigma \, \delta_S[\alpha_S v_3(S) + (1 - \alpha_S) v_1(S)]$$
$$\leq \max \Sigma \, \delta_S \alpha_S v_3(S) + \max \Sigma \, \delta_S (1 - \alpha_S) v_1(S)$$
$$\leq \alpha_I v_3(I) + (1 - \alpha_I) v_1(I) = v_2(I). \quad \text{QED.}$$

This result shows that if the intermediate market M_2 is well bounded by its extremes M_1 and M_3 in the sense of (19) and (20) then it has a nonempty core, that is, $C(M_2) \neq \varnothing$.

As mentioned at the beginning of this section, some of the results with concave preferences are relevant to the study of exchange when the owners are producers of a good. Thus, if the A's are producers, then $v(A_i)$ represents the opportunity cost of production.

$$v(A_i) = \sum_{s=1}^{m_i} a_{is}$$

is the total cost of producing m_i units of the good and a_{m_i} is the marginal cost of the (m_i)th unit. Thus concave preferences correspond to decreasing average cost of production and convex preferences to increasing average cost. With this interpretation theorem 3 allows us to conclude that if there are a number of producers who all have increasing average cost and a number of potential buyers with insatiable concave preferences then the core will not be empty.

A more interesting application of these ideas is to the case of competing producers who all have decreasing average cost discussed above in section I.8. In particular, suppose that all of the producers have a given fixed cost and no running expenses. If the producers have no capacity limit, then

$$v(A, \cup_S B_j) = \max \{a, \sum_S b_j\},$$

assuming that each buyer desires at most one unit. It is not hard to show that under these conditions the core will not be empty. However, if the producers have capacity limitations, then the core can be empty. Thus, consider the example of an airplane that can seat only two passengers and suppose there are no running costs. Let

A_1	A_2	B_1	B_2	B_3
0	0			
		$5 + c$	$5 + c$	$5 + c$
10	10			

$v(A_1) = v(A_2) = 10; \quad v(A_1, A_2) = 20;$
$v(I) = 10 + (10 + 2c), \qquad \qquad \text{if } c < 5;$
$v(A_1 A_2 B_1 B_2) = 10 + (10 + 2c), \quad v(A_1 B_1 B_2 B_3) = 10 + 2c,$
$v(A_1 A_2 B_1 B_3) = 10 + (10 + 2c), \quad v(A_2 B_1 B_2 B_3) = 10 + 2c,$
$v(A_1 A_2 B_2 B_3) = 10 + (10 + 2c).$

The core of this market is empty because

$$\frac{3[10+(10+2c)]+2[10+2c]}{4} > v(I),$$

and not all of the core constraints can be satisfied by a feasible set of imputations. Therefore, there is no competitive equilibrium, or, more familiarly, a natural monopoly. If there were one more customer or one less plane, the core would not be empty. This example can easily be generalized to the case in which the buyers desire an arbitrary number of units and have convex preferences. As long as there are sellers with increasing returns and capacity limits, the core can be empty.

These remarks may appear to suggest that a market can have a nonempty core for a more general class of trader preferences if the traders use money than if they barter. However, it can be shown that the means of trade is immaterial for the emptiness of the core of a market provided every trader has a continuous utility function that is a strictly increasing function of his bundle of goods, which are assumed to be continuously divisible. Moreover, under these conditions only balanced markets have nonempty cores so that it is not only true that balance is sufficient but also it is necessary for a non-empty core (Telser 1971). The present analysis is more general in so far as it covers goods that are not continuously divisible, allows for flats in the utility function, and allows us to use linear programming techniques to study the status of the core of a market.

In the next section we study when there are feasible trades for implementing the money imputations of the core.

4. The Feasibility of Trade

The feasibility of trade concerns the problem of determining an allocation of the initial stock of goods among the traders so that their imputations can satisfy the core constraints. In the case of single-unit trade there are no complications because for any set of imputations satisfying the core constraints there is a feasible set of trades. However, this is not always true for multiunit trade as shown by the following example:

Consider the market in which the traders' preferences are given as follows:

$A_1^{(2)}$	$A_2^{(2)}$	B_1	B_2
20	14	25	26
15	12	24	10
0	0	23	0

Each of the A's begin with 2 units of the stock, and each of the B's begin with none. The core constraints for the imputations are as follows:

$$x_1 \geqslant 35, \quad x_2 \qquad \geqslant 26,$$
$$x_1 + y_1 \geqslant 49, \quad x_1 + x_2 + y_1 \geqslant 92,$$
$$x_1 + y_2 \geqslant 46, \quad x_1 + x_2 + y_2 \geqslant 75,$$
$$x_2 + y_1 \geqslant 49, \quad x_1 + y_1 + y_2 \geqslant 51,$$
$$x_2 + y_2 \geqslant 40, \quad x_2 + y_1 + y_2 \geqslant 51,$$
$$x_1 + x_2 + y_1 + y_2 \leqslant v(I) = 98.$$

The minimum is 95, which can be attained with the following two core constraints

$$x_1 + y_2 = 46,$$
$$x_2 + y_1 = 49,$$

by the choice of

$$x_1 = x_2 = 44, y_1 = 5, \quad \text{and} \quad y_2 = 2.$$

We can readily verify that these choices satisfy all of the core constraints.

There is not necessarily group rationality. However, there is a set of trades that can implement these imputations and thereby satisfy the core constraints given as follows:

B_1 buys 2 units and pays $49 - 5 = 44$;
B_2 buys 1 unit and pays $26 - 2 = 24$;
A_1 sells 1 unit and receives 24 so that A_1's final imputation is $20 + 24$;
A_2 sells 2 units and receives 44.

Now the problem is to see whether, given any set of imputations x_1, x_2, y_1, and y_2, there is a set of feasible trades to implement them. To show that in fact such a set of feasible trades does not always exist, let us consider the imputations

$$x_1 = x_2 = 44, y_1 = 6 \quad \text{and} \quad y_2 = 3.$$

It is readily verified that these imputations satisfy all of the core constraints and fall short of $v(I)$ by 1. However, there are no trades that can implement these imputations. To show this in detail, we need only consider the five cases in which each of the B's buy something, for otherwise, there would not be the required positive values of the y's. There are 4 units of the good to allocate among the four traders. Suppose that each of the B's buys 2 units.

B_1 buys 2 and pays $49 - 6 = 43$;
B_2 buys 2 and pays $36 - 3 = 33$.

Hence their total payments are 76, which is less than the total receipts of the two sellers, $44 + 44 = 88$. Consider a second case in which

B_1 buys 3 units and pays $72 - 6 = 66$;
B_2 buys 1 unit and pays $26 - 3 = 23$.

In this case the total payments of the B's are 89, which exceeds the postulated receipts of the two A's by 1. Hence these trades are also inconsistent with the postulated imputations that satisfy the core constraints. Similarly, one can verify that in the three other cases in which the B's buy 3 units in the aggregate, the implied imputations are not consistent with the given set. Hence this example demonstrates that there may not exist a set of trades for implementing a given set of imputations already satisfying the money core constraints.

To proceed with this analysis, it is convenient to adopt a somewhat more general point of view than has hitherto been used. First, let us drop the distinction between owners and nonowners and simply assume that every trader begins with an initial stock of goods denoted by k. Thus, there are m types of traders, $A_i, i = 1, ..., m$, such that A_i has k_i units of the good initially. Second, we can encompass the cases in which the units are either continuous or discrete by using a Stieltjes integral to represent the gains from trade. Thus

$$(1) \qquad\qquad F(q) = \int_0^q dF(r)$$

represents the maximum amount that a trader is willing to pay for the acquisition of q units of the good rather than to go without. Similarly, if the trader held q units initially, then $F(q)$ gives the least acceptable amount of money for selling them all. It is assumed that

$$(2) \qquad\qquad dF(r) \geqslant 0, \qquad\qquad \text{for all } r \geqslant 0.$$

DEFINITION. *The characteristic function of a coalition S is given by*

$$(3) \qquad\qquad v(S) = \sup_{\{q_i,\, i \text{ in } S\}} \sum_S \int_0^{q_i} dF_i(r)$$

subject to

$$(4) \qquad\qquad \sum_S q_i = K_S,$$

where

$$(5) \qquad\qquad \sum_S k_i = K_S.$$

In fact the assumptions about F and the constraint set imply the existence of a unique maximum at points of continuity of F.

DEFINITION. *The imputation of the trader A_i is denoted by x_i and is given by*

$$(6) \qquad\qquad x_i = F(q_i) + z_i,$$

where z_i denotes the money payments or receipts of z_i so that $z_i > 0$ denotes a receipt and $z_i < 0$ denotes a payment. Since by definition $v(A_i) = F(k_i)$, it follows that $q_i < k_i$ implies $z_i > 0$, and $q_i > k_i$ implies $z_i < 0$.

We are now ready for a rigorous definition of the core of a market with multiunit trade taking account of the requirement that the trades for implementing the money imputations must be feasible. In a single-unit trade market the one-to-one correspondence between the money imputations and the exchanges of goods make it unnecessary to check the feasibility of the trades required to implement the money core constraints.

DEFINITION. *The core of a market consists of a set of money imputations* $\{x_i\}$ *and a set of commodity allocations* $\{q_i\}$ *such that*

   i. $\Sigma_I q_i = K_I$, *where* $K_I = \Sigma_I k_i$;
   ii. $\Sigma_I z_i = 0$;
   iii. $\Sigma_S x_i \geqslant v(S)$ *for all S such that* $S \subset I, S \neq I$;
   iv. $\Sigma_I x_i \leqslant v(I)$.

The definition of a balanced game given in section 2 above remains in force and is easily adapted to a market.

DEFINITION. *A market with n traders equipped with a characteristic function as defined above is said to be balanced if there exists a set of nonnegative numbers* $\{\delta_S\}$ *with*

(7) $$\sum_{i \text{ in } S} \delta_S = 1$$

such that

(8) $$\max_{\delta_s} \sum \delta_S v(S) \leqslant v(I).$$

Next we require the definition of a Pareto optimal allocation of goods.

DEFINITION. *A Pareto optimal allocation of goods, denoted by* $\bar{q}_i$, *is a set of quantities such that*

$$\sum_I \bar{q}_i = K_I \quad and \quad \sum_I F_i(q_i) \leqslant \sum_I F_i(\bar{q}_i) = v(I).$$

We are now ready to prove lemma 2.

LEMMA 2. *Let* $x_i = F(\bar{q}_i) + z_i$.
If the market is balanced, then a set of z's can be found to satisfy ii and iii.

Proof. By hypothesis

(9) $$\max \sum \delta_S v(S) \leqslant v(I) = \sum_I \bar{v}_i, \; \delta_S \geqslant 0 \quad and \quad \sum_i \delta_S = 1.$$

where, for convenience, we write $F_i(\bar{q}_i) = \bar{v}_i$. Hence, the imputations, x_i, satisfy $x_i = \bar{v}_i + z_i$. The core constraints ii and iii may be written as follows:

$$\sum_S z_i + \sum_S \bar{v}_i \geqslant v(S)$$

(10) $$\sum_I z_i \qquad \geqslant 0$$

$$-\sum_I z_i \qquad \geqslant 0$$

Gale's theorem 2.7 (1960) applies to this system of inequalities. This theorem asserts that either the system of inequalities (10) has a solution or the following system of equations has a nonnegative solution, $\beta_S \geqslant 0$:

$$(11) \qquad \sum \beta_S = 0 \Leftrightarrow \sum_{i \text{ in } S} \beta_S = (\beta_N - \beta_{N-1}),$$

$$(12) \qquad \sum \beta_S(v(S) - \bar{v}_i) = 1,$$

where β_{N-1} and β_N refer to the last two inequalities in (10).

Equation (12) can be simplified by means of the following identity

$$(13) \qquad \sum_{i \text{ in } S} \beta_S \bar{v}_i = (\beta_N - \beta_{N-1}) \sum_I \bar{v}_i.$$

Therefore, we are led to study the expression given by

$$(14) \qquad \sum \beta_S v(S) - (\beta_N - \beta_{N-1}) \sum_I \bar{v}_i.$$

We now appeal to the hypothesis given in (9) which implies that for all $\beta_N - \beta_{N-1} > 0$, we can choose β_S according to the expression as follows:

$$(15) \qquad \delta_S = \beta_S / (\beta_N - \beta_{N-1}),$$

and it will be true that

$$(16) \qquad \sum \beta_S v(S) \leqslant (\beta_N - \beta_{N-1}) \sum_I \bar{v}_i.$$

Therefore, for all β_S that satisfy (15), it is impossible to satisfy (12) because

$$\sum \beta_S v(S) - (\beta_N - \beta_{N-1}) \sum \bar{v}_i \leqslant 0$$

for all $\beta_N - \beta_{N-1} > 0$. Moreover, it is clearly impossible to satisfy (11) and (12) with $\beta_N - \beta_{N-1} = 0$ for then all of the β_S's would have to be zero. Hence (11) and (12) cannot have a nonnegative solution so that the system of inequalities (10) must have a solution. QED.

THEOREM 5. *Let* $q' = \{q_i'\}$ *be a feasible set of allocations such that*

$$(17) \qquad v(I) = \sum_I F_i(\bar{q}_i) \geqslant \sum_I F_i(q_i').$$

If the market is balanced and if

$$(18) \qquad \sum_i F_i(q_i') \geqslant \max \sum \delta_S v(S),$$

then there exists a set of money transfers $z = \{z_i\}$ *to satisfy the money core constraints i, ii, and iii.*

Proof. Let $v_i' = F_i(q_i')$ and apply the same argument as in lemma 2. For if there must be equality in (17) then lemma 2 applies directly to give the desired conclusion, and if there is inequality in (17) then the hypothesis that the market is balanced together with Gale's theorem 2.7 yields the desired result. QED.

We can summarize these results conveniently in the following theorem.

THEOREM 6. *A necessary condition for a nonempty core is that the money core constraints (10) can be satisfied by a Pareto allocation of goods. In a balanced market the Pareto allocations satisfy (10) so that the core is not empty.*

Proof. The first assertion is virtually immediate by the definition of a core. The second assertion follows from lemma 2 and theorem 5. QED.

There is also the following interesting corollary.

COROLLARY. *Let* $x = \{x_i\}$ *be a set of imputations satisfying the money core constraints* $\Sigma_S x_i \geqslant v(S)$ *for all* $S \subset I$, $S \neq I$. *If* $\min \Sigma_I x_i = v(I)$, *then there is a uniquely determined set of trades, namely, the Pareto optimal allocations that satisfy the core constraints.*

Proof. By the duality theorem of linear programming,

$$\min \sum_I x_i \quad \text{subject to} \quad \sum_S x_i \geqslant v(S)$$

is equal to

$$\max \sum \delta_S v(S) \quad \text{subject to} \quad \sum_{i \text{ in } S} \delta_S = 1 \qquad (\delta_S \geqslant 0).$$

Therefore, the market is balanced, and the core is not empty by theorem 5. Since the minimum amount acceptable to all traders equals the maximum attainable amount, the allocation of goods must be Pareto optimal. By lemma 2 these can be implemented with money exchanges. QED.

This is an important result that generalizes the familiar property of a competitive equilibrium. Moreover, it can be shown that under these conditions there are nonnegative prices that can clear the market. Hence, if the most the traders can get is the least they will accept, then there is a classical competitive equilibrium. It should be noted that convex preferences are not necessary for this conclusion. Moreover, the so-called money core constraints shown in (10) are the analogues in core theory for the intersection of supply and demand schedules.

It is convenient at this point to provide some intuitive understanding of the concept of balance because this is closely related to feasible trades. A coalition S can be regarded as constituting a separate market. If there is group rationality within this market, then the traders belonging to this coalition would receive quantities of goods that would maximize the valuation of the coalition. Trader A_i in coalition S would receive q_i^S, where

$$\sum_S q_i^S = \sum_S k_i^S \quad \text{and} \quad v(S) = \sum_{i \text{ in } S} F_i(q_i^S).$$

However, not enough goods are available for every trader to receive q_i^S from all coalitions simultaneously, and it is impossible for trader i to obtain

(19) $$q_i = \sum_{i \text{ in } S} q_i^S$$

because were (19) to hold for all A_i we would have

$$\sum_I q_i = \sum_I \sum_{i \,in\, S} q_i^S > K_I.$$

Nevertheless, the final allocation any trader receives depends on the fact that the trader can threaten to join a coalition S where he would receive q_i^S, although not all traders can carry out their threats simultaneously. Now here is where balance enters. It is readily verified that for a balanced collection of coalitions there is a sense in which the threat to form all possible coalitions *is* feasible. This is because

$$\sum_{i \,in\, S} \delta_S = 1$$

allows every trader to receive a fraction $\delta_S q_i^S$ of his allocation in the coalition S, and it will be true that

(20) $$\sum_{i \,in\, I} \sum_{i \,in\, S} \delta_S q_i^S = K_I.$$

In other words, it is true that

$$\sum_i \delta_S \sum_i q_i^S = \sum_i \delta_S \sum_S k_i^S = K_I.$$

To convince oneself of this fact, it is helpful to write out the algebra in detail for a special case, say, a market with four traders. For a balanced collection of coalitions, the traders have a credible threat to block a proposed allocation of goods.

It may be helpful for some readers to see another interpretation that is closer to a linear programming framework. Imagine that every coalition S has a manager who must pay the members of the coalition the imputation x_i, which is as if the manager's receipts are $v(S)$ and his costs are $\Sigma_S x_i$. Hence δ_S refers to the activity level of a coalition. If $\delta_S \Sigma_S x_i > v(S)$, then the manager of this coalition would incur a loss so he would be unwilling to operate this coalition, that is, $\delta_S = 0$. But if $\Sigma_S x_i = v(S)$, then he can just break even; say this means he obtains the competitive return, so he is willing to have the coalition run at a level δ_S. Thus the dual variables determine the coalitions it just pays for their managers to operate.[6]

5. Group Rationality with Multiunit Trade

In addition to the problem of determining when a market has a nonempty core that can be implemented by a feasible set of trades, two other problems arise with multiunit trade. First, when is there group rationality and, second, when is there a common unit price? In studying these problems we assume feasible imputations and a nonempty core.

---

6. I am grateful to Lionel McKenzie for pointing this interpretation out to me.

Group rationality is of considerable importance, a point stressed in chapter I, because it is easier to accept the proposition that individuals are rational than that groups are. To describe an individual as rational is to say little or nothing since only the person himself can be the judge, and virtually any kind of individual behavior can be rational. In a market a "rational" trader may refuse trades that well-meaning outsiders think he ought to accept. Nor are arbitrarily ordered limit prices irrational no matter how bizarre this may seem. At most, rationality requires an individual to behave consistently with respect to his own goals, which need not be explicit or even conscious to the individual. Nor is that all. The goals may change with the passage of time. As a result individual rationality amounts to the weak assertion that the individual is the best judge of his own behavior regardless of how this may appear to others. In market exchange rationality means that the participants *have* limit prices but that these do not necessarily obey any regularities except for nonnegativity.

Group rationality raises a new and more difficult set of problems. For a group to be rational entails that its members are in communication and capable of reaching a common decision compatible with their individual preferences. Even casual reflection quickly reveals the sources of possible conflict. For example, individuals would have an incentive to misrepresent their true preferences to each other in order to improve their final imputations. Even honest traders encounter difficulties. The larger the size of the group the more complex is the communications network necessary to achieve group rationality for even the best-intentioned individuals. Messages can become distorted or may be incomplete, instructions ambiguous, delegations of authority inadequate to cover unforeseen events, and so forth. To assume that a large group can be collectively rational is naïve. In many economic settings such an assumption would give no upper limit to the size of a decision-making unit. Hence very large firms could manage as well as very small ones, and, indeed, central planners could manage the whole economy as well as a collection of economic units operating through a free market system. For these reasons it is desirable to investigate when a market will exhibit group rationality.

In a market, group rationality has a precise meaning. There is said to be group rationality if the imputations of the traders are such that the sum of their imputations achieves the upper bound given by $v(I)$. The approach here is to see whether group rationality results when trade is conducted among overlapping subsets of traders. In particular we shall study the possibility of achieving group rationality with pairs of traders (A_i, B_j) under multiunit trade.

With multiunit trade A_i can sell at most μ_i units and B_j will buy at most σ_j units. Let the reservation prices for A_i and the limit prices for B_j obey the order as follows:

$$a_{i1} \leqslant a_{i2} \leqslant \dots \leqslant a_{i\mu_i}$$

(1) $(i, j = 1, \dots, m).$

$$b_{j1} \leqslant b_{j2} \leqslant \dots \leqslant b_{j\sigma_j}$$

Thus it is assumed that both kinds of traders have convex preferences. This ensures the existence of a nonempty core and allows us to study the problems of group rationality free of these complications. As with single-unit trade it is convenient and innocuous to retain the assumption of an equal number of A's and B's. However, unlike single-unit trade, with multiunit trade it is no longer possible to order the traders with respect to their reservation prices, which may overlap. An intramarginal trader is simply one who engages in active exchange and is not otherwise distinguished by his valuations from the other participants in the market. The market is essential if and only if some trade actually occurs. Otherwise, the initial allocation of goods is optimal so that it does not pay for anyone to buy or to sell. Hence a necessary condition for active trade is that

(2) $$\min_i \{a_{i1}\} < \max_j \{b_{j\sigma_j}\}$$

(see section I.4). For the remainder of this section it is assumed that the market is essential so that condition (2) is satisfied.

Consider a coalition $A_i \cup B_j$.

(3) $$v(A_i \cup B_j) = \max \sum_1^{\mu_i} z_h \quad \text{with} \quad z_h \quad \text{in} \quad \{a_{is}\} \cup \{b_{jt}\}.$$

This expression assumes an especially simple form with convex preferences. There is a number $k_{ij} \leqslant \min \{\mu_i, \sigma_j\}$ giving the maximum number of units of the good that the two parties A_i and B_j can trade to their mutual advantage.

(4) $$v_{ij} = \sum_1^{k_{ij}} b_{j, \sigma_j+1-s} + \sum_{k_{ij}+1}^{\mu_i} a_{is},$$

where v_{ij} denotes $v(A_i \cup B_j)$.

As before, let x_i and y_j denote the imputations of A_i and B_j, respectively. We wish to examine the properties of the solution of the system as follows:

$$x_i \quad \geqslant v(A_i) = \sum_s a_{is},$$

(5) $$y_j \quad \geqslant v(B_j) = 0,$$

$$x_i + y_j \geqslant v_{ij}.$$

The primal linear programming problem is

(6) $\min [\Sigma \, x_i + \Sigma \, y_j]$ subject to (5).

Since

$$v(A_i, B_j) \geqslant v(A_i), \, y_j \geqslant 0 \quad \text{and} \quad x_i + y_j \geqslant v_{ij},$$

the dual problem may be restated as follows:

(7) $$\max \Sigma \, \delta_{ij} v_{ij}$$

subject to $\delta_{ij} \geqslant 0$ and

(8) $$\sum_i \delta_{ij} \leqslant 1 \quad \text{and} \quad \sum_j \delta_{ij} \leqslant 1.$$

In this case it is more illuminating to begin with the dual problem.

LEMMA 3. *A solution of the dual problem is given by a collection of m pairs* $\{(A_i, B_j)\}$ *such that every* A_i *appears exactly once, and every* B_j *appears exactly once. The trades necessary for this solution are feasible.*

Proof. The dual problem is an example of the well-known assignment problem in linear programming literature (Koopmans and Beckmann 1957). If $v_{ij} \geqslant 0$, then in fact the constraints of (8) are equalities and the solution of the assignment problem applies. This solution has the properties stated in the lemma. It remains only to show that the solution does in fact require feasible trades. However, every A_i and B_j appears exactly once in the optimal solution. Moreover, by the definition of v_{ij} all of the necessary trades can be carried out by the members of the pairs. Since the pairs in the solution do not overlap, the necessary trades are indeed feasible. QED.

The lemma does not assert that the collection of pairs $\{(A_i, B_j)\}$ is balanced. To prove this it is necessary to show that the solution of the linear programming problem satisfies all of the core constraints. This is not always true. Before pursuing this topic, let us settle some preliminaries.

Let the A's and B's in the solution of the dual linear programming problem be labelled such that the pairs appear on the diagonal of the matrix $[v_{ij}]$ defined by $i+j = m+1$. The given solution asserts that

$$\begin{aligned} \delta_{i,\, m+1-i} &= 1, &&\text{if} \quad i = 1, \ldots, m, \\ \delta_{ij} &= 0, &&\text{otherwise.} \end{aligned}$$

Therefore, by the duality theorem of linear programming it follows that

(9) $$\begin{aligned} x_i + y_{m+1-i} &= v_{i,\, m+1-i}, &&\text{if} \quad i = 1, \ldots, m, \\ x_i + y_j &\geqslant v_{ij}, &&\text{otherwise.} \end{aligned}$$

To prove that the collection $\{(A_i, B_j)\}$ balances the market, one can show that for every coalition $S \subset I$, the imputations that satisfy (9) also satisfy

(10) $$\sum_{i,\, j \in S} (x_i + y_j) \geqslant v(S).$$

However, it is not true in general that the conditions in (9) imply those in (10), nor, for that matter is the converse generally true. One avenue of analysis would be to determine properties of a characteristic function that would ensure that (9) does imply (10). This would be rather complicated. Fortunately, there is another approach that gives a simple sufficient condition for determining when the collection of pairwise coalitions $\{(A_i, B_j)\}$ balances the market without the necessity of a detailed examination of the characteristic function. This result is contained in theorem 7.

THEOREM 7. *Let* $T = \{(A_i, B_j)/i, j = 1, ..., m\}$ *and let* $V = [v_{ij}]$, *so that* V *is an* $m \cdot m$ *matrix with entries in row i and column j given by the values of the characteristic function* $v(A_i, B_j) = v_{ij}$. *If max* $\Sigma \delta_{ij} v_{ij} = v(I)$ *subject to* $\delta_{ij} \geqslant 0$ *and*

$$\sum_i \delta_{ij} \leqslant 1, \quad \sum_j \delta_{ij} \leqslant 1,$$

then the market is balanced and, therefore, has a nonempty core.

Proof. To prove this theorem, we shall show that the solution of the dual linear programming problem that is restricted to the pairwise coalitions $\{(A_i, B_j)\}$ is actually a solution of the general dual problem defined as follows:

$$\max \sum \delta_S v(S) \quad \text{subject to} \quad \delta_S \geqslant 0$$

and

$$\sum \delta_S \leqslant 1, \qquad\qquad \text{for all} \quad S \subset I.$$

Since it is assumed throughout this section that preferences are convex, the core is not empty and

$$\max \sum \delta_S v(S) \leqslant v(I).$$

By lemma 3 the solution of the dual linear programming problem restricted to the pairwise coalitions is of the form

$$\sum_1^m v_{i, m+1-i} \geqslant \sum \delta_{ij} v_{ij}.$$

It follows that the dual variables given as follows:

$$\begin{aligned} \delta_{i, m+1-i} &= 1, &\qquad \text{if} \quad i = 1, ..., m, \\ \delta_{ij} &= 0, &\qquad \text{otherwise,} \end{aligned}$$

also solve the general dual problem since by hypothesis

$$\sum_i^m v_{i, m+1-i} = v(I).$$

Therefore, by the duality theorem of linear programming we have

$$\sum_{i, j \in S} [x_i + y_j] \geqslant v(S), \qquad\qquad \text{for all} \quad S \subset I,$$

where the imputations $\{(x_i, y_j)\}$ satisfy (9). QED.

    This theorem shows that pairwise group rationality is sufficient for balance. Is this condition also necessary? The answer is no for the following reasons. Even if

$$\sum v_{i, m+1-i} < v(I),$$

it may still be true that

(11) $$\max \sum \delta_{ij} v_{ij} = \max \sum \delta_S v(S) < v(I).$$

The last inequality holds provided

(12) $$\min \sum_I (x_i + y_j) < v(I)$$

and

$$\sum_S (x_i + y_j) \geqslant v(S), \qquad \text{for all } S \subset I.$$

This is to say that, while no collection of coalitions achieves the upper bound, the collection of pairwise coalitions can do as well as any other collection. It follows that there can be balance for the pairwise coalitions without group rationality.

For the pairwise coalitions to attain the upper bound $v(I)$ means that the total market can be decomposed into m disjoint submarkets such that within every submarket the Pareto optimal trade occurs. Let q_{ij} denote the largest quantity that the pair (A_i, B_j) agrees that it would be mutually advantageous to trade. Similarly, let Q denote the maximum quantity that must be exchanged to attain $v(I)$. Q is uniquely determined by $v(I)$. For pairwise group rationality it is necessary and sufficient that

(13) $$\sum_i q_{i, m+1-i} = K.$$

It is natural to study the properties of markets when the coalitions are confined to pairs (A_i, B_j) because then every trader represents only himself and need not consult others before reaching a decision which presumably makes negotiation easier. Even so there is not necessarily group rationality, raising the question of whether or not it would be possible to attain $v(I)$ with larger trading units short of forming a coalition of all traders.

To increase the size of the trading units means to form small teams of A's and B's that will trade with each other. The team members can join other teams and all possible teams can form. Hence the final imputations are forced into the core and the indeterminacy is minimized. Does a non-empty core imply that teams larger than pairs result necessarily in group rationality? That is, if the imputations are in the core, must it follow that

(14) $$\min \sum_I (x_i + y_j) = v(I)?$$

The answer to this question is no. The example at the beginning of section 4 illustrates the possibility of satisfying all of the core constraints, which gives the minimum value of the total imputations, without reaching $v(I)$.

Let us now continue the analysis of the outcome of trade for a collection of pairwise coalitions $\{(A_i, B_j)\}$. With single-unit trade we have seen that a necessary and a sufficient condition for a common price per unit is non-unique solutions of the linear programming problem. This suggests the desirability of studying the uniqueness of the solutions of the linear programming problems with multiunit trade to find necessary conditions for a common price per unit.

Any solution of the assignment problem described in lemma 3 can be expressed as a convex combination of basic solutions, a basic solution being a set of pairs $\{(A_i, B_j)\}$ such that every A_i appears exactly once and every B_j appears exactly once in the set.[7] Equivalently, a basic solution consists of exactly one element from every row and column of the matrix $[v_{ij}]$. Therefore, if there are several solutions of the assignment problem, then there must be at least two distinct basic solutions each with the properties asserted in lemma 3. There would be two sets of elements $\{v_{ij}\}$ and $\{v_{i'j'}\}$ such that

$$(15) \qquad\qquad \Sigma\, v_{ij} = \Sigma\, v_{i'j'} = M,$$

where M denotes the maximum value of the objective function $\Sigma\, \delta_{ij} v_{ij}$. Therefore,

$$(16) \qquad\qquad \alpha\, \Sigma\, v_{ij} + (1-\alpha)\, \Sigma\, v_{i'j'} = M \qquad\qquad (0 \leqslant \alpha \leqslant 1).$$

It follows that one can construct an infinite number of solutions according to the following procedure. Let every coalition (A_i, B_j) in the first basic solution be assigned a number α and let every coalition in the second basic solution $(A_{i'}, B_{j'})$ be assigned a number $1-\alpha$ with $0 \leqslant \alpha \leqslant 1$. Let $\delta_{ij} = \alpha$ for all (i, j) in the first basic solution and let $\delta_{i'j'} = 1-\alpha$ for all (i', j') in the second basic solution. Let all other δ's be zero. With these choices of δ, the constraints (8) are satisfied and the maximum is achieved. Hence by the duality theorem of linear programming there must be equality in at least $2m$ of the pairwise core constraints, (9).

In the general case with N distinct basic solutions, there must be equality in at least mN of the corresponding core constraints and there must be N equations connecting sets of m v's such as (16). Given that there are more equations among the pairwise core constraints than unknowns $\{(x_i, y_j)\}$, does this uniquely determine the imputations? The answer is no. In general even with $2m$ equations among the pairwise core constraints, the imputations are indeterminate. This is because one can add a constant $c \neq 0$ to every x_i and subtract c from every y_j while satisfying all of the $2m$ equations. Equivalently, one can show that the matrix of the coefficients for the $2m$ equations in the two distinct basic solutions has rank less than $2m$.

---

7. In the assignment problem the constraint set is a convex polyhedral cone whose vertexes are represented by permutation matrixes. Hence the maximum of the objective function is attained on a vertex. If there is no unique solution, then the value of the objective function equals the maximum on an edge or face of the cone. An edge or a face is a convex combination of two or more distinct permutation matrixes. These facts support the statements in the text. For good expositions, see Koopmans and Beckmann (1957) and Gale (1960, chap. 5). For an explicit treatment of a market in terms of the assignment problem see Shapley and Shubik (forthcoming).

It is worth mentioning another property of the pairwise solution. A basic solution of the dual problem satisfies

$$\Sigma \, v_{i,\,m+1-i} \geqslant \Sigma \, \delta_{ij} v_{ij}.$$

Sometimes $v_{i,\,m+1-i}$ is the largest element in its row so that the solution consists of pairs such that every A_i trades with that B_j for whom there is the largest mutual advantage from trade. If so, the solution would have the property that

(17) $$v_{i,\,m+1-i} = \max_{j} \, \{v_{ij}\}.$$

However, this will not be true if at least two row maxima are in the same column of the matrix $[v_{ij}]$ so that a given buyer is the best trading partner for two different sellers, with a similar argument for columns instead of rows.

Let us now consider when the core imputations establish a common price per unit. A standard proposition of economic theory asserts that in a competitive market there is a common price per unit that is independent of the purchases and sales of any individual trader. Actually a rigorous proof of when this is true had to await the development of core theory. We shall divide this proposition in two parts and consider here only the question of when there will be a common price per unit, leaving to the next section a study of how the number of traders affects the outcome.

With single-unit trade the core constraints imply both group rationality and a common price per unit. With multiunit trade even if the core is not empty neither proposition need be true. If the traders have convex preferences, then theorem I.6 applies and permits us to conclude that there is a single-unit trade market that includes all possible constraints of the multiunit market. If in fact the multiunit traders act as if they were coalitions of independent single-unit traders who compete with each other, then multiunit and single-unit trade would have precisely the same equilibrium. However, it is not in the interest of a multiunit trader thus to act, for it would dissipate his advantage of having several units to trade. Although a common price per unit *can* satisfy the core constraints with multiunit trade, because there is a single-unit trade market that encompasses the multiunit trade market, this is not sufficient grounds for concluding that the two markets would have the same equilibrium. Other conditions must be present in addition to a nonempty core before concluding that there must be a common unit price in the market.

A formal restatement of these propositions is now helpful. Let A_{is} denote a single-unit owner with a reservation price a_{is} and let B_{jt} denote a single-unit nonowner with a limit price b_{jt}. The relation between these single- and multiunit traders is shown by

$$A_i = \cup_s A_{is},$$

(18)

$$B_j = \cup_t B_{jt}.$$

Let x_{is} denote the imputation of A_{is} and let y_{jt} denote the imputation of B_{jt}:

$$x_i = \sum_s x_{is},$$

$$y_j = \sum_t y_{jt}.$$

The study of single-unit trade shows that there is a common price per unit, denoted by p, such that

(19)
$$x_{is} \begin{cases} = p, & \text{for all intramarginal } A\text{'s,} \\ = a_{is}, & \text{for all extramarginal } A\text{'s.} \end{cases}$$

(20)
$$y_{jt} \begin{cases} = b_{jt} - p, & \text{for all intramarginal } B\text{'s,} \\ = 0, & \text{for all extramarginal } B\text{'s.} \end{cases}$$

Let q_i denote the quantity sold by A_i and let q_j denote the quantity bought by B_j in this regime. It follows that

(21) $$x_i = pq_i + \sum_{q_i+1}^{\mu_i} a_{is} = \sum_1^{q_i} (p - a_{is}) + v(A_i),$$

(22) $$y_j = \sum_1^{q_j} (b_{j,\sigma_{j+1-s}} - p),$$

and

(23) $$\sum_i q_i = \sum_j q_j, \quad q_i, q_j \leqslant \min \{\mu_i, v_j\}.$$

Since $\{x_{is}\}$ and $\{y_{jt}\}$ are in the core, it follows that x_i and y_j would also be in the core under this regime. However, one must realize that with multi-unit trade nothing *forces* every trader A_i and B_j to regard himself as a union of independent traders as is shown in (18). For the classical competitive equilibrium it is not large numbers that are decisive. What is decisive is the assumption that every unit of the good competes with every other unit so that a market with multiunit traders is equivalent to one with as many single-unit traders as there are units of the good. If these conditions, which are *sufficient* for the classical results, were also *necessary*, then the case for the classical analysis would be very weak. This is because the core constraints in terms of A_{is} and B_{jt} conflict with the interests of the traders. For a trader to regard himself as a coalition of independent single-unit traders each seeking the best possible terms and ignoring the consequences for their "fellows" is unreasonable. Put differently, it is generally not in a multi-unit trader's interest to tender offers one unit at a time. This raises three questions. First, can other sufficient conditions yield the classical results? Second, do we actually observe the contract terms implied by the classical theory? Third, what kind of prices are implied by the core imputations with multiunit trade?

Consider the first question. Assume there is multiunit trade. Let every seller have the same reservation price and let this price be constant for all quantities thereby giving the following characteristic function:

$$(24) \qquad\qquad v(A_i) = a\mu_i,$$

where, as above, μ_i denotes the maximum quantity that A_i has available for sale. These assumptions are equivalent to supposing that all sellers have the same constant average cost and that they differ only with respect to how much they can sell. The next chapter examines this case assuming also that each buyer wants at most one unit. For these conditions theorem III.1 proves that there will be a common price per unit equal to a given certain other conditions. Moreover, this conclusion remains valid for more general demand conditions where the buyers want different amounts of the good. Therefore, there are situations where multiunit trade implies a common unit price. Moreover, these situations require large numbers of traders in the sense of having several essentially interchangeable traders of each type so that a given trader can obtain precisely the same terms from several other traders.

The second question pertains to the prevalence of a common unit price in the real world. It cannot be doubted that this practice is common in many retail markets, but it can be explained without recourse to the state of competition in these markets. The theory herein neglects transaction costs and studies the outcome of trade while assuming that traders can costlessly join any coalition. In reality there are costs due to the time and trouble of negotiating and seeking information. To avoid or merely reduce such costs, the traders may well announce constant unit prices and be willing to forgo the chance of obtaining better terms because this raises the net gain, allowing for transaction costs. These considerations weigh heavier the more numerous the traders. Since competition is commonly supposed to increase with the number of suppliers and since a common constant unit price is supposed to characterize competition, this conjuncture of circumstances may explain the prevalent belief that a common unit price is *necessary* for competition. The preceding analysis does prove that convexity, implying a nonempty core, also implies the existence of a common unit price capable of satisfying all of the core constraints, but this is not necessarily the only possible manifestation of competition.

In fact we do not observe a common unit price in all competitive markets in the real world, which is implied by core theory for multiunit trade. Moreover, a complex set of terms is consistent with the core in these cases. For example, quantity discounts are common so that buyers of large amounts pay lower average prices than buyers of small amounts. Sellers may quote different terms to different customers depending on the buyer characteristics. Core theory as applied to multiunit trade determines the quantity exchanged and the amount spent, but it does not imply that the average price per unit is necessarily the same for all buyers and sellers.

According to core theory competition is the *process* by which the traders reach agreement on the terms, and many kinds of contracts are compatible with competition.

6. Competition and Numbers

In this section we study how the number of traders affects the outcome of exchange assuming throughout that the imputations are in the core. We shall begin with a general proposition about nonzero-sum n-person games.

LEMMA 4. *Denote player i by A_i and let*

$$(1) \qquad\qquad I_n = \cup_{i=1}^{n} A_i.$$

Let a characteristic function be defined for all possible coalitions of $\{A_i\}$ and assume a nonempty core for the n-person nonzero sum game defined by this characteristic function. Let $x = \{x_i\}$ denote the imputations in the core of this game. If the imputations satisfy group rationality, then

$$(2) \qquad\qquad v(A_n) \leqslant x_n \leqslant v(I_n) - v(I_{n-1}),$$

where

$$(3) \qquad\qquad I_{n-1} = I_n - A_n.$$

Proof. By hypothesis

$$(4) \qquad\qquad \sum_{1}^{n} x_i = v(I_n),$$

and since x is in the core,

$$(5) \qquad\qquad \sum_{1}^{n-1} x_i \geqslant v(I_{n-1}).$$

In addition

$$v(I_n) = v(I_{n-1} \cup A_n) \geqslant v(I_{n-1}) + v(A_n).$$

Without loss of generality we may assume that $v(A_n) \geqslant 0$ from which it follows that $v(I_n) \geqslant v(I_{n-1})$. From (4) and (5)

$$x_n = v(I_n) - \sum_{1}^{n-1} x_i \leqslant v(I_n) - v(I_{n-1}).$$

Since x is in the core, $x_n \geqslant v(A_n)$. QED.

If we regard $v(I_n) - v(I_{n-1})$ as the marginal contribution of A_n to the market, then the lemma asserts that the imputation of A_n cannot exceed his marginal contribution, if there is group rationality, but A_n may receive less than his marginal contribution. We now study when every trader does in fact obtain his marginal contribution.

We begin by defining a certain collection of coalitions. Let

$$A'_i = \cup_1^m A_h - A_i,$$
$$B'_j = \cup_1^m B_h - B_j,$$
(6) $$T = \{(4'_i, \cup_j B_j), (\cup_i A_i, B'_j)\},$$

define the collection we shall study. The coalitions in T are of particular interest because of lemma 4. Thus, if there is group rationality, then

(7) $$x_i \leqslant v(I) - v(A'_i, \cup_j B_j),$$

and

(8) $$y_j \leqslant v(I) - v(\cup_i A_i, B'_j).$$

The problem is to find sufficient conditions for equality in (7) and (8) so that the imputations of the traders attain the upper bounds.

We first analyze the collection T for single-unit trade. Assume there are k units exchanged and that A_i are the owners, B_j the nonowners. The intramarginal traders are A_i and B_{m+1-i} for $i = 1, ..., k$, while the extramarginal traders are those for which $i = k+1, ..., m$. Consequently,

(9) $$v(I) = \sum_1^k b_{m+1-i} + \sum_{k+1}^m a_i.$$

It is easy to calculate the values of the characteristic function for the coalitions in T given as follows:

(10) $$\begin{cases} v(A'_i, \cup B) = v(I) - b_{m+1-k} & (i = 1, ..., k), \\ v(A'_i, \cup B) = v(I) - a_i & (i = k+1, ..., m), \\ v(\cup A, B'_j) = v(I) - (b_j - a_k) & (j = m, m-1, ..., m+1-k), \\ v(\cup A, B'_j) = v(I) & (j = 1, ..., m-k). \end{cases}$$

Let

(11) $$x^* = \sum_1^m x_i \quad \text{and} \quad y^* = \sum_1^m y_j.$$

Corresponding to the coalitions in T there are the core constraints as follows:

(12) $$\begin{cases} x^* + y^* - x_i \geqslant v(I) - b_{m+1-k} & (i = 1, ..., k), \\ x^* + y^* - x_i \geqslant v(I) - a_i & (i = k+1, ..., m), \\ x^* + y^* - y_j \geqslant v(I) - (b_j - a_k) & (j = m, m-1, ..., m+1-k), \\ x^* + y^* - y_j \geqslant v(I) & (j = 1, ..., m-k). \end{cases}$$

Let $\{\delta_i / i = 1, ..., m\}$ be the set of dual variables for the first m constraints and let $\{\sigma_j / j = 1, ..., m\}$ be the set of dual variables for the second set of m

constraints. Also let

(13) $$\Delta = \sum_1^m \delta_i \quad \text{and} \quad \Omega = \sum_1^m \sigma_j.$$

The dual problem is defined by

(14) $$\max \left\{ \sum_1^k \delta_i[v(I) - b_{m+1-k}] + \sum_{k+1}^m \delta_i[v(I) - a_i] \right.$$
$$\left. + \sum_1^k \sigma_{m+1-i}[v(I) - (b_{m+1-i} - a_k)] + \sum_{k+1}^m \sigma_{m+1-i}v(I) \right\}$$

subject to

(15) $$\Delta + \Omega - \delta_i \leqslant 1 \quad \text{and} \quad \Delta + \Omega - \sigma_j \leqslant 1$$

with $\delta_i, \sigma_j \geqslant 0$. As long as there are extramarginal traders this problem has a trivial solution given by $\delta_i = 0$ for all i, and one of the $\sigma_{m+1-i} = 1$ for $i = k+1, ..., m$, with all other σ's set equal to zero. A more interesting situation is where there are no extramarginal traders. If so, only the first and the third set of constraints in (12) are relevant and the objective function in (14) includes only the first and third members. The closest approach to the upper bound $v(I)$ is by choosing the following values of the dual variables:

$$\delta_i = 0, \qquad\qquad \text{for all} \quad i,$$
$$\sigma_1 = 1,$$
$$\sigma_{m+1-i} = 0, \qquad \text{for} \quad i = 1, ..., m-1.$$

With these choices the objective function attains its maximum given by

(16) $$v(I) - (b_1 - a_m).$$

In both of these cases there is equality in only one of the core constraints shown in (12). In the first case it is trivial to show that the collection is balanced since there is group rationality. In the second case the collection T is balanced if and only if $b_1 = a_m$. Moreover, from the previous work on single-unit trade it is known that the collection of pairwise coalitions

$$\{(A_i, B_{m+1-i})/i = 1, ..., m\}$$

is not only balanced but also achieves group rationality. Finally, in both cases it is false that the traders receive the value of their marginal contribution. This analysis provides the clues for the general case of multiunit trade. We now prove theorem 8.

THEOREM 8. *Let a multitrade market consist of m A_i's and m B_j's. Let the imputations be in the core and satisfy group rationality. Every trader's imputation equals his marginal contribution if and only if the collection T is balanced with the choice of*

(17) $$\delta_i = \sigma_j = 1/(2m-1).$$

Proof. As in (11) let

$$x^* = \sum_1^m x_i \quad \text{and} \quad y^* = \sum_1^m y_j.$$

Sufficiency. Assuming (17) is true and that T is a balanced collection, it follows that there is equality in the core constraints given by

(18) $\qquad x^* + y^* - x_i = v(A'_i, \cup B) \qquad (i = 1, ..., m),$

(19) $\qquad x^* + y^* - y_j = v(\cup A, B'_j) \qquad (j = 1, ..., m).$

Since there is group rationality and T is balanced, $x^* + y^* = v(I)$. Therefore,

(20) $\qquad x_i = v(I) - v(A'_i, \cup B),$

and

(21) $\qquad y_j = v(I) - v(\cup A, B'_j).$

Necessity. Assume (20) and (21) hold. Summing over i and j gives

(22) $\qquad x^* + y^* = 2mv(I) - \sum_i v(A'_i, \cup B) - \sum_j v(\cup A, B'_j).$

By group rationality $x^* + y^* = v(I)$. Hence (22) implies

(23) $\qquad \sum v(A'_i, \cup B) + \sum v(\cup A, B'_j) = (2m - 1)\, v(I).$

Therefore, (17) does in fact yield a balanced collection T. QED.

Observe that when the hypotheses of theorem 8 are satisfied then the imputations in the core are uniquely determined so that there is no range of indeterminacy, as is typically true for either single- or multiunit trade. For instance, traders receive the value of their marginal contribution with single-unit trade if

(24) $\qquad a_k = b_{m+1-k}.$

This result can be verified by applying theorem 8 to single-unit trade.[8] Moreover, with single-unit trade and the choice of weights given in (17), the value of the objective function in the dual problem is

(25) $\qquad v(I) - [k/(2m-1)](b_{m+1-k} - a_k) \leqslant v(I).$

Hence for the collection T to be balanced with single-unit trade it is sufficient that either (24) is satisfied or that

---

8. With single-unit trade and k intramarginal traders the value of the objective function is

(i) $\qquad (\Delta + \Omega) v(I) - \sum_1^k \delta_i b_{m+1-k} - \sum_{k+1}^m \delta_i a_i - \sum_1^k \sigma_{m+1-i}(b_{m+1-i} - a_k).$

If

$$\delta_i = \sigma_j = 1/(2m-1),$$

then

$$\Delta + \Omega = 2m/(2m-1)$$

and (i) reduces to (25) in the text.

(26) $$k/(2m-1) = 0.$$

Since $k > 0$ for an essential market (one in which some trade is advantageous), the latter condition suggests the role of numbers in a single-unit market. However, if the ratio of k to $2m-1$ remains constant as m increases, then condition (24) is pertinent for the determinacy of the equilibrium.

Another way to show how numbers affect the result is to show the consequences of adding a trader in a single-unit market. Let us begin with mA's and mB's such that k units are exchanged. Assume that A_{m+1} with a reservation price a_{m+1} enters the market. The quantity traded remains the same if $a_{m+1} > a_k$ though the indeterminacy of the price may be reduced if $a_{m+1} < a_{k+1}$. However, if $a_{m+1} < a_k$, then one more unit is traded, but whether the common price decreases depends on the relation between a_k and b_{m+k}, which is the limit price of the most eager of the extramarginal buyers. Plainly, the common price cannot *increase* as a result of entry of A_{m+1}. A similar analysis applies to the effects of an additional B, say, B_{m+1}. If $b_{m+1} < b_{m-k}$, then there can be no effect on either the quantity traded or the price. However, if $b_{m+1} > b_{m+1-k}$, then the quantity traded must rise by one unit and the effect on the price depends on the relation between b_{m+1-k} and a_{k+1}. Notice that everything depends on the relations between the limit prices at the margin and not on the number of traders.

7. The Number of Traders and the Emptiness of the Core

In this section we derive some general properties of a market core by studying a certain sequence of markets generated by increasing the number of traders within a prescribed class of trader types which are characterized by their preferences and initial stocks. In this work we shall not distinguish between initial owners and initial nonowners and shall instead use the more general framework of section 4. The present investigation is inspired by a simple example to which we now turn.

Let a trader start with one unit of the good and limit prices as follows:

$$A^{(1)}: \quad 3 \quad 6 \quad 0.$$

This trader would not accept less than 3 for one unit nor pay more than 6 for an additional unit. Finally, he would be willing to acquire at most two units. We now construct a sequence of markets $\{M_m\}$ such that market M_m contains m identical traders of type A, where $m = 1, 2, 3, \dots$. Let $v_m(I)$ denote the value of the characteristic function for the coalition of all of the m identical traders in M_m. We can readily verify that

(1) $$v_m(I) = v_{m-1}(I) + \begin{cases} 3, & \text{for odd } m, \\ 6, & \text{for even } m, \end{cases}$$

and, plainly,

(2) $$v_0(I) = 0.$$

Since the traders are identical, all must receive the same imputation or the core is empty. In fact the core is empty for odd m and is nonempty for even m.

The core constraints are of the form as follows:

(3) $$|S|x \geqslant v(S),$$

where x denotes the common imputation of the traders and $|S|$ denotes the number of members of the coalition S. To ascertain the emptiness of the core, we solve min mx subject to (3), for all $S \subset I$. Let

$$x = v_{m-1}(I)/(m-1), m > 1.$$

Clearly, if $m-1$ is even so that m is odd, then the core must be empty since all traders must receive an imputation of 4.5 and $v_m(I)$ is too small. However, if $m-1$ is odd, so that m is even, then $v_m(I)$ is big enough to give every trader an imputation of 4.5, thereby satisfying all core constraints. This proves that the core is empty for m odd and nonempty for m even.

The example contains the essential property of a sequence of markets generated so that the number of trader types is constant while the number of each type of trader rises. Let there be n classes of traders, $A_j, j = 1, \ldots, n$, with m members of each class giving a total of mn traders. We shall examine the effect of increasing m.

DEFINITION. *Let M_m denote a market with m traders of each type A_j, $j = 1, \ldots, n$.*

(4) $$P = \{M_m/m = 1, 2, 3, \ldots\}$$

denotes a countable set of markets.

We now prove theorem 9.

THEOREM 9. *Let*

(5) $$P = P_c \cup P_{\bar{c}},$$

where P_c denotes the subset of P containing the markets with nonempty cores and $P_{\bar{c}}$ denotes the subset of P containing the market with empty cores. For every trader class A_j assume that

(6) $$dF_j(r) \geqslant 0, \qquad \qquad for\ all\ r \geqslant 0,$$

so that every trader class has nonnegative limit prices. Also assume that for every trader class A_j there is an r_{ju} such that

(7) $$dF_j(r) = 0, \qquad \qquad for\ all\ r \geqslant r_{ju}.$$

Then P_c is a countably infinite nonempty set, and one cannot have $P = P_{\bar{c}}$. However, if $P_{\bar{c}}$ is not an empty set, then it must be countably infinite.

Observe that none of the traders are insatiable so that eventually their excess demand functions become and remain zero. Also no conditions are imposed on the nature of their excess demands save that the limit prices are nonnegative. Before proving the theorem we must determine the nature of the allocations of goods consistent with core imputations. As in section 4, let k_{ij} denote the initial holdings of goods by trader i in class j. Since every trader in a class is identical and has the same initial holdings,

(8) $$k_{ij} = k_j > 0, \qquad\qquad \text{for all } j.$$

Let

(9) $$K = \sum_j k_j$$

so that

(10) $$mK = \sum_{i,\,j} k_{ij}.$$

By definition

(11) $$v_m(I) = \max_{q_{ij}} \sum_j \sum_i F_j(q_{ij})$$

subject to

(12) $$q_{ij} \geqslant 0 \quad \text{and} \quad \sum q_{ij} = mK,$$

where

$$F(q) = \int_0^q dF(r)$$

as in section 4.

Form the Lagrangian expression

(13) $$L(q, w, \lambda) = \sum_{j,\,i} F_j(q_{ij}) + (q, w) + \lambda(mK - \sum q_{ij}),$$

where

(14) $$(q, w) = \sum_{ij} w_{ij} q_{ij}$$

denotes the scalar (inner) product, and both w and λ are Lagrangian multipliers. A rather mild regularity condition on F permits the application of the Kuhn-Tucker condition to this problem in order to obtain the necessary condition for a maximum (Kuhn and Tucker 1950). The regularity condition is that

(15) $$F(q) = \int_0^q f(r)\, dr$$

with $f(r)$ a continuous function of the continuous variable q thereby making $F(q)$ a continuously differentiable function. The necessary conditions for a maximum give the following results:

(16) $$\frac{\partial L}{\partial q} = f_j(q_{ij}) - \lambda + w_{ij} \leqslant 0,$$

(17) $$q\frac{\partial L}{\partial q} = 0,$$

(18) $$w_{ij} \geqslant 0, \quad \text{and} \quad w_{ij}q_{ij} = 0.$$

Consequently,

$$[f_j(q_{ij}) - \lambda]q_{ij} = 0$$

so that

(19) $\qquad q_{ij} > 0 \quad \text{implies} \quad q_{ij} = q_j \quad \text{and} \quad f_j(q_j) = \lambda > 0.$

Therefore, all traders in a class A_j receiving a positive quantity must all receive the same positive quantity, but not all traders in A_j need receive a positive quantity of goods. That is, within a class A_j there may be two groups of traders such that members of one group receive the same positive amount while members of the second group receive none of the good. For the latter

$$0 \leqslant w_{ij} \leqslant \lambda - f_j(0)$$

implying that

(20) $$f_j(0) \leqslant \lambda = f_j(q_j).$$

The Langrangian multiplier λ in this calculation resembles the classical competitive price, but the analogy is imperfect because not all the identical traders in a given class receive the same quantity of the good. For example, this can happen for demand schedules like UV shown in figure 2.1 and demand schedules like XY in the same figure are also admissible. Therefore, in class A_j: α traders each get $q_j > 0$; $m - \alpha$ traders each get 0. It follows that

$$F_j(q_{ij}) = \begin{cases} F_j(q_j), & \text{for } \alpha \text{ traders,} \\ 0, & \text{for } m - \alpha \text{ traders.} \end{cases}$$

The existence of a maximum is guaranteed by the continuity of the objective function $\Sigma\, P_j(q_{ij})$, and the boundedness of the constraint set as expressed by (6) and (7). Hence we may conclude that $v_m(I)$ is given by

(21) $$v_m(I) = \Sigma\, \alpha_j F_j(q_j),$$

where both the α's and q's may depend on m. Although the Pareto optimal allocation does not imply that all of the traders in a given class receive the same amount of *goods*, for a nonempty core all traders must receive the same *imputation* x_j, where, as in section 4,

(22) $$x_j = F_j(q_j) + z_j.$$

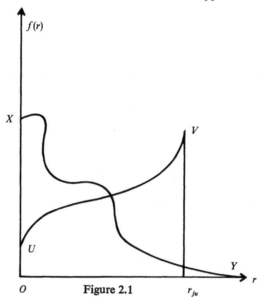

Figure 2.1

A coalition S in M_m is defined by the number of traders it contains from each class. Let β_j denote the number of traders of class A_j in S. Then,

$$(23) \qquad\qquad S = \langle \beta_1, \beta_2, ..., \beta_n \rangle, \qquad\qquad \text{with } 0 \leqslant \beta_j \leqslant m.$$

The term $v(S)$ is derived by the same procedure as the one for $v_m(I)$. Consequently, for coalition S and class A_j,

$$\alpha^{S}_{j,m}, \qquad\qquad \text{traders each get } q^{S}_j > 0,$$
$$\beta^{S}_j - \alpha^{S}_{j,m} \qquad\qquad \text{traders each get } 0,$$

where $0 \leqslant \alpha^{S}_{j,m} \leqslant \beta^{S}_j$. Hence,

$$(24) \qquad\qquad v(S) = \sum \alpha^{S}_{j,m} F_j(q^{S}_j),$$

and

$$(25) \qquad\qquad \sum_S q^{S}_{ij} = \sum \alpha_j k_j.$$

We may now begin proving theorem 9. From the previous results if all traders have convex preferences so that $f(r)$, the excess demand function, is a decreasing function of r, then $P = P_c$ and $P_{\bar{c}}$ is empty establishing that the theorem is true for this case, and we need only consider the case where not all traders have convex preferences. Moreover, by theorem 6 we need only consider whether the money core constraints can be satisfied with the Pareto allocations. As we shall see, we may specialize even more since it suffices to take the case in which only one trader class, say A_1, has concave preferences, or, more precisely, preferences requiring satiation of at least some traders of the class, provided the total stock is adequate. To prove these assertions consider the Pareto allocations for a market where some traders have concave preferences as shown by XZ in figure 2.2 and others have convex preferences as shown by WC. Assume that the area $OXZE$ exceeds the area OWC and that the area $OXYD$ equals the area OWC. If a coalition S forms with the two classes of traders, then $v(S)$ would be attained by dividing the total stock among the traders with convex preferences, WC, when the total stock is below OD, but when the total stock is above OD then it would be necessary to give it all to the trader with concave preferences until he is satiated, that is, until he has OE. Since the total stock rises with m, this argument shows that in a market with a mixture of preferences such that at least one class includes traders who will become satiated due to their nonconvex preferences, the characteristic function will assume the shape given by the arc OC_1 in figure 2.5. The maximum $v_m(I)/m$ occurs for those values of m permitting the largest number of class 1 traders to be satiated and leaving the smallest possible remaining stock to be divided among the other trader classes.

Next we can convince ourselves that without loss of generality we may assume that the market has only one class of traders with absorbing preferences like A_1. Suppose on the contrary that there were two such classes with the preferences as depicted in figure 2.3. For both classes to be absorbing, the areas $OR_1S_1Q_1$ and $OR_2S_2Q_2$ would have to be equal. But then clearly $v_m(I)$ is attained by satiating the class with the smaller upper bound, OQ_1. Therefore, to have two absorbing classes, necessarily, at the Pareto optimal allocations both the quantities q_j and the values $F_j(q_j)$ must be the same for the two classes. This would require a relation between the preferences of the two absorbing classes as shown in figure 2.4. Consequently, for all practical purposes we may as well assume that there is only one absorbing class.

Let the only absorbing class in the market be A_1 and assume that q_1 units satiate an A_1 trader so that

$$(26) \qquad mK = \alpha q_1 + r,$$

where K is defined in (9), m and α are integers, and r, the remainder, satisfies

$$(27) \qquad 0 \leqslant r < q_1.$$

This means that α of the A_1 traders are "full" and the remaining stock, r is either used to begin filling another A_1 trader or it is divided among the other trader types.

We lose no generality by assuming that $K < q_1$. This simplifies the exposition without essentially affecting the behavior of $v_m(I)$ as a function of m. The function $v_m(I)$ will be a periodic function of m so that there will be identical remainders r for prescribed multiples of m if and only if q_1/K is a

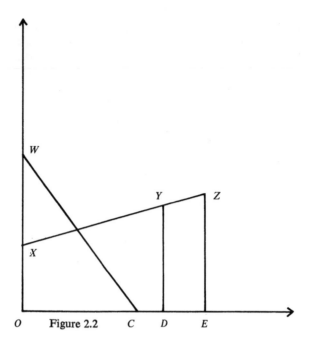

Figure 2.2

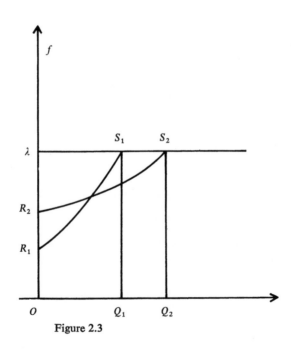

Figure 2.3

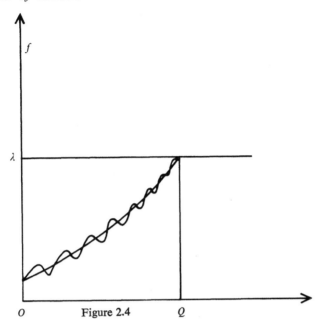

Figure 2.4

rational number. But if q_1/K is not rational, then $v_m(I)$ will be arbitrarily close to a periodic function of m because there are rational numbers arbitrarily close to q_1/K.

Initially, assume that q_1/K is rational so that $v_m(I)$ is periodic and there is a smallest m, denoted by m', such that $v_{m'}(I)/m'$ is a maximum. Moreover, for all integer multiples of m', hm' with $h = 1, 2, ...$,

$$v_{hm'}(I)/hm' = v_{m'}(I)/m',$$

because one can have precisely the same Pareto optimal allocations in the sequence of markets $\{M_{hm'}/h = 1, 2, ...\}$.

We now show that the empty core markets are the ones for which $m \neq hm'$ and $m > m'$. Consider an $m \neq hm'$ with $m > m'$ so that there is a coalition of an appropriate multiple of m', say, $m = gm' + p$, where g is an integer and $0 \leqslant p < m'$, such that any complete collection of all of the trader types having gm' traders of each type can guarantee their members an amount $v_{gm'}(I)/gm'$ so that

$$v_{gm'}(I)/gm' > v_m(I)/m.$$

All such collections are balanced and with appropriate choices of dual variables δ_S it would follow that

$$\max \Sigma \, \delta_S v_m(S) > v_m(I).$$

In carrying out this calculation we have all possible coalitions of the trader types taken gm' at a time and use the familiar methods from section 3 and the beginning of this section to establish the emptiness of the core for all m such that $m \neq hm'$, $h = 1, 2, \ldots$.

Retaining the assumption that q_1/K is rational, we must now show that all core constraints can be satisfied for all $m = hm'$ so that

$$P_c = \{M_{hm'}/h = 1, 2, \ldots\}.$$

In this case all traders can receive the same imputation as follows:

(28) $x_j = x = v_m(I)/mn,$

and all core constraints can be satisfied with this choice of x_j since

$$\Sigma \ \alpha_j x_j \geqslant v(\alpha_1, \ldots, \alpha_n)$$

and

(29) $v_m(I)/mn > v(\alpha_1, \ldots, \alpha_n)/\displaystyle\sum_j \alpha_j.$

There remains only the case where q_1/K is irrational. If so, there is a rational number arbitrarily close to q_1/K, and for each of the rational numbers in a neighborhood of q_1/K there is an m' having the property that the cores of all markets $M_{hm'}$ are all nonempty for integral h. Moreover, the characteristic functions for these markets will also be arbitrarily close to the characteristic functions for the original market. Hence the same properties hold for the sequence of markets M_m when q_1/K is irrational as when it is rational.

Since P_c contains a countable sequence of markets, whether $P_{\bar c}$ is empty depends on the value of m'. If $m' = 1$ so that the core of the market is not empty, then there is one trader of each type, and all markets having an equal number of each trader type will have nonempty cores. But if $m' > 1$, then only markets with integral multiples of m' have nonempty cores so that $P_{\bar c}$ contains a countably infinite sequence of markets. This concludes the proof of theorem 9.

Theorem 9 characterizes markets in general in a manner capable of graphical illustration as shown in figure 2.5. If $v_m(I)$ is a periodic function of m, its graph is composed of a succession of arcs $OC_1C_2C_3 \ldots$, so that the only markets with nonempty cores are those for which m is an integral multiple of ON_1 where ON_1 corresponds to m' in theorem 9. If $m' = 1$, then $v_m(I)$ is a straight line through the origin connecting the points O, C_1, $C_2 \ldots$. In this case $P_{\bar c}$ is empty and all markets with an equal number of each trader type have nonempty cores. When ON_1 is irrational, there are rational numbers arbitrarily close to it, and for each of these a corresponding sequence of markets with a periodic function $v_m(I)$, which approximates the given function $v_m(I)$ as closely as desired. Hence the conclusion of the theorem remains valid whether or not ON_1 is rational because there are rational numbers in an arbitrarily small neighborhood of ON_1. The frequency of nonempty cores depends on the period m', which may be a very large number.

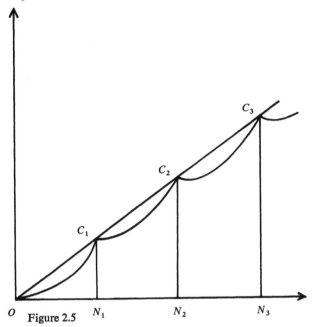

Figure 2.5

Therefore, given a set of trader types such that m' is large and proportionate growth in the number of each trader type, it may well be that almost all markets will have empty cores although a countably infinite number will have nonempty cores.

8. Conclusions

As shown above in section I.9 the classical competitive equilibrium is characterized by a set of positive prices such that every person maximizes his utility $U_i(x^i)$ subject to a budget constraint $(p, x^i) = (p, \sigma^i)$, where the m-vector x^i denotes his bundle of goods, σ^i is his initial endowment, and p is the competitive price vector. Since everyone's budget constraint is satisfied, the competitive price vector p clears all of the markets so that $\Sigma_i x^i = \Sigma_i \sigma^i$.

Shapley's theorem asserts that if a competitive equilibrium exists then the associated set of imputations of goods $\{x^i\}$ must be in the core. We also know that if all individual utility functions are concave so that the indifference curves are convex to the origin then a classical competitive price equilibrium does in fact exist. Therefore, a nonempty core is necessary for the existence of a classical competitive price equilibrium but not sufficient. This is to say that the core may be nonempty without there existing a set of positive prices capable of clearing the market at which the individual traders maximize their utilities subject to their budget constraints. For instance, we know that every balanced market has a nonempty core, but this does not mean that there are constant unit prices for the goods such that the implied trades can satisfy all of the core constraints.

The salient property of the classical competitive equilibrium is the existence of a set of prices independent of the individual traders' actions that is capable of clearing the market. Mathematically, this is expressed by the linearity of the budget constraint. Each individual's budget constraint is a hyperplane which in the case of concave utility functions supports a convex set of the preferred commodity bundles. The individual can be thought of as minimizing his outlay over the latter set. However, in the more general situation encompassed by core theory in which the individuals are not restricted to having convex indifference curves there may not be a set of linear budget constraints capable of generating commodity bundles in the core. In other words, in the general case where the individual preference sets need not be convex a more complicated set of terms of trade is required to generate allocations of goods in the core. Linear budget constraints no longer support the individual preference sets. Essentially, the classical concept of a competitive equilibrium refers to the possibility of having a decentralized economy of a simple sort, namely, one in which constant unit prices can satisfy the individual demands. But this is not the only possible kind of competition. Competition as represented by the freedom of contracting may require more complicated terms of trade among the traders than that represented by linear budget constraints.

III

Applications of the Core to Oligopoly

1. Introduction

Oligopoly refers to a market in which a "few" producers sell a homogeneous product to "many" buyers. However, this terminology is misleading because some aspects of oligopoly appear even when there are two sellers and one buyer. The theory of the core is well suited for analyzing those situations treated in oligopoly theory not only for its new results but also because it forces a rigorous examination of several often neglected aspects of oligopoly. For example, with core theory it is necessary to prove in every case whether or not there will be group rationality (Pareto optimality) and whether or not there will be price discrimination. The absence of price discrimination is often taken for granted, but it should not be, because a seller may maximize his return by charging every buyer the highest price he would be willing to pay rather than go without the good altogether. Why does this not always happen?

These considerations pose the general problem of distinguishing between the assumptions and the deductions about behavior in oligopoly. Most economists develop theories which assume that individuals attempt to maximize something, be it utility, profit, the value of the firm or what not, subject to given constraints. One may regard the constraints as equivalent to assumptions about behavior. Of course, assumptions are indispensable elements of theory, but they are not proper subjects of dispute. The best guide for choosing among theories is to see which have useful empirical implications and not whether their assumptions are distasteful. For example, in the classical theory of monopoly it is assumed that all buyers pay the same price. This is an assumption about behavior. Presumably, some firms can obtain a higher return by price discrimination. Hence the absence or presence of price discrimination should be deduced and not assumed. The Cournot theory of duopoly also illustrates this thesis. This theory is sometimes attacked on the ground that it assumes certain things about the behavior of the duopolists. These assumptions about behavior are easily restated as constraints in a maximum problem. Seen in this light, the Cournot theory is no less plausible than the simple theory of monopoly, which assumes no price discrimination. The point is that assumptions about behavior do not constitute grounds for rejecting a theory, a moral to be kept in mind when reading theories of oligopoly.

In the following analysis it is assumed that all sellers have the same production cost schedules such that average and marginal cost both equal a constant, a. Let m_i denote the number of units available for sale by A_i. Therefore, the imputation of seller A_i must satisfy

(1) $$x_i \geqslant am_i.$$

Every buyer B_j is assumed to desire at most one unit of the good at a limit price b_j. The buyers are indexed so that

(2) $$b_{j+1} \geqslant b_j.$$

To illustrate the problems of collusion among sellers, consider the case in which there are two sellers, $m_1 = m_2 = 1$, and there is one buyer. If the imputations are in the core, then they must satisfy the inequalities as follows:

(3) $$x_i \geqslant a,$$

(4) $$x_i + y \geqslant v(A_i, B) \qquad\qquad (i = 1, 2),$$

(5) $$x_1 + x_2 + y \leqslant v(I) \qquad\qquad \text{(feasibility)}.$$

We first prove that there must be group rationality, that is, equality in (5). For $i \neq i'$ we have

$$v(I) = \max \{a + \max \{a, b\}\} \leqslant a + \max \{a, b\} = v(A_i) + v(A_{i'}, B).$$

Therefore,

$$x_i + y + x_{i'} \geqslant v(A_i) + v(A_{i'}, B) \geqslant v(I),$$

giving an implication of equality in (5), which becomes the following:

(6) $$x_1 + x_2 + y = v(I).$$

Next we prove that if the two sellers do not collude, then they will obtain the same imputation, a. Equation (6) implies that

(7) $$x_2 = v(I) - (x_1 + y) \leqslant v(I) - v(A_1, B) = a.$$

Therefore, together with (3), we obtain

(8) $$x_2 = a,$$

and, similarly, $x_1 = a$. However, if the sellers collude, then the two inequalities of (4) do not apply, and the imputations satisfy $a \leqslant x_i \leqslant b$, which are the limits for bilateral monopoly. In this model we cannot determine whether or not the sellers collude although their potential gain from collusion can be large.

Because there is only one buyer in this example, plainly there cannot be price discrimination. With two buyers having different price limits, price discrimination becomes possible if and only if the two sellers collude while the buyers do not collude. Thus, if the sellers compete, both buyers would pay the same price. However, if the buyers collude, then regardless of what the sellers do, price discrimination becomes impossible. We now prove these assertions.

Let there be two buyers B_1 and B_2 such that $b_1 < b_2$. Let the sellers be as described above. The feasibility constraint is

$$(9) \qquad \sum_{i=1}^{2} x_i + \sum_{j=1}^{2} y_j \leqslant v(I),$$

and the basic constraints of the core are

$$(10) \qquad x_i + y_j \geqslant v(A_i, B_j) = \max \{a, b_j\} \qquad (i, j = 1, 2).$$

Obviously,

$$v(I) = v(A_1, B_2) + v(A_2, B_1),$$

so that the core imputations are group rational. Given group rationality, namely,

$$(11) \qquad \Sigma x_i + \Sigma y_j = v(I),$$

together with the basic core constraints, (10), it follows that

$$(12) \qquad x_i + y_j = v(A_i, B_j) \qquad (i, j = 1, 2).$$

Group rationality and the equalities of (12) readily yield the result that there must be a common price, but it does not follow that this common price necessarily equals the lower limit, which is given by a, though this is true in the preceding case.

If the two sellers collude then none of the constraints involving a single seller and one or more of the buyers apply. In particular, the imputations need not satisfy the inequalities, (10). Hence there need not be group rationality for the core imputations nor need (12) hold. As a result it is no longer true that both buyers necessarily pay the same price. It is worth stating explicitly the constraints for this case. Let

$$(13) \qquad x = x_1 + x_2.$$

This represents the hypothesis that the two sellers collude and act as one.

Thus

$$x \geqslant \qquad 2a \qquad = v(A_1, A_2),$$

(14)
$$x + y_1 \geqslant v(A_1, A_2, B_1) = \max \{2a, a + b_1\},$$
$$x + y_2 \geqslant v(A_1, A_2, B_2) = \max \{2a, a + b_2\},$$

$$x + \sum_1^2 y_j \leqslant v(I) \qquad \qquad \text{(feasibility)}.$$

The price paid by the first buyer is $p_1 = b_1 - y_1$, and the price paid by the second buyer is $p_2 = b_2 - y_2$. Since

$$y_j \geqslant 0 \Rightarrow b_j - p_j \geqslant 0,$$

it follows that

(15)
$$p_j \leqslant b_j.$$

However, it is readily verified that $p_1 \neq p_2$ can satisfy the system of inequalities (14). Hence price discrimination is consistent with (14). The buyers can prevent price discrimination by cooperating with each other if the buyer who obtains the good at a lower price can resell it to the other in competition with the original sellers. In terms of the inequalities, such collusion among the buyers is represented by a system in which no one buyer treats individually with one or more sellers. The collective imputation of the buyers satisfies $y = y_1 + y_2$. Assuming that the buyers collude with each other and that the sellers also collude with each other, the pertinent inequalities are as follows:

$$y \geqslant v(B_1, B_2) = 0,$$
(16)
$$x \geqslant v(A_1, A_2) = 2a,$$
$$x + y \leqslant v(I).$$

This system represents bilateral monopoly. In (16) neither the buyers' method of cost sharing nor the sellers' method of revenue sharing is determinate, and there need not be group rationality.

One can *assume* group rationality in this case by the following argument. First, if trade is not mutually beneficial, it will not occur. In this case

$$x = 2a, \quad y = 0, \quad v(I) = 2a,$$

implying

$$x + y = v(I)$$

so that there is group rationality. Next, *assume* that if trade *is* mutually beneficial then it will occur. There are two possibilities: either one unit is traded because $b_1 < a < b_2$; or it is mutually beneficial to trade two units because $a < b_1 < b_2$. In the first case the sellers' receipts are $x - a$ (since only one unit is sold), and the buyers' payment is $b_2 - y$. Since payments must equal receipts, $x - a = b_2 - y$, and we can conclude that

$$x + y = a + b_2 = v(I).$$

In the second case it is mutually beneficial for two units to be sold. The sellers' receipts are x and the buyers' payments are $\Sigma\, b_j - y$. Payments must equal receipts, $x = \Sigma\, b_j - y$, and, therefore,

$$x + y = \Sigma\, b_j = v(I).$$

Hence, there is group rationality in all cases. The assumption of group rationality is seen to be equivalent to the assumption that whenever trade would be mutually beneficial, it would always happen. Since this seems so reasonable, why should we not assume group rationality?

However, group rationality would also predict the impossibility of simple monopoly, the situation where one seller sets a common price for all buyers in order to maximize his net revenue. This is a consequence of the well-known fact that simple monopoly is not Pareto optimal (equivalent to group rationality). The assertion is readily demonstrated in an instructive way. Let k' denote the quantity sold by the monopolist and let p' denote the monopoly price, where k' maximizes the monopoly net revenue given by $(p' - a)\,k'$, and the monopoly price satisfies

(17)
$$
\begin{aligned}
a < p' &< b_{n-(j-1)}, && \text{for } j = 1, \ldots, k', \\
p' > b_{n-(j-1)} &\geqslant a, && \text{for } j = k' + 1, \ldots, k.
\end{aligned}
$$

Under competition k units would be sold with $k > k'$. The nonowners $B_{n-(j-1)}$ for $j = k' + 1, \ldots, k$ would be willing to pay at least the marginal cost, a, and the seller would be willing to accept a unit price between a and p'. But then the monopolist would have to prevent sales in the lower price market from being resold in the higher price market thereby lowering his return in the latter more lucrative market. If it is too costly for the monopolist to stop the leakage between the markets, then simple monopoly would be more profitable although the imputations are not group rational. A similar argument explains the rationale for the Cournot-Nash solution of oligopoly, as we shall see.

These examples illustrate the many possibilities arising in even the simplest market structures. The following analysis elaborates this theme. There are three main topics. First, the previous results on the core are extended to a situation in which the sellers make a homogeneous product under constant returns to scale. Next, the core theory is applied to noncooperative duopoly for one time period in order to derive the main results of the Cournot-Nash theory. Finally, the Cournot-Nash theory is extended to a situation where sales occur over a sequence of time periods. This makes seller collusion enforceable for an infinite horizon but not for a finite horizon. We shall return to the latter problem in chapter V and again in chapter VI.

2. Properties of the Core under Constant Returns

It is convenient to repeat the basic assumptions of the model representing constant returns. Let there be s sellers with $s \geqslant 2$. Let each seller have the same constant cost per unit equal to the marginal cost per unit, a. Hence the imputation of seller A_i, denoted x_i, is required to satisfy

$$(1) \qquad\qquad x_i \geqslant v(A_i) = am_i \qquad\qquad (i = 1, ..., s),$$

where m_i is the maximum number of units available for sale by A_i. There are n buyers, B_j, who each desire to buy at most one unit. B_j's limit price is b_j, which are ordered as follows: $b_j \leqslant b_{j+1}$.

LEMMA 1. *Let the imputations of the sellers satisfy (1),*

$$(2) \qquad\qquad y_j \geqslant v(B_j) = 0,$$

and

$$(3) \qquad\qquad v(A_i, B_j) = max \{am_i, a(m_i - 1) + b_j\}.$$

Let the imputations be feasible and in the core. Then the market is essential if and only if there is a $k \geqslant 1$ such that

$$(4) \qquad\qquad b_{n-(j-1)} \geqslant a, \qquad\qquad for\ all \quad j \leqslant k.$$

Proof. If the imputations are in the core and the market is essential, then it pays for coalitions to form. Hence trade must be mutually beneficial. For the latter to be true, it is necessary and sufficient that (4) is satisfied. Conversely, if (4) is satisfied then trade is mutually beneficial, coalitions can form with benefit to their members so that the market must be essential. QED.

Before giving the properties of the core for this market we should see how it differs from the market structures analyzed in the preceding chapter. Chapter I considers two main kinds of structures. In the first, every potential seller A_i owns one unit of a good that he would be willing to sell at a price not less than a_i while the demand conditions are the same as in the present case. The properties of this market structure are given in theorems I.1, I.2, and I.3. These properties coincide with the well-known conclusions of the neoclassical theory. The second main market structure in chapter I works with multiunit traders so that the potential buyers and sellers are each interested in more than one unit of the good. The properties of this case, given in theorem I.6, differ from the neoclassical results. In particular there are wider price limits and there is no longer a one-to-one correspondence between units of the good and the traders. The present case constitutes an intermediate situation. While the buyer preferences are as in the first case, the seller constraints resemble those for the second case. However, because the sellers have the same cost functions, the properties of the core are akin to the neoclassical theory, as we shall see.

To facilitate the analysis, we shall use some game theoretic concepts introduced in chapter I. A game is said to be *decomposable* if there is a collection of exhaustive and disjoint coalitions, namely a partition,

$$\{S_k/S_k \subset I, S_k \neq \varnothing \quad \text{and} \quad S_k \neq I, \cup_k S_k = I, k = 1, \dots, K\}$$

such that

(5) $$\Sigma \, v(S_k) = v(I).$$

The sets in this partition are called *splitting sets.*[1] In chapter I, for example, theorem 5 asserts that the imputations of a decomposable game with a nonempty core satisfies group rationality. The pairs of traders defined in equation I.5 (11) constitute a splitting collection or partition. The present application shows how a market can be regarded as a collection of nonoverlapping submarkets in order to study the properties of its core. Notice that every inessential game is decomposable but not conversely. Decomposability means that there is at least *one* partition to satisfy (5); it definitely does not mean that *every* partition satisfies (5). Hence essential games can be decomposable.

THEOREM 1. *Let the market be essential so that*

$$a < b_{n-(j-1)}, \qquad\qquad \text{for} \quad j = 1, \dots, k.$$

If the imputations are feasible, in the core, and satisfy (1) and (2), then
  i. *there is group rationality;*
  ii. *there is a common price p such that*

(6) $$p = b_{n-(j-1)} - y_{n-(j-1)} \qquad\qquad (j = 1, \dots, k);$$

  iii. $\Sigma \, m_i > k$ *implies* $p = a$;
  iv. $\Sigma \, m_i \leqslant k$ *implies* $p \geqslant a$.

Proof. The idea of this proof is simple though the mechanics are complicated. To prove group rationality one can use theorem I.5 by displaying a collection of splitting sets. In fact there are many possible collections of splitting sets formed by mating a group of buyers with a seller. This decomposes the market into a collection of nonoverlapping submarkets plus the extramarginal traders. To establish a common price, the intramarginal buyers are permuted among the sellers. This directly demonstrates how the competition among the sellers establishes a common price. The last assertion of the theorem raises the delicate issue of the intensity of competition and shows that when the quantity offered by the sellers is large enough, the price is forced down to the lowest level they will accept.

We begin by constructing a collection of splitting sets. By hypothesis the

---

1. The concepts of a decomposable game and splitting sets are due to von Neumann and Morgenstern (1947, esp. chap. 9).

maximum quantity that can be sold is k units, where $k \geqslant 1$. Let the intra-marginal buyers be ordered according to their limit prices and assembled into the sets as follows:

$$W_1 = \cup_1^{r_1} B_{n-(j-1)},$$

$$W_2 = \cup_{r_1+1}^{r_2} B_{n-(j-1)},$$

$$\cdots$$

$$W_J = \cup_{r_J-1+1}^{r_J} B_{n-(j-1)},$$

where $J \leqslant s$, s being the number of sellers, and $r_J \leqslant k$. The extramarginal buyers are assembled in the set W_M so that

$$W_M = \cup_{k+1}^{n} B_{n-(j-1)}.$$

Let $r_0 = 0$. Then the r's must satisfy the constraints as follows;

$$(7) \qquad\qquad 0 < r_i - r_{i-1} \leqslant m_i, \qquad\qquad \text{for} \quad i = 1, ..., s.$$

W_1 contains the r_1 most eager buyers, W_2 the next $r_2 - r_1$ most eager buyers, and so on. Join these sets of buyers with sellers so that

$$(8) \qquad\qquad S_i = A_i \cup W_i \qquad\qquad (i = 1, ..., J).$$

If $k \geqslant s$, then it is possible to join every seller with at least one intramarginal buyer. However, this cannot be done if $k < s$. Hence let

$$(9) \qquad\qquad S_{J+1} = \cup_{J+1}^{s} A_i,$$

and let

$$(10) \qquad\qquad S_{J+2} = W_M.$$

S_{J+1} is the set of all sellers who are not joined with buyers, and the set S_{J+1} can be empty. The set S_{J+2} contains all of the extramarginal buyers, and it may also be empty. It is claimed that the collection of disjoint sets

$$S = \{S_i / i = 1, ..., J+2\}$$

splits the market. To prove this, one must show that

$$(11) \qquad\qquad v(I) = \sum_1^{J+2} v(S_i).$$

By hypothesis and by the definition of $v(I)$ it follows that

$$(12) \qquad\qquad v(I) = \sum_1^{k'} b_{n-(j-1)} + \left[\sum_1^{s} m_i - k' \right] a,$$

where

$$k' = \min \{k, \Sigma m_i\}.$$

To verify that (11) actually holds is easy given this expression for $v(I)$. Since (11) is true, theorem I.5 implies there must be group rationality so that

(13) $$\Sigma\, x_i + \Sigma\, y_j = v(I).$$

Next, we must show that there is a common price p satisfying (6). Since the imputations are in the core, it follows that for any splitting set S_i

(14) $$\sum_{S_i} x_i + \sum_{S_i} y_j \geq v(S_i), \qquad\qquad \text{for} \quad i = 1, ..., J;$$

(15) $$\sum_{S_{J+1}} x_i \geq v(S_{J+1});$$

(16) $$\sum_{S_{J+2}} y_j \geq v(S_{J+2}).$$

Sum these inequalities over any of the $J+1$ i-subscripts and if the remaining i-subscript is in the range from 1 to J, it follows from (13) that

(17) $$\Sigma\, x_i + \Sigma\, y_j \leq v(S_i).$$

If the remaining subscript $i = J+1$, we have

(18) $$\Sigma\, x_i \leq v(S_{J+1});$$

or finally if the remaining subscript is $i = J+2$, we have

(19) $$\Sigma\, y_j \leq v(S_{j+2}).$$

In all three cases there must be equality in (14) to (16) so that

(20) $$\sum_{S_i} x_i + \sum_{S_i} y_j = v(S_i), \qquad\qquad \text{for} \quad i = 1, ..., J;$$

(21) $$\sum_{S_{J+1}} x_i = v(S_{J+1});$$

(22) $$\sum_{S_{J+2}} y_j = v(S_{J+2}).$$

One result is immediate, for consider (22). For S_{J+2} we have

(23) $$\sum_{k'}^{n} y_{n-(j-1)} = v(\cup_{k'}^{n} B_{n-(j-1)}) = 0.$$

Since $y_{n-(j-1)} \geq 0$ for all j, it follows that

(24) $$y_{n-(j-1)} = 0, \qquad \text{for all} \quad j = k'+1, ..., n.$$

It is convenient to prove ii and iii simultaneously when $\Sigma\, m_i > k$. In this case we can choose a set of r's such that S_{J+1} is not empty and then (21) implies

(25) $$\sum_{S_{J+1}} x_i = v(\cup_{S_{J+1}} A_i) = a \sum_{S_{J+1}} m_i.$$

The coalition S_{J+1} can itself be further partitioned as shown by the following:

(26)
$$\sum_{S_{J+1}} v(A_i) = v(S_{J+1}).$$

Therefore, instead of (25) one may write

(27) $x_i = am_i,$ for all A_i in S_{J+1}.

implying that

(28) $x_i/m_i = a.$

But any A_i in S_{J+1} can be exchanged for an A_i in S_i with $i = 1, ..., J$ for an appropriate choice of a set of r's. Therefore, (22) is true for all $i = 1, ..., s$. This proves that if $\Sigma\, m_i > k$, then there is a common price $p = a$ satisfying (6).

However, if $\Sigma\, m_i \leq k$, then, for any set of r's, it follows that S_{J+1} is empty, and a different argument is necessary to prove the existence of a common price p to satisfy (6). In this case a set of r's can be chosen so that $J = s$. Consider the first two splitting sets, $S_1 = A_1 \cup W_1$ and $S_2 = A_2 \cup W_2$. By hypothesis neither A_1 nor A_2 is empty. Moreover,

(29)
$$v(A_1 \cup W_1) = \sum_1^{r_1} b_{n-(j-1)} + (m_1 - r_1)a;$$

$$v(A_2 \cup W_2) = \sum_{r_1+1}^{r_2} b_{n-(j-1)} + [m_2 - (r_2 - r_1)]a.$$

Equation (20) implies that

$$x_1 + \sum_1^{r_1} y_{n-(j-1)} = v(A_1 \cup W_1)$$

and that

$$x_2 + \sum_{r_1+1}^{r_2} y_{n-(j-1)} = v(A_2 \cup W_2).$$

These two equations can be written as follows:

(30)
$$x_1 = \Sigma[b_{n-(j-1)} - y_{n-(j-1)} - a] + am_1;$$
$$x_2 = \Sigma[b_{n-(j-1)} - y_{n-(j-1)} - a] + am_2.$$

Now let two buyers be interchanged between W_1 and W_2. In particular let $B_{n-(j_1-1)}$ be taken from W_1 and placed in W_2 and let it be replaced by $B_{n-(j_2-1)}$ from W_2. This generates two new W's that can be mated with A_1 and A_2 to form new splitting sets. For this new partition we can obtain equations like (30), one for x_1 and another for x_2. These new equations together with (30) imply that

(31) $b_{n-(j_1-1)} - y_{n-(j_1-1)} = b_{n-(j_2-1)} - y_{n-(j_2-1)} = p.$

By permuting the intramarginal buyers in this way among all of the W's, it follows that (31) holds for all of the intramarginal buyers. QED.

It is worth noting that the quantity sold by an individual is generally indeterminate in this model though the total sales are determinate. This is the same conclusion as the neoclassical theory when there are competing sellers with constant returns to scale. However, even if there were given amounts sold by each A_i, there would still be a common price if competition among the sellers allows the buyers to deal with whom they please. Hence for price discrimination, the sellers must agree both on how much to sell and on to whom they will sell.

In the market assumed in theorem 1 there are also coalitions that generate basic primary constraints analogous to those discussed in section I.5. These are generated by the *minimal splitting sets*, which are, as their name suggests, the smallest, nonoverlapping and exhaustive collections that decompose the market. By appropriately permuting the participants, one can construct certain collections of minimal splitting sets that constitute basic primary constraints among buyers and sellers that imply a common price and the equilibrium amount of sales. However, since an explicit derivation of these gives no new ideas, it is pointless to pursue this topic.

The market structure assumed in theorem 1 represents the maximum degree of competition given the potential supply, $\Sigma\, m_i$, and it is natural to describe this as a state of *maximum competition*. However, the potential supply is a parameter in this market entering the core constraints; and, since there is production, we should determine the potential supply given by $\Sigma\, m_i$. Theorem 1 shows that if there are at least two sellers with identical average cost equal to marginal cost and if the potential buyers each desire at most one unit, then, provided the buyers can deal with any of the sellers, there will be a common price such that any buyer who is willing to pay at least the marginal cost can obtain the good. Moreover, with the offer of a large enough quantity so that $\Sigma\, m_i > k$, the common price would equal the marginal cost. The price exceeds the marginal cost only if the quantity offered is below the quantity demanded at the price a. The classical theory would assert that, at any price above the marginal cost, an individual seller would have an incentive to expand his output. As a result, more would be offered until at a price equal to marginal cost there is no longer the incentive to expand output. The classical theory does not describe a market as being in a state of maximum competition if the price is above the marginal cost. A supply theory for oligopoly is one of the topics in the next section; the theory of the core as applied to market exchange leaves open the question of what determines the maximum quantity that could be offered by the producers.

There is a lesser degree of competition if either the buyers or the sellers cooperate among themselves. If the sellers restrict their offers so as to force prices above marginal cost, then they must agree on their total offer; but, if they cannot assign themselves specific customers, then they cannot price discriminate. Hence there is less competition though still a common price if there are at least two sellers. Without price discrimination there would not be group rationality. In fact the situation is equivalent to maximum competition and a constraint on supply given by $\Sigma\, m_i \leqslant k$. To describe this we may use the term *reduced market*, which is a market derived from the initial one by the removal of potential buyers. There is a trivially reduced market if the extramarginal buyers are removed since these would be the ones who are unwilling to pay as much as the marginal cost. A nontrivial reduced market results with the removal of intramarginal buyers. In a reduced market there is a state of reduced competition which is well illustrated by simple monopoly.

Consider a market served by one seller who charges all of the buyers the same price.[2] At the output maximizing the seller's net revenue, marginal cost equals marginal revenue which is the output Ok' in figure 3.1. Now

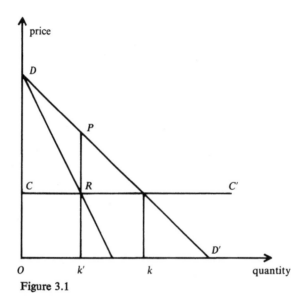

Figure 3.1

assume that the demand curve were $DPRk'$ instead of DD'. This corresponds to the removal of all potential buyers whose limit prices are below $k'P$. With the new demand curve and a single seller, the seller's output that maximizes

2. In figure 3.1 the demand conditions are represented by a continuous relation between price and quantity. If every buyer desires at most one unit and is willing to pay at most b_j, then the aggregate demand function cannot be continuous throughout (unless the demand is perfectly elastic). Hence the curve drawn in the figure must be regarded as a tolerably good approximation to the true demand relation. A fairly intricate mathematical analysis would be necessary to describe the conditions that would permit such a continuous approximation, but this would not advance the economic understanding of the problem so it is omitted.

net revenue remains Ok', and the monopoly price remains $k'P$. However, with the reduced demand curve, even if there were many sellers and maximum competition, the quantity sold would remain Ok'. If instead the supply schedule were CC', which means that $\Sigma\, m_i > k$, then the price would equal the marginal cost, a. However, the price would be in the interval RP and, therefore, can exceed marginal cost, if $\Sigma\, m_i = k'$. Hence if the sellers agree not to sell more than k', they can force the price above a.

This device of a reduced market can be used to prove that there would be a common price and group rationality in a reduced market if the sellers agree to restrict their total output to an amount below k, which would be demanded in the original market with maximum competition. There is group rationality in the reduced market but not in the original market because some intra-marginal buyers cannot obtain the good. Since the proof of these statements is straightforward, it is omitted.

3. The Cournot-Nash Theory of Duopoly for Finite Horizons

Theorem 1 shows that even with only two sellers, the imputations can be forced into the core of a maximum competitive market. This seems at variance with reality in so far as we observe an inverse relation between the number of competitors on the one hand and either prices or rates of return on the other.[3] These observations have inspired the construction of theories

---

3. Actually, an empirical investigation of the relation between the number of competitors and either prices or rates of return is beset with difficulties. To illustrate these, I shall describe two studies concerning the relation between prices and the number of bidders. The first study is reported in a Senate committee investigation of administered prices in the drug industry (1961)—it shows how *not* to proceed. The second study is a highly sophisticated analysis of the effects of the number of bidders on the underwriting cost of bonds of local governments by Reuben Kessel (1971). Kessel's study shows how such an investigation ought to proceed.

The government study is both simple and naïve. It relates two variables. The first is the ratio of the lowest price paid for drugs by the Military Medical Supply Agency to the major drug companies' price to retailers. The second variable is the number of bidders. There is an inverse relation between these two variables shown in a scatter diagram on p. 95, and the data are given on p. 262 (1961). Unfortunately, this evidence is far from conclusive because other pertinent variables are ignored such as the quantity bought by the government agency. It may well be that the larger the amount bought, the lower the price, because of scale economies and also the larger the amount bought, the greater the number of bidders. Hence the observed inverse relation between the price ratio and the number of bidders might reflect scale economies instead of increased competition, or, at the very least, the effect of competition can be exaggerated.

In contrast, the Kessel study is painstaking in its efforts to control for the effects of other pertinent variables on the price besides the number of bidders. Indeed the details are far too complicated to report here, and the reader should consult the study itself. The results leave no doubt that there is an inverse relation between the dependent variable, the difference between the buying and selling prices of underwriters of tax-exempt bonds, and the number of bidders.

In a different vein are those studies that relate rates of return to the strength of competition. However, competition is not usually measured by the number of firms but by the relative size of firms. Many of these studies find a positive association between the relative size of the leading firms in an industry and the rate of return. Still in doubt is the extent to which this relation reflects competition or other factors. Chapter VIII gives the results of an empirical investigation of this relation.

that would not only embrace these phenomena but would also yield other verifiable propositions. Perhaps the earliest of these, Cournot's theory (1838), has continued to attract considerable attention. More recently and independently, Nash (1950b and 1951) has proposed a more general theory originating in game theory that includes Cournot as a special case.[4] My task is to describe this theory rigorously so as to clarify the distinction between what it assumes and what it deduces.

Let there be two sellers with identical cost functions subject to constant returns to scale and let both have access to the same potential buyers. In short, assume precisely that market structure which is postulated in theorem 1 with $s = 2$. A crucial assumption of Cournot's model, correctly interpreted, is that the sales are made on a *single* occasion. At least at the outset of the analysis we rule out actions for a sequence of periods. On a given occasion there is one set of market transactions; in a dynamic setting there is a sequence of sets of market transactions. I shall gradually remove the assumption of a one-period model in the course of the analysis. Much of the criticism of the Cournot model seems to have repeated trade in mind and, even so, misunderstands the correct strategy. The assumption that transactions occur on a single occasion must be fully appreciated. It means there cannot be retaliation in the future for anything done in the present by either of the sellers. It also means that tacit collusion signalled by the nature of the sales over time is ruled out. Moreover, the Cournot model assumes that the sellers do not cooperate explicitly. Therefore, in a one-period model they cannot agree on a collusive strategy before trade begins. In terms of the core this means that the sellers individually can deal with the buyers.

The nature of the characteristic function suggests that the sellers have an advanage over the buyers because $v(A_i) = am_i$, while $v(B_j) = 0$. How can the sellers exploit their advantage? Cournot postulates that each seller chooses a rate of sales to maximize his net return, that is, to maximize his imputation, given the rate of sales chosen by his rival. This has the effect of making m_i a decision variable instead of a given parameter as is assumed above.

Before analyzing the Cournot-Nash model in terms of the core, it is helpful to give a conventional description of the model. Let us suppose the aggregate demand can be represented by a continuous function as follows:

(1) $$p = f(q)$$

such that the price p and the aggregate quantity q vary inversely. This assumes

4. Nash's work was independent of Cournot's theory. It was not until after Nash published his main results that the connection between his equilibrium point and the Cournot solution became evident. Nash was inspired by game theory and developed the theory of n-person games while Cournot, the economist, was one of the first to provide a mathematical theory of the effects of competition on price (Cournot 1960, chap. 7; Nash 1950b and 1951).

that all buyers pay the same price. Let q_i denote the sales of seller A_i. Hence,

$$(2) \qquad q = q_1 + q_2.$$

Let total cost of seller i be a function of his rate of sales, represented by the function as follows:

$$(3) \qquad c_i = g(q_i).$$

Thus both sellers are assumed to have the same cost function. In the preceding analysis the cost function is assumed to be linear in q_i so that

$$(4) \qquad c_i = a\, q_i$$

is a special case of (3). The net revenue of A_i, denoted by r_i, is defined by

$$(5) \qquad r_i = pq_i - c_i.$$

This corresponds to the imputation of seller A_i, namely, x_i. Because of the demand function, the net revenue of a seller depends not only on his rate of sales but also on his rival's rate of sales. The Cournot theory postulates that a rate of sales is chosen by each seller that maximizes his net revenue given the competitor's rate of sales. Hence the Cournot solution has the property that there exists a pair (q_1^0, q_2^0) such that

$$(6) \qquad \begin{aligned} r_1(q_1, q_2^0) &\leqslant r_1(q_1^0, q_2^0), \\ r_2(q_1^0, q_2) &\leqslant r_2(q_1^0, q_2^0). \end{aligned}$$

The existence of such a pair is guaranteed by the demand and cost functions, if $q\, f(q)$ is a bounded function of q and $g(q)$ is an increasing function of q.[5] If the net revenue is a differentiable function of the q's, then the pair (q_1^0, q_2^0)

---

5. The conditions stated in the text would imply the existence of a supremum for each player and not necessarily a maximum. Moreover, one would also require $q_i \geqslant 0$ so that (7) should be written as an inequality as follows:

$$\frac{\partial r_i(q_1, q_2)}{\partial q_i} \leqslant 0 \qquad\qquad (i = 1, 2).$$

Actually, if both firms have identical cost functions and if there is a strictly positive quantity demanded at a price equal to marginal cost, then (7) will have a strictly positive solution.

Nash has shown that every n-person nonzero-sum game has an equilibrium point in terms of mixed strategies. Moreover, this Nash point is the same as the minimax solution for zero-sum games. Therefore, the Nash point generalizes the celebrated minimax theorem to nonzero sum games. The proof that the Nash equilibrium point is also a saddlepoint for a zero-sum game is simple. Since the game is zero-sum,

(i) $\qquad\qquad\qquad r_1 + r_2 = 0.$

Therefore, the second inequality in (6) becomes

(ii) $\qquad\qquad\qquad -r_1(q_1^0, q_2) \leqslant -r_1(q_1^0, q_2^0).$

Together with the first inequality in (6), this implies

(iii) $\qquad\qquad\qquad r_1(q_1^0, q_2) \geqslant r_1(q_1^0, q_2^0) \geqslant r_1(q_1, q_2^0).$

Hence, the pair (q_1^0, q_2^0) is a saddlepoint.

that satisfies (6) can be found by solving the two equations as follows:

(7)
$$\frac{\partial r_i(q_1, q_2)}{\partial q_i} = 0 \qquad (i = 1, 2).$$

Cournot himself is responsible for much of the confusion surrounding his model. This is because he gives it the appearance of applying to a sequence of periods instead of to one period. Thus instead of writing (7), Cournot writes

(8)
$$\frac{\partial r_1(q_{1,t}, q_{2,t-1})}{\partial q_{1,t}} = 0,$$

$$\frac{\partial r_2(q_{1,t-1}, q_{2,t})}{\partial q_{2,t}} = 0,$$

where the t subscript refers to period t. He justifies these relations by the argument that each firm expects his rival to offer the same quantity for sale in the current period as he did in the preceding period so that A_i chooses $q_{i,t}$ to maximize r_i with $q_{i'} = q_{i',t-1}$. Cournot shows that the sequence of solutions to (8) converges to the solution of (7). That is,

(9)
$$(q_{1,t}^0, q_{2,t}^0) \rightarrow (q_1^0, q_2^0), \qquad \text{as} \quad t \rightarrow \infty.$$

However, if in fact sales are assumed to occur for an infinite number of periods, as is implicitly assumed by (8), then it is by no means obvious that the sequence of sales implied by the Cournot postulate does in fact satisfy the pair of equations (8). Moreover, to interpret $q_{i',t-1}$ as representing the expectations by A_i regarding its rival's sales in the current period seems strange because in fact $q_{i,t} \neq q_{i,t-1}$, and one would think that firm's would not have expectations that persist in error. These problems will occupy our attention in chapter IV and again in chapter VI.

The correct interpretation of the static situation is best illustrated in a diagram. Consider figure 3.2 where the sales of the first firm are plotted on the horizontal axis and of the second firm on the vertical axis. The line $R'R$ represents equation (7) for A_1, and the line $P'P$ represents (7) for A_2. Each firm can choose only one rate of sales that may not subsequently be altered. This is the meaning of a one-period model. If A_2 chooses to sell q_2^0, then A_1 can do no better than to choose to sell q_1^0. Conversely, if A_1 chooses to sell q_1^0, then A_2 can do no better than to choose to sell q_2^0. Hence the pair of sales represented by the point Q^0 is the only mutually consistent and profit maximizing pair given that the firms cannot cooperate, can choose only one rate of sales, and must charge all customers a common price. If the price were equal to marginal cost, then the total quantity sold would be on the locus $R'P$, while if the price were chosen to maximize the monopoly return, the total quantity sold would be on the locus $P'R$. To satisfy group rationality including the buyers, the price must equal marginal cost so that both the

Cournot solution as well as the monopoly solution fail to satisfy group rationality though as the diagram clearly shows, Q^0 is closer to Q^1, which is an acceptable total output under maximum competition, than is Q^{-1}, which is a possible monopoly output. However, leaving the buyers out of account, the Cournot solution implies a lower return to the sellers than collusion which is capable of yielding them the monopoly return. For example, with simply monopoly the two sellers could choose the point Q^{-1} and share equally the net return of simple monopoly. The trouble with this solution is that it cannot be enforced. Thus suppose A_2 chooses to sell q_2^{-1} hoping that A_1 would choose q_1^{-1} so that in the aggregate the total imputation, $r = r_1 + r_2$ would be larger than under the Cournot solution. Given that A_2 would choose to sell q_2^{-1}, then A_1, acting alone can obtain a

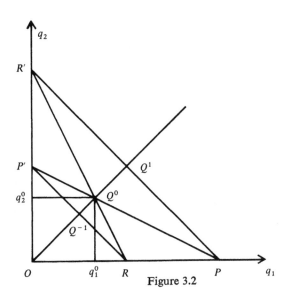

Figure 3.2

larger net revenue by offering that amount of q_1, determined by the locus $R'R$ that corresponds to q_2^{-1}. By hypothesis there is nothing to stop A_1 from doing so, and he can thereby obtain a larger net revenue than from the joint profit maximum of simple monopoly. Of course, the same argument applies to A_2. As a result both remain at Q^0. Therefore, the inability to enforce cooperation by penalizing departures from it together with the other assumptions of the model gives the Cournot solution. The argument is like the one used to explain why simple monopoly violates group rationality despite the fact that price discrimination is more profitable to the seller than charging all of the buyers the same price.

Among the more interesting aspects of the Cournot model is its prediction of a precise relation between the number of firms and the difference between the price and marginal cost. For simplicity we may work with the simple cost function given by (4). Let the demand function (1) be twice continuously differentiable. Assume there are s sellers who all have the same cost function given by (4). The total quantity q is given by

$$(10) \qquad q = \Sigma \, q_i,$$

and the net revenue of firm i satisfies

$$(11) \qquad r_i = (p-a)q_i.$$

The Cournot-Nash point is a set $\{q_i^0\}$ that satisfies the set of s equations as follows:

$$(12) \qquad \frac{\partial r_i}{\partial q_i} = (p-a) + q_i \frac{\partial f}{\partial q} \frac{\partial q}{\partial q_i} = (p-a) + q_i \frac{\partial f}{\partial q} = 0.$$

The latter equation is a consequence of $(\partial q/\partial q_i) = 1$. For q_1^0 to maximize r_i given the values of the other q's, it is sufficient that

$$(13) \qquad \frac{\partial^2 r_i}{\partial q_i^2} = 2 \frac{\partial f}{\partial q} + q_i \frac{\partial^2 f}{\partial q^2} < 0.$$

Summing (12) over all of the q's gives the result that

$$(14) \qquad (p-a)\, s + q \frac{\partial f}{\partial q} = 0.$$

Notice that $q_i = 0$ for all i unless $p > a$, and that the latter condition is necessary for an essential market. Equation (14) defines q as an implicit function of s and a. To find the relation between p and s, we use

$$(15) \qquad \frac{\partial p}{\partial s} = \frac{\partial f}{\partial q} \frac{\partial q}{\partial s}$$

and derive the value of $(\partial q/\partial s)$ from (14) which implies that

$$\frac{\partial(p-a)}{\partial s} + (p-a) + \frac{\partial f}{\partial q} \frac{\partial q}{\partial s} + q \frac{\partial^2 f}{\partial q^2} \frac{\partial q}{\partial s} = 0.$$

After some simple reductions and collecting terms, this gives

$$(16) \qquad \frac{\partial q}{\partial s} = -(p-a) \bigg/ \left[2 \frac{\partial f}{\partial q} + \frac{\partial^2 f}{\partial q^2} q \right].$$

Since (13) implies the denominator is negative and since the numerator is positive in an essential market, it follows that

$$(17) \qquad \frac{\partial q}{\partial s} > 0.$$

Therefore, because price and quantity vary inversely, (15) and (17) imply

(18)
$$\frac{\partial p}{\partial s} < 0.$$

This analysis has several other implications. First, all of the firms will have the same market share given by $1/s$. However, the time path of market share is indeterminate because the rate at which new firms enter the industry is indeterminate. This model says only that the more firms there are, the closer is the price to marginal cost. In the limit with an infinite number of firms, the price becomes equal to marginal cost. We can derive the relation between sales per firm and the number of firms. Equation,

(19)
$$\bar{q} = q/s,$$

defines the sales per firm. Equation (14) implies that

(20)
$$\bar{q} = -(p-a)/\left(\frac{\partial f}{\partial q}\right).$$

It is a simple exercise to calculate the slope of sales per firm with respect to the number of firms. The result is given as follows:

(21)
$$\frac{\partial \bar{q}}{\partial s} = -\left(\frac{\partial q}{\partial s}\right)\left[\frac{\partial f}{\partial q} + \bar{q}\frac{\partial^2 f}{\partial q^2}\right]\bigg/\left(\frac{\partial f}{\partial q}\right).$$

One might think that since the market share decreases as the number of firms increases, it would follow that the sales per firm would also decrease as the number of firms increases. This is not necessarily true. To see why, examine the signs of the factors in (21). Since the denominator is positive and the term

$$-\left(\frac{\partial q}{\partial s}\right) < 0$$

because of (17), the sign of $(\partial \bar{q}/\partial s)$ will be negative if the sign of the terms within the square brackets of (21) is positive. Let us examine the sign of this term. Summing (13) over all i implies that

$$2s\frac{\partial f}{\partial q} + q\frac{\partial^2 f}{\partial q^2} < 0,$$

which becomes

(22)
$$2\frac{\partial f}{\partial q} + \bar{q}\frac{\partial^2 f}{\partial q^2} < 0.$$

Now s and $\bar{q}$ vary inversely if and only if

(23)
$$\frac{\partial f}{\partial q} + \bar{q}\frac{\partial^2 f}{\partial q^2} < 0,$$

but this is certainly not a necessary consequence of (22). A sufficient but not a necessary condition for (23) is that the price is a concave function of q so that

$$(24) \qquad\qquad \frac{\partial^2 f}{\partial q^2} < 0.$$

But this is not always true. Thus, if the quantity demanded becomes highly elastic as the price approaches marginal cost, it would not follow that the average sales would vary inversely with the number of firms because the entry of firms so lowers the price that the total quantity demanded increases more than in proportion to the number of firms. As a result the market share falls while sales per firm rises. Hence the Cournot-Nash theory predicts an inverse relation between the market share and the number of firms and derives a relation between sales per firm and the number of firms, given by (21), which depends on the properties of the demand function.

Now consider the Cournot-Nash theory in the light of the core. In a pure exchange model of the kind discussed in chapter I, the maximum amount available for sale is a given constant and the analysis centers on the conditions of sale. These include the determination of who sells and who buys, how much each individual obtains as his final imputation, and whether or not there will be a common price and group rationality. However, with production possible one should not assume a given upper bound on sales but should deduce one from a suitable theory. In fact the Cournot-Nash theory does determine the output. Once this is determined, if the imputations are in the core then theorem 1 applies and proves there must be a common price. What remains to be shown is that the assumptions of the Cournot-Nash theory are consistent with imputations in the core of a reduced market.

Before doing this it is helpful to review the general theme. All along the analysis assumes that every participant tries to make himself as well-off as possible by maximizing his imputation. However, the imputations must satisfy certain constraints because exchange requires cooperation at least to the extent of agreement on the terms of sale. If all coalitions are feasible, then competition is at its maximum, and the imputations are forced into the core. This means that the constraints are so numerous that the individuals' power over their imputations is substantially weakened. The maximization problem reduces to one of determining a set of imputations to satisfy the inequalities implied by the possibility of forming coalitions. The core constraints applicable to pure exchange do not suffice to determine the imputations if there is production. However, if for any given output the resulting imputations must satisfy all possible constraints of a reduced market, then there must be a common price, which justifies the assumption of a common price for all buyers in the Cournot-Nash model. Let us prove these assertions. Suppose an essential market so there is a k such that

$$b_{n-(j-1)} - a \geqslant 0, \qquad\qquad \text{for all} \quad j = 1, \ldots, k,$$

with strict inequality for $j = k$. The quantity exchanged under maximum competition with two sellers having constant average cost would be k units and the price would be a. These are the consequences of theorem 1. However, in the Cournot-Nash model each seller determines a rate of output, m_i^0 for A_i, with the property that

(25) $\qquad x_1(m_1, m_2^0) \leqslant x_1(m_1^0, m_2^0), \quad x_2(m_1^0, m_2) \leqslant x_2(m_1^0, m_2^0),$

assuming that all buyers pay the same price. There results a combined output $m^0 = m_1^0 + m_2^0$ such that $m^0 < k.$[6]

Hence none of the intramarginal buyers $B_{n-(j-1)}$, with $j = m^0, \ldots, k$, can obtain the product. For any seller to expand his output in order to satisfy this demand, given the output of his rival and a common price for all buyers, would only lower his imputation. Hence any coalition between the extramarginal buyers and one seller is dominated by the Cournot-Nash solution. Moreover, price discrimination is not possible so long as all coalitions can form. Hence no group of these intramarginal buyers can combine with one of the sellers to obtain the good at a price below the one established by the Cournot-Nash theory and above the marginal cost. As a result, the imputations satisfy group rationality for the reduced market but not for the original market because in the reduced market

$$y_{n-(j-1)} = 0, \qquad\qquad \text{for} \quad j = k, \ldots, m^0,$$

while in the original market with maximum competition, these imputations would be

$$y_{n-(j-1)} = b_{n-(j-1)} - p > 0, \qquad \text{for} \quad j = m^0, \ldots, k.$$

We conclude that the assumptions of the Cournot-Nash model are consistent with the imputations being in the core of the reduced market for which sales are m^0.

It is remarkable that the Cournot-Nash theory retains its validity in a T-period model for finite T despite the fact that now implicit collusion between the two sellers is feasible. Implicit collusion means that the sellers have an informal agreement such that violations in one period can be punished by increasing the rate of sales in later periods. As a result it would appear that the sellers have an incentive to adhere to an implicit agreement. For example, they can agree to sell an amount that would maximize the sum of their imputations in every period given that all buyers pay the same price which is equivalent to maximizing the return under simple monopoly. The total return exceeds the sum of the imputations obtainable under the Cournot-

---

6. First, by the assumption made in section I.1 the price must not exceed the limit price of the least eager of the included buyers so that this buyer must obtain some positive gain from trade. Second, it may be that $m^0 = 1$ so that one of the sellers is inactive, meaning that he makes and sells nothing.

Nash theory, and it would presumably be preferred by the two sellers. This is the problem to be considered.

Let x_{it} denote the imputation of A_i in period t and let y_{jt} denote the imputation of B_j in period t. Let the output of A_i in period t be m_{it}. Hence,

$$(26) \qquad\qquad x_{it} \geqslant a\, m_{it}, \qquad\qquad \text{for all } i \text{ and } t.$$

Let the limit price of B_j in period t be b_j, which means that the demand remains constant over time so that it is the same in every period as it is for a one-period horizon. Every potential buyer wants at most one unit for which he will pay at most his limit price, implying

$$(27) \qquad\qquad y_{it} \geqslant 0, \qquad\qquad \text{for all } j \text{ and } t.$$

It is assumed that neither the buyers nor the sellers can store the goods.

Under one set of assumptions this situation reduces to the one-period static model, and the logic of the Cournot-Nash theory clearly applies. Thus assume that trade for all periods is arranged by means of binding contracts on the initial day. This is analogous to forming a super-game whose constituent games are played in sequence and such that one must lay down a strategy in advance for playing the whole sequence. The coalitions that buyers and sellers form at the beginning determine the imputations for all of the T period. This is equivalent to forward contracting arrived at by the Edgeworth recontracting mechanism operating only at the outset. Once the imputations are determined for all future periods by the negotiations at the outset, no one can change. These conditions effectively destroy the distinction between present and future. Hence, it is hardly surprising that, as a result, the solution is the same as in a one-period model. Given forward contracting, the Cournot-Nash model requires the sellers to choose sequences $\{m_{it}\}$ that maximize their total imputation, which is $\sum_{t=1}^{T} x_{it}$, given the complete sequence of sales chosen by their rival. It is assumed that while the price can vary over time, all buyers pay the same price in the same time period. Formally, the Cournot-Nash equilibrium is the solution of the following maximum problem:

$$\max_{\{m_{it}\}} \sum_{t=1}^{T} x_{it} \quad \text{given} \quad \{m_{i't}\} \qquad\qquad (i' \neq i).$$

It is readily verified that the solution in any period is a rate of sales that is the same as it would be in a one-period horizon. The simple monopoly equilibrium fails here for the same reasons as in the one-period case. However, the solution may acquire stronger credentials if the buyers and sellers are uncommitted in advance by a set of forward contracts. We now study this case.

Assume that the sellers can determine their sales afresh in every period and that there are only spot contracts. Hence each seller can guard himself against a rival's departure from the tacitly agreed upon output and can respond accordingly. No one is committed in advance to actions for all future periods. Although the firms make plans, nothing compels the carrying out of their plans if this is not in their interest. Nevertheless the solution is the same whether or not the firms are free to determine their sales in each period sequentially, provided there is a *finite* number of periods. The argument is due to Luce and Raiffa (1957, sec. 5.5). Consider the last period T. In this period each firm can choose its rate of sales, m_{iT}, without fear of reprisals in future periods.

Hence for this period the situation is the same as in the static one-period case. Since each firm can take its rival's sales as given, it can choose its own rate of sales to maximize its own imputation given the rival's sales. For any given rate of sales chosen by its rival, including the output necessary for the maximum joint return, the firm can do no better than to make its own imputation as large as possible, given the constant rate of sales of its rival. Moreover, even with a tacit agreement, the firm cannot be punished in the last period for violations because the punishment must follow the crime, and this is impossible by hypothesis. Hence the solution in the last period must be the same as in the static one-period case. Given this outcome for the last period, consider the determination of the best strategy for period $T-1$.

Nothing done in period $T-1$ can affect the choice in period T because, by the nature of the cost and demand conditions, the strategy for period T described above is best given any strategy in all of the preceding periods. Therefore, given the choice in period T, whatever choice is made in period $T-1$ is the same as if $T-1$ were in fact the last period. Hence the problem reduces to a $T-1$ period horizon, and by exactly the same reasoning we conclude that the Cournot-Nash solution holds in period $T-1$. Proceeding by induction it follows that the Cournot-Nash solution is the same in every period as it is in the last period. Hence the solution for a one-period horizon is the same as the solution for a T period horizon for all finite T. These results complete the proof of the theorem as follows:

THEOREM 2. *Let there be two sellers and n buyers such that seller i offers up to* m_{it} *units for sale in period t with* $t = 1, \ldots, T$ *and T finite. Each buyer desires at most one unit in period t at a limit price* b_j. *The characteristic function for this market is defined as follows:*

$$v(A_{it}) = a\, m_{it},$$
$$v(B_{jt}) = 0,$$
$$v(A_{it}, B_{jt}) = \max \{am_{it}, b_j + (m_{it} - 1)a\},$$

and, in general,

$$v(\cup_i A_{it}, \cup_j B_{jt}) = \max z_t,$$

where

$$z_t = \sum_{h=1}^{m_{it}} z_{ht}$$

with z_{ht} chosen from the set $\{a, \cup b_j; b_j$ corresponding to $B_j\}$. Intertemporal coalitions are not allowed because it is assumed that the good is not storable and because the sellers cannot explicitly cooperate. Assume the market is essential so that there is a $k \geqslant 1$ such that

$$b_{n-(j-1)} - a > 0, \qquad\qquad \text{for all} \quad j = 1, \ldots, k.$$

Thus k would be the output under maximum competition. Let each firm choose its sequence of sales to maximize its imputation given by

$$\sum_{t=1}^{T} x_{it}$$

subject to the constraints that for every t, $\{x_{it}\}$ and $\{y_{jt}|j = 1, \ldots, m_{jt}\}$ are in the core of a reduced market.

Then there is a common price p and an output m^0 such that

$$m^0 = m_1^0 + m_2^0, \; |m_1^0 - m_2^0| \leqslant 1$$

with the property that

$$x_{it} = x_i = (p-a) \, m_i^0,^7$$

and

$$x_1(m_1^0, m_2^0) \geqslant x_1(m_1, m_2^0),$$
$$x_2(m_1^0, m_2^0) \geqslant x_2(m_1^0, m_2).$$

With an infinite horizon the situation is entirely different. This is the topic of the next section.

4. The Cournot-Nash Theory of Duopoly for Infinite Horizons

Although there are many interpretations of an infinite horizon, one of these is especially relevant for the present problem. Until now we have assumed that trade continues for a definite number of periods known to all of the participants. Instead, one may assume that the continuance of the demand is uncertain. If it is unlikely that the demand will last forever but it is unknown when the demand will end, then this is equivalent to assuming an infinite horizon. This can be demonstrated mathematically.

Let the probability of continuing trade for one more period be α so that the probability of trade lasting for exactly T periods is $(1-\alpha)\,\alpha^T$. Thus the continuance of trade is a random event independent of the previous history. A seller's imputation is x_{it} if there is trade in period t and is zero otherwise.

---

7. These conditions cover the possibility that the sales of the two firms can be unequal; if so, the sales cannot differ by more than one unit.

Because of the independence assumption about the continuance of trade, it is easy to calculate the expected imputation of seller A_i. If there were trade for exactly one period, the imputation of A_i would be x_{i1}, and the probability of this event would be $\alpha(1-\alpha)$. With trade for exactly two periods the imputation would be $x_{i1}+x_{i2}$ with probability $\alpha^2(1-\alpha)$. In general, trade for exactly T periods would have a probability of $\alpha^T(1-\alpha)$ and the imputation would be $\sum_1^T x_{it}$. Hence the expected value of the imputations would be

$$(1) \qquad \sum_1^\infty \alpha^T(1-\alpha)\left(\sum_1^T x_{it}\right) = \sum_1^\infty \alpha^t x_{it}.$$

Formally, this expected imputation resembles the present value of an income stream $\{x_{it}\}$ discounted by the constant discount factor $\alpha = (1+\rho)^{-1}$, where ρ is the discount rate per period. The problem now is to determine for which values of the probability α the Cournot-Nash solution will be valid.

Assuming there is a common price for all buyers, it is easy to derive the Cournot-Nash solution on the hypothesis that each seller chooses his rate of sales in every period to maximize his expected imputation, (1), given the sequence of sales of its competitor. Formally, the problem is to find $\{m_{it}\}$ that maximizes

$$(2) \qquad P_i = \sum_1^\infty \alpha^t x_{it}.$$

The solution to this problem is precisely the same as in the static one-period situation. To verify this assertion, the easiest procedure postulates a continuous demand relation such as 3(2), the cost function 3(4), and a net revenue function 3(5). The net revenue corresponds to the seller's imputation. The problem is one of finding the sequence of sales, $\{q_{it}\}$, to maximize

$$(3) \qquad P_i = \sum_1^\infty \alpha^t r_{it} = \sum \alpha^t [f(q_{1t}+q_{2t})q_{it}-aq_{it}].$$

The solution must satisfy

$$(4) \qquad \frac{\partial P_1}{\partial q_{1t}} = 0, \quad \frac{\partial P_2}{\partial q_{2t}} = 0, \qquad \text{for all } t.$$

However, it now seems unreasonable to attribute significance to a result that assumes each seller pursues a strategy that is independent of his rival's. This unreasonableness is due in particular to the fact that the sellers are not compelled to choose their strategies in advance for all time. Instead they can decide what to do period by period as events develop.

The simplest case that poses these objections to the Cournot-Nash solution is known as the "prisoner's dilemma" though it is better described herein as the "cartel's dilemma."[8] Assume that each firm confines its choice of

---

8. For an analysis of the prisoner's dilemma, see Luce and Raiffa (1957, sec. 5.4).

output to only two alternatives. The first alternative is to offer a "small" quantity, S, equal to one half the output that maximizes the combined monopoly return. The second alternative is to offer a "large" quantity, L, that is the Cournot-Nash output. The possible imputations for a given period are represented in the matrix as follows:

$$
\begin{array}{cc}
 & \begin{array}{cc} S & \qquad\qquad L \end{array} \\
\begin{array}{c} S \\ L \end{array} &
\left[\begin{array}{cc}
(u, u) & (u-\delta_1-\delta_3, u+\delta_2) \\
(u+\delta_2, u-\delta_1-\delta_3) & (u-\delta_1, u-\delta_1)
\end{array} \right],
\end{array}
$$

where $\delta_j > 0$. If one firm chooses S while the other does not, then it suffers a larger loss than if it were also to choose L. The weakness of collusion is that it exposes the loyal party to the risk of a smaller return than he could obtain from a constant choice of L. The L choice combines the virtue of safety with the chance of profit if the opponent happens to choose S. With only one play, this argues for the choice of L by both. With a sequence of plays that can be unending, both can choose S and gain, since such implicit collusion is enforced by the possibility of punishing current violations in later periods.

By hypothesis the combined return is a maximum if both firms choose S. Their implicit cooperation is broken if one should choose L. Prudence then dictates that the choice of L must be punished for some time afterwards. Otherwise the gain from cheating would be irresistable. Let us study the expected return to a firm that contemplates the choice of L. The situation is as follows:

    i. both firms choose S in the first $T-1$ periods;
    ii. the first firm chooses L in the Tth period while the second firm, caught by surprise, continues to choose S;
    iii. for k periods thereafter the second firm punishes the cheater by choosing L;
    iv. both firms resume collusion.

The problem is to determine for which range of values of the probability of continuance, α, if any, this policy is more profitable to A_1 than continuous collusion.

The return to A_1 from this policy minus the return from perpetual collusion is given by the expression as follows:

$$
(5) \qquad \phi(k, T, \delta_1, \delta_2, \alpha) = \left[\sum_{1}^{T-1} \alpha^t u + (u+\delta_2)\alpha^T + \sum_{T+1}^{T+1+k} (u-\delta_1)\alpha^t \right.
$$

$$
\left. + \sum_{T+1+k}^{\infty} u\alpha^t \right] - \sum_{1}^{\infty} \alpha^t u
$$

$$
(6) \qquad\qquad = \alpha^T [\delta_2 - \alpha\,\delta_1(1-\alpha^k)/(1-\alpha)].
$$

Cheating is profitable if and only if

$$
(7) \qquad\qquad\qquad\qquad \phi > 0.
$$

Several facts are obvious. If it pays to cheat at all, it pays to do so immediately. This is because ϕ is larger, the smaller is T. Second, the temptation to cheat is greater, the smaller is k. This is obvious intuitively and is verified formally in (6) by the presence of the term $(1-\alpha^k)$. Thus the larger the penalty of cheating, as expressed by the number of periods for which one period of cheating is punished, the less is the expected gain and the less is the temptation to cheat. The temptation is least if a single violation of the implicit agreement is followed by a relentless choice of L thereafter by the aggrieved party. This corresponds to k set equal to infinity. To study the relation between α and the δ's, it is convenient to assume that $k = \infty$. Thus for $T = 1$ and $k = \infty$, ϕ becomes

(8) $$\phi(\infty, 1, \delta_1, \delta_2, \alpha) = \alpha[\delta_2 - \alpha\delta_1/(1-\alpha)].$$

Therefore, cheating has a positive expected return for all α that satisfy

(9) $$\alpha < \delta_2/(\delta_1+\delta_2).$$

For example, if the aggregate demand function is linear in price and quantity, $p = c_0 + c_1 q$, and both firms have the same constant average cost, $C = aq$, then the expected return from cheating once is positive if $\alpha < 1/2$.[9] In general given δ_1 and δ_2, which depend on the cost and the demand conditions, for all probabilities of continuance that satisfy (9), the Cournot-Nash solution yields a higher return than implicit collusion that punishes violations of the agreement as described above.

Summarizing, we see that if there is a sufficiently high probability of continuing, then the two firms will collude in order to obtain the joint profit maximum. Otherwise, despite the infinite horizon, they will find it to be more profitable to compete.

---

9. Given the demand function (8) and the cost function (9), one can derive the Cournot-Nash output, L, and the joint profit maximum output, S. Let $\sigma = -(c_0-a)/c_1$. Then the L output is as follows: $q_i = (1/3)\,\sigma$. The S output is given by $q_i = (1/4)\,\sigma$.

For these two alternatives, the pay-off matrix is as follows:

	S	L
S	$((18/144)c_1\sigma^2, (18/144)c_1\sigma^2)$	$((15/144)c_1\sigma^2, (20/144)c_1\sigma^2)$
L	$((20/144)c_1\sigma^2, (15/144)c_1\sigma^2)$	$((16/144)c_1\sigma^2, (16/144)c_1\sigma^2)$

Actually, given that one of the firms chooses S, the most profitable action for the other firm is to choose not L but that output rate which maximizes its net return given the choice of S by its rival. This output is given by $q_1 = (3/8)\,\sigma$. In this case the pay-off table is as follows:

	S	L
S	$((72/576)c_1\sigma^2, (72/576)c_1\sigma^2)$	$((54/576)c_1\sigma^2, (81/576)c_1\sigma^2)$
L	$((81/576)c_1\sigma^2, (54/576)c_1\sigma^2)$	$((64/576)c_1\sigma^2, (64/576)c_1\sigma^2)$

Thus the return to the firm that chisels is slightly larger than in the preceding case. In this case it pays to chisel if $\alpha < 9/17 = .53$. In the preceding case it pays to chisel if $\alpha < 1/2$. As one would expect, the probability of continuance must be slightly larger to penalize violations of the implicit collusive agreement if the chiselling firm follows its optimal chiselling strategy. Of course, these results depend on the nature of the cost and the demand conditions.

IV

Theories of Expectations for N Competing Firms

1. Introduction

The theory of oligopoly discussed so far ignores the consequences of the fact that information is a costly good. It is time to start remedying this neglect. If information is costly, then some ignorance is optimal. The behavior of the market's participants, who lack omniscience, depends on how they learn, what they know, and how they use their knowledge. Neglect of these considerations in many theories of oligopoly is more a testimony to the difficulty of analyzing these aspects of reality than to widespread agreement among scholars that these are unimportant problems. It is prudent, therefore, to proceed from the simpler to the more complicated cases. In particular some simplifying assumptions about demand are necessary. Let both the price and aggregate quantity be continuous variables and let the demand condition be represented by a continuously differentiable function relating price and aggregate quantity such that these two variables vary inversely. Assume that the firm offers the same product to all customers at a uniform price per unit thereby ruling out price discrimination among the customers of a given firm. However, even these assumptions do not imply a uniform price in the market if information is a costly good. This important point deserves some explanation.

If information is costly so that market participants limit the number of their contacts, then, as shown in sections I.6 and I.7, different sellers can obtain different prices for the same product. We can represent this by assuming that the products of the various firms are imperfect substitutes in the short run that become increasingly better substitutes in time. Thus suppose the effect of a price change by one seller is spread out over time as the news spreads among the market participants. With unchanging conditions there would eventually be a uniform price throughout the market. The speed with which the ultimate equilibrium is approached depends on the cost of and

146

the returns to information. The more expensive is information, the slower is its dissemination and the slower is the approach to equilibrium. In the short run the existence of information costs results in a price distribution among firms such as is shown in figure 4.1. However, in a market with the same group of customers the price range AB should be related to the cost of obtaining price information. In particular it should depend on the cost of a few searches or shopping expeditions because even the least diligent customers will eventually discover the lowest prices under stable conditions. Therefore, if all sellers are alike and offer the same product, then we should expect a distribution of sales among them of the same form as the price distribution shown in figure 4.1. Such a price distribution is hardly distinguishable from the kind that would result from the sellers' offering different kinds of products. A skewed sales distribution such as the one shown in figure 4.2 might result if old customers leave and new ones enter the market and if all customers do not have the same net return from information. In this case the higher priced sellers would obtain the patronage of those newer customers who search the least while the lower priced sellers would attract those customers who have the most information. Either case is consistent with the conclusion that the costs of information imply a price and sales distribution analogous to those that would appear in a market with product heterogeneity.

These considerations apply to the behavior of buyers. However, sellers of the competing products have as much an incentive to discover the going prices of goods as do the buyers. This is especially true if the products are close substitutes. Under these circumstances a seller whose price differs from his competitors' sacrifices profits. To focus on the implications of the cost of information to the sellers, it is convenient to assume that the products are perfect substitutes and that all sellers obtain the same price per unit at a given time.

A seller's policy variable is the quantity that he should offer for sale. This depends on his expected price, which is a function of the total quantity offered by all of the sellers thereby giving each seller an incentive to forecast the amount offered by all of his competitors. The problem becomes one of relating the current decision of each seller to his predictions of the future and of studying the consequences of different kinds of predictions.

Predictions of the future often depend on past history because experience has shown that the past is a more or less reliable guide to the future in that the nearer the future the more reliable a guide is the past. We shall study the implications of different sorts of predictions using past observations. Cournot originally postulated a simple kind of expectations. He assumed that each seller would expect his rival's sales to be the same in the current period as it was in the preceding period. If there are two firms whose products are perfect substitutes, then the sequence of actual sales implied by the Cournot postulate does converge to an equilibrium. However, it fails to converge

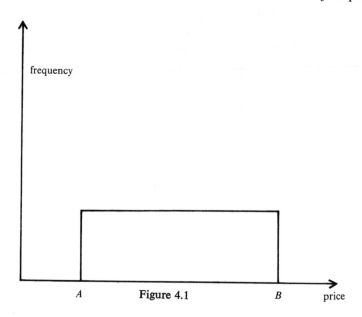

frequency

A   **Figure 4.1**   B   price

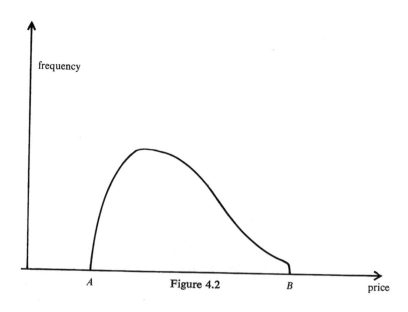

frequency

A   **Figure 4.2**   B   price

if there are three or more firms. Section 2 studies this expectations model as well as some others. We shall see that some expectations models imply convergence for any number of firms while in others whether there is convergence depends on the number of firms. For instance, if the firm's expectation is the arithmetic average of all past sales, then there is convergence for any number of firms; but with adaptive expectations convergence depends on how rapidly the expectations coefficient decreases with a rising number of firms.

The third section studies expectations theories in a situation in which the products of the firms are not perfect substitutes so that the firms do not necessarily sell their products at a common price. Their price policy depends on the prices they expect of rival goods. For n competing products, an economically meaningful representation of the demand conditions imposes a particular structure on the matrix of price coefficients, which implies that Cournot's expectations model leads to a convergent solution for any number of firms. A fortiori, arithmetic mean expectations and adaptive expectations converge for any number of firms.

Throughout this chapter, I shall assume that the firms compete with each other. Hence they do not collude and maximize their combined net revenue. To assume that the firms compete means that every firm supposes that its actions do not invite explicit retaliation by its rivals. Therefore, it can regard the choices of prices or quantities by its competitors as given on a par with the state of nature. These assumptions are equivalent to the Cournot-Nash postulate of noncooperation discussed in chapter III.

The next chapter studies the net return to collusion where I shall argue that collusion occurs whenever it is the most profitable course. This assumes that collusion is not always more profitable than competition because the former includes certain costs absent under competition that can exceed the larger gross revenue obtainable by a collusive price policy.

2. Expectation Models with Quantity as the Policy Variable

The basic model specifies the demand and cost conditions. The continuous function

$$(1) \qquad\qquad p = f(q)$$

represents the demand conditions so that the price p varies inversely with the aggregate quantity demanded q. In addition assume that the demand function has a continuous derivative such that

$$(2) \qquad\qquad -\infty < \frac{\partial f}{\partial q} < 0.$$

Let the demand function remain constant over time.

There are n sellers, A_i, $i = 1, ..., n$. The quantity sold by each is q_i so that aggregate sales satisfy

$$(3) \qquad q = \sum_{1}^{n} q_i.$$

Assume that average cost is constant and independent of the quantity sold and in addition assume that every firm has the same average cost so that A_i's total cost function is given by

$$(4) \qquad C_i = cq_i.$$

These assumptions imply a uniform price for every seller who can control the quantity that he offers but not the price. Given the total quantity offered by all of the sellers, the price adjusts to clear the market. For this to be true, it is necessary to assume that the sellers do not hold inventories, or, more generally, that they hold a constant amount of inventories. For a long-run equilibrium there must be a price and an aggregate quantity offered capable of persisting indefinitely. Since in every period the price and quantity satisfy the demand function, it is the sellers' reactions that determine whether or not a given one-period equilibrium can persist.

The incomplete knowledge of the sellers is introduced via random variables. Each firm knows only its own rate of sales but does not know in advance the aggregate sales of its competitors. This is equivalent to saying that no firm knows the market price in advance. If the firms regard the aggregate quantity as a random variable, then they also perceive the market price as a random variable. Given the assumption of constant inventories, the price must adjust so that the purchasers become willing to buy the total quantity offered for sale by the n firms. Let Q_i^* denote the total sales as expected by A_i. Since A_i knows his own sales,

$$(5) \qquad s_i^* = Q_i^* - q_i$$

gives A_i's forecast of the total quantity that will be offered by his $n-1$ competitors. However, every firm makes its own prediction without advance knowledge of the plans of the other firms, and there is nothing to prevent every firm from making a different prediction of the total sales. This means that it is possible for $Q_i^* \neq Q_j^*$. If so, the market cannot be in long-run equilibrium. To prove this, let us suppose that every firm expects a different total quantity to be sold. As a result every firm expects a different price. After the market is cleared every firm would observe a discrepancy between the actual and its expected price. If this caused no change in their forecast, then these discrepancies would persist, and, eventually, some firms would discover how to increase their return by offering a different quantity for sale. Consequently, differences in the predictions of aggregate sales eventually lead to actions that change the amount offered. As long as there are discrepancies between the actual and the expected prices in this model, there is an incentive to change expectations and the quantity offered. Therefore, for a long-run equilibrium there must be a total rate of sales q and an expected Q^* such that

$$(6) \qquad\qquad Q_i^* = Q^* = q, \qquad\qquad \text{for all } i = 1, \dots, n.$$

Under the stationary demand conditions represented in (1) the actual price must eventually coincide with the expected price of every firm in order that a long run equilibrium exist.

The relation between the expected price and the expected rate of sales depends on the shape of the demand function. A firm that regards the actual sales of its rivals as if it were a random variable will also regard its net return as a random variable. Hence if the actual return of A_i is given by

$$(7) \qquad\qquad r_i = q_i[f(q) - c],$$

then the expected net return to A_i is given by

$$(8) \qquad\qquad Er_i = Eq_i[f(q) - c].^1$$

In this expression the symbol E is used to denote the expectation of a random variable. Since A_i does not regard his own sales as a random variable, q_i can be moved outside the expectation operator E and the expression for A_i's expected net return becomes

$$(9) \qquad\qquad Er_i = q_i E[f(q) - c].$$

Because every firm regards the combined sales of its competitors as a random variable, this also makes the price, $p = f(q)$, a random variable. However, in general

$$(10) \qquad\qquad Ep = Ef(q) \neq f(Eq) = f(Q_i^*).$$

For there to be equality in (10), it is necessary and sufficient that the price is a linear function of the total quantity. Let us consider some implications of nonlinearity.

Denote the combined actual sales of the $n-1$ competitors of A_i by s_i so that

$$(11) \qquad\qquad s_i = q - q_i.$$

Since A_i assumes his forecast error is a random variable, he considers the difference

$$(12) \qquad\qquad u_i = s_i - s_i^*$$

as if it were generated by a random drawing from a probability distribution. If A_i's expectation is unbiased then the mean of the distribution that gene-

---

1. E is the expectations operator which applies to a random variable and yields the mean value. For more information see any standard statistics textbook (Mood and Graybill, 1963).

rates u_i is zero so that $Eu_i = 0$. How does the shape of the demand function influence the expected price? To answer this question let us expand $f(q)$ around Q_i^* in a Taylor Series up to second degree terms as follows:

$$(13) \quad f(q) - f(Q_i^*) = \frac{\partial f}{\partial q}\bigg|_{q=Q_i^*} (q - Q_i^*) + (1/2)\frac{\partial^2 f}{\partial q^2}\bigg|_{q=Q_i^*} (q - Q_i^*)^2,$$

$$q - Q_i^* = q - q_i - (Q_i^* - q_i)$$

together with (11) and (12) implies that

$$(14) \qquad\qquad\qquad q - Q_i^* = s_i - s_i^* = u_i.$$

Taking expectations in (13), it follows that

$$(15) \qquad\qquad p_i^* = Ef(q) = f(Q_i^*) + (1/2)\frac{\partial^2 f}{\partial q^2}\sigma_i^2,$$

where $Eu_i^2 = \sigma_i^2$.

Before going on to the supply conditions, notice that it is harder to infer the true state of the market if the demand itself is subject to random shocks. For instance, suppose that the demand function has an additive random component v that is independent of q so that the demand function becomes

$$(16) \qquad\qquad\qquad p = f(q) + v.$$

In this case the actual price would differ from the expected price not only because of the firm's forecast error regarding the quantity offered by its competitors but also because of the random variable v. Under these conditions it would be more difficult for the firm to identify the source of its forecast errors. However, if the expected value of v is zero, then, even with the demand function (16) instead of (1), it remains true that the expected price is given by (15) up to terms of the second degree.

To complete the model, one must derive the supply conditions. These describe the relation between q_i and the state of the firm's knowledge. Assume that A_i chooses q_i to maximize its expected net return given in (9). If there is a maximum, then q_i satisfies

$$(17) \qquad\qquad p_i^* - c + q_i \frac{\partial p_i^*}{\partial q_i} = 0,$$

where

$$(18) \qquad \frac{\partial p_i^*}{\partial q_i} = \left[\frac{\partial f}{\partial q}\bigg|_{q=Q_i^*} + (1/2)\sigma_i^2\frac{\partial^3 f}{\partial q^3}\bigg|_{q=Q_i^*}\right]\frac{\partial Q_i^*}{\partial q_i}.$$

Next one must postulate or derive a relation between Q_i^* and q_i since (17) and (18) imply the presence of the term $(\partial Q_i^*/\partial q_i)$. Such a relation has come to be called a "conjectural variation," and there are many more or less ad hoc suggestions about the form of this relation, but only two are of particular importance.[2] The first is the Cournot-Nash hypothesis which asserts that

(19)
$$\frac{\partial Q_i^*}{\partial q_i} = 1.$$

The second is the monopoly solution whereby the firms agree to maximize their joint net return so that

(20)
$$\frac{\partial Q_i^*}{\partial q_i} = q/q_i.$$

Without additional assumptions about the cost of obtaining information and about the state of the firm's knowledge, one can hardly say anything useful about conjectural variations. Nevertheless for any theory of conjectural variations there is a general result contained in the following theorem.

THEOREM 1. *Let (1) represent the demand conditions and let (4) represent every firm's cost function. Let every firm choose q_i to maximize its expected net return given in (9). For the existence of a long-run equilibrium it is necessary that the point $q = (q_1, \ldots, q_n)$ with $q = \Sigma_1^n q_i$ have the properties as follows:*

> i. $q_{it} \rightarrow q_i$
>
> ii. $Q_{it}^* \rightarrow q$ *for all i as t $\rightarrow \infty$,*

where Q_{it}^* and $q_t = (q_{1t}, \ldots, q_{nt})$ satisfy

(21)
$$f(Q_{it}^*) - c + q_{it} \frac{\partial p_{it}^*}{\partial q_{it}} = 0, \qquad \textit{for all i and t.}$$

> iii. *q and q satisfy*

(22)
$$f(q) - c + q_i \frac{\partial f}{\partial q} = 0, \qquad \textit{for all i.}$$

Proof. A long-run equilibrium is defined as a set of quantities offered by the n sellers and a price determined by the demand function corresponding to the total quantity offered that can persist indefinitely. By hypothesis every seller chooses q_{it} to maximize his expected net return. Hence in every period t the q_{it}'s and the Q_{it}^*'s must satisfy (21). Both i and ii follow from the definition of a long run equilibrium while iii asserts that the equilibrium values must also satisfy the necessary conditions for a maximum expected net revenue. Condition iii asserts that the space of admissible strategies is closed. Therefore, if an equilibrium exists, then it must also satisfy (21). QED.

---

2. For a thorough discussion of conjectural variations see (Fellner 1949).

This theorem directs our attention to the question of determining how expectations depend on past observations. Cournot answered this question in his original analysis by postulating the following expectations equation:

(23) $$Q_{it}^* = q_{it} + s_{i, t-1}.$$

This asserts that every firm expects the combined sales of its competitors to be the same in the current period as it was in the preceding period. Some other theories of expectations are more plausible than that of Cournot. We now study a class of expectations theories in which the expectations depend on past observations to forecast the future in order to derive sufficient conditions concerning the properties of the expectations for the existence of a long-run equilibrium. In this analysis let us agree at the outset to accept the basic postulate of the Cournot-Nash theory, given in (19), which is equivalent to the assumption that there is no formal cooperation among the n firms. The conditions determining whether or not the firms will cooperate depend on the net return; these are considered in detail in chapter V. Therefore, on the basis of the Cournot-Nash postulate the necessary conditions for the maximum net revenue shown in (21) become

(24) $$f(Q_{it}^*) - c + q_{it} \frac{\partial p_{it}^*}{\partial Q_{it}^*} = 0.$$

Recall that

(25) $$s_{it} = q_t - q_{it}.$$

Now assume that s_{it}^* is a linear function of all past s_{it}'s so that

$$s_{it}^* = w_{i0} s_{i, t-1} + w_{i1} s_{i, t-2} + \dots$$

(26) $$= \sum_{k=0}^{\infty} w_{ik} s_{i, t-1-k}.$$

Equation (26) is the basic assumption about the nature of expectations to be analyzed in the present work. Assuming the existence of an equilibrium, we seek the implied properties of the weights $\{w_{ik}\}$, which thereby constitute the necessary conditions for the existence of an equilibrium.

The first property is simple. Suppose $s_{i, t-1} = s_i$, for all t. Then

(27) $$s_i^* = s_i \sum_{k=0}^{\infty} w_{ik} = (q - q_i) \sum_{0}^{\infty} w_{ik}.$$

By definition $s_i^* = Q_i^* - q_i$. If there is an equilibrium then theorem 1 implies that $Q_i^* = q$, for all i. Hence (27) becomes

$$(q - q_i) = (q - q_i) \sum_{0}^{\infty} w_{ik}.$$

If $(q - q_i) \neq 0$, we obtain

(28) $$\sum_{k=0}^{\infty} w_{ik} = 1,$$

completing the proof of lemma 1. QED.

LEMMA 1. *If a long-run equilibrium exists and if the expected quantity satisfies (26), then the weights must satisfy (28).*

Another property of these weights can be deduced by considering other possible sequences of sales. Now assume that

(29) $$s_{it} = M_i z^{-t}.$$

Hence, s_{it} grows (or declines) geometrically. In fact such a sales path can be the solution of the system if the equations (24) are linear difference equations. Substituting the value of s_{it} given in (29) into the general form of (26), we obtain

(30) $$s_{it}^* = M_i \sum_{k=0}^{\infty} w_{ik} z^{-(t-1)+k} = M_i z^{-(t-1)} \sum_{k=0}^{t-1} w_{ik} z^k.$$

Define the generating function $w_i(z)$ as follows:

(31) $$w_i(z) = \sum_{k=0}^{\infty} w_{ik} z^k.^3$$

Thus, $w_i(z)$ is a power series in z whose coefficients are the weights w_{ik}. Let

$$w_{i(t)}(z) = \sum_{k=0}^{t} w_{ik} z^k$$

denote the partial sum from 0 to t of the series defined in (31). With this notation, equation (30) becomes

(32) $$s_{it}^* = M_i z^{-(t-1)} w_{i,(t-1)}(z).$$

It follows from (28) that $w_i(1) = 1$. Since $w_i(z)$ is a power series in z, a standard theorem on power series applies and gives the result that $w_i(z)$ represents a convergent power series for all $|z| < 1.^4$ ($|z|$ denotes the modulus of z.)[5] This proves the following lemma.

---

3. Generating functions are power series used to study many kinds of problems involving linear differential equations, linear difference equations, and stochastic time series. For an excellent exposition with applications to probability closely related to the usage in the text see (Feller 1962, chap. 11).

4. Knopp (1951, chap. 5, sec. 18, thm. 1).

5. If z is a complex number so that $z = x + iy$, where x and y are real and $i = \sqrt{-1}$, then the complex conjugate of z, denoted by $\bar{z}$, is $\bar{z} = x - iy$ so that $z\bar{z} = x^2 + y^2$, and $|z| = \sqrt{(z\bar{z})}$. Therefore, the region of convergence is the interior of the unit circle in the complex plane.

LEMMA 2. *Under the hypotheses of lemma 1, the generating function of the weights, $w_i(z)$ defined in (31), converges for all $|z| < 1$ and for $z = 1$.*

These two lemmas give necessary conditions for the weights of a system for which an equilibrium exists. A weighting system that satisfies these two lemmas is *admissible*.

Let us now study a few important kinds of expectations theories to see whether or not their weights are admissible. In the original Cournot expectations model the weights satisfy

$$(33) \qquad w_{ik} = \begin{cases} 1, & \text{if } k = 0, \\ 0, & \text{otherwise.} \end{cases}$$

Obviously, these weights are admissible.

In a first-order adaptive expectations model,

$$(34) \qquad s_{it}^* - s_{i,t-1}^* = \gamma_i(s_{i,t-1} - s_{i,t-1}^*).^6$$

This equation implies that the weights satisfy

$$(35) \qquad w_{ik} = \gamma_i(1-\gamma_i)^k.$$

Hence (dropping the subscript i for convenience) the generating function is

$$(36) \qquad w(z) = \gamma/[1-(1-\gamma)z].$$

It readily follows by substituting $z = 1$ in (36) that these weights are admissible.

An analytically more complicated though conceptually simpler case than adaptive expectations is that in which the expected sales are the arithmetic mean of all past observations. Thus,

$$(37) \qquad s_t^* = (s_{t-1} + s_{t-2} + \ldots + s_0)/t \qquad (t = 1, 2, \ldots).$$

In contrast with adaptive expectations, where the weights decrease at a geometric rate, (37) gives every past observation the same weight, namely $1/t$. Moreover, the weights in (37) depend on calendar time in the sense that the generating function for period t is as follows:

$$(38) \qquad w_t(z) = (1 + z + z^2 + \ldots + z^t)/(t+1) \qquad (t = 0, 1, \ldots).$$

The generating function for the arithmetic mean is the limit of $w_t(z)$ as t approaches infinity. Now

$$w_t(z) = (1 - z^{t+1})/[(1-z)(1+t)]$$

---

6. Adaptive expectations as used herein were introduced into economics by Nerlove (1958). Cagan earlier (1956) used closely related models in the form of differential equations. For an excellent discussion of the statistical properties of such models see Whittle (1963, chap. 3, sec. 3).

and

$$\lim_{t \to \infty} w_t(z) = \begin{cases} 0, & \text{if } z < 1, \\ 1, & \text{if } z = 1. \end{cases}$$

In fact, $w_t(1) = 1$, for all t.

The expectations model given by the arithmetic mean, (37), is a limiting case of the adaptive expectations model shown in (34). To prove this, consider the adaptive expectations model for a t-period horizon. In this case the weights are as follows:

$$(39) \qquad w_k = (1-\gamma)^k / \sum_0^t (1-\gamma)^k = \gamma(1-\gamma)^k / [1-(1-\gamma)^{t+1}].$$

It is obvious from (39) that the choice of $\gamma = 0$ gives precisely the weights of the arithmetic mean — namely, $1/(1+t)$. In addition the weights are a continuous function of γ. For the adaptive expectations generating function given in (36)

$$\gamma/[1-(1-\gamma)z] \begin{cases} = 0, & \text{for all } z < 1 \text{ and } \gamma = 0, \\ = 1, & \text{for } z = 1 \text{ and all } \gamma < 1. \end{cases}$$

These results agree with the generating function for the arithmetic mean and show that adaptive expectations for $\gamma = 0$ are equivalent to arithmetic mean expectations. The arithmetic mean also illustrates the point that weights can be admissible without the generating function being a continuous function of z. Thus, for $z < 1$ the generating function is zero, while for $z = 1$, the generating function is 1.

We continue our analysis of the model including expectations by considering the implications of (24). This equation represents the policy of the firm since it relates the firm's choice of q_{it} to its perception of its environment. To derive some properties of the system, take a linear approximation to the demand function evaluated at the equilibrium levels. Thus

$$(40) \qquad f(Q_{it}^*) - f(q) = \left. \frac{\partial f}{\partial q} \right|_{Q^* = q} (Q_{it}^* - q).$$

Let

$$b_0 = f(q) - \frac{\partial f}{\partial q} q \quad \text{and} \quad b_1 = \frac{\partial f}{\partial q}.$$

$$(41) \qquad f(Q_{it}^*) = b_0 + b_1 Q_{it}^*$$

is the linear approximation to the demand function. The implied linear approximation to the policy equation (24) now becomes

$$(42) \qquad b_0 + b_1 Q_{it}^* - c + b_1 q_{it} = 0.$$

But since $Q_{it}^* = s_{it}^* + q_{it}$, (42) simplifies to

(43) $$(b_0 - c)/b_1 + 2q_{it} + s_{it}^* = 0 \qquad (i = 1, ..., n).$$

For there to be an equilibrium solution of the original set of policy equations, (24), the set of n linear difference equations, (43), must have a convergent solution. In other words, the difference equations implied by (43) and the expectations hypothesis (26) together must represent a stable system.[7]

It is easiest to study the solution of (43) with the help of generating functions. Multiply every term of (43) by z^t and sum over all $t = 0, 1, ...,$ giving

(44) $$b(z) + 2\, q_i(z) + s_i^*(z) = 0,$$

where

$$b(z) = [(b_0 - c)/b_1]/(1 - z),$$

$$q_i(z) = \sum_0^\infty q_{it} z^t,$$

and

$$s_i^*(z) = \sum_0^\infty s_{it}^* z^t.$$

The latter expression depends on the generating function of the weights as follows:

(45) $$s_i^*(z) = z w_i(z) s_i(z),$$

where

$$s_i(z) = \sum_0^\infty s_{it} z^t,$$

and, it will be recalled,

$$s_{it} = \sum_{j \neq i}^n q_{jt}.$$

To verify (45), observe that, dropping the subscript i for convenience,

$$s_t^* = w_0 s_{t-1} + w_1 s_{t-2} + w_2 s_{t-3} + ...$$

so that

$$z^{t-1} s_t^* = w_0 s_{t-1} z^{t-1} + w_1 z s_{t-2} z^{t-2} + w_2 z^2 s_{t-3} z^{t-3} + ...$$

Therefore, if $s_t = 0$, for all $t < 0$, so that $q_0^* = 0$,

$$\sum_0^\infty z^t s_t^* = z w(z) s(z)$$

---

7. The assertion in the text is subsumed by the more general proposition about nonlinear dynamic systems to the effect that such a system is stable only if its linear approximation is stable. For a rigorous discussion see Vainberg (1964, chap. 8).

as required. Hence (44) becomes

(46) $$b(z) + 2q_i(z) + zw_i(z)s_i(z) = 0.$$

However, $s_i(z) = \Sigma_{j \neq i}^n q_j(z)$ so that (46) has a concise matrix representation as follows:

(47) $$b(z) + W(z)q(z) = 0,$$

where

$$b(z) = (b_0 - c)/[b_1(1-z)] \begin{bmatrix} 1 \\ 1 \\ \cdots \\ 1 \end{bmatrix};$$

$$q(z) = \begin{bmatrix} q_1(z) \\ \cdots \\ q_n(z) \end{bmatrix};$$

and

$$W(z) = \begin{bmatrix} 2 & zw_1(z) & \cdots & zw_1(z) \\ zw_2(z) & 2 & \cdots & zw_2(z) \\ & & \cdots & \\ zw_n(z) & & \cdots & 2 \end{bmatrix}$$

Hence the formal solution of (47) is

(48) $$q(z) = -W(z)^{-1}b(z).$$

Our task now is to discover necessary and sufficient conditions in terms of the expectations coefficients for the existence of convergent solutions to the linearized policy equations (43). This is equivalent to studying the stability of the linearized policy equations. Stability of the linearized policy equations is necessary, but is not sufficient for the stability of the original policy equations (24). To this end it is convenient to express the generating matrix $W(z)$ somewhat differently in a form that provides an explicit representation of the inverse $W(z)^{-1}$. Every element of $W(z)$ is a power series in z. Hence we may collect the coefficients of like powers of z in a matrix and represent $W(z)$ as follows:

(49) $$W(z) = W_0 + W_1 z + W_2 z^2 + \cdots.$$

For example, W_0 is a diagonal matrix with 2's on the diagonal. The definition of $W(z)$ implies that the $n \times n$ matrixes, $W_j, j = 1, \ldots,$ are all equal so that for all nonnull W_j's

(50) $$W_j = W.$$

For instance, for the Cournot expectations model W_1 is an $n \times n$ matrix whose elements are all 1 and $W_j = 0$, for $j = 2, \dots$. In general we may write $W(z)$ as follows:

(51) $W(z) = W_0 + zW/(1-z).$

With the change of variable

(52) $y = z/(1-z),$

the power series $W(z)$ becomes

(53) $W(y) = W_0(I + yW_0^{-1}W),$

and

(54) $W(y)^{-1} = (I + yW_0^{-1}W)W_0^{-1}.$

Let $M = W_0^{-1}W$. By means of (54) one may express $W(y)^{-1}$ as a power series in powers of My. For (54) to represent a convergent power series in y with a positive radius of convergence, it is necessary and sufficient that the spectral radius of M be less than 1. (The spectral radius of a matrix is the largest modulus of the proper values of the matrix.)[8]

We may also express the inverse of $W(z)$ differently as follows: $W(z)^{-1} = [w^{ij}(z)]$. Because of the simple nature of $b(z)$, we can write the solution of (48) in a more explicit form. The ith component of $q(z)$ is given by

(55) $q_i(z) = -\sum_{j}^{n} w^{ij}(z)[(b_0 - a)b_1](1-z)^{-1}.$

The expression

(56) $w^{ij}(z) = \sum_{0}^{\infty} v_{it}z^t$

is a power series in z as is indicated by the right-hand term in (56), while

$$(1-z)^{-1} = 1 + z + z^2 + \dots.$$

But

$$\sum v_{it}z^t(1 + z + z^2 + \dots) = v_0 + (v_0 + v_1)z + (v_0 + v_1 + v_2)z^2 + \dots$$

so that

(57) $q_{it} = [-(b_0 - a)/b_1] \sum_{k=0}^{t} v_{ik} \rightarrow [-(b_0 - a)/b_1] \sum_{0}^{\infty} v_{ik}.$

However, for the convergence in (57) it is necessary and sufficient that the spectral radius of M be less than one. Assuming convergence,

(58) $\sum_{t=0}^{\infty} v_{it} = \sum_{j=1}^{n} w^{ij}(1).$

---

8. Varga (1962, chap. 1, sec. 3) discusses the spectral radius.

That is, the series on the left-hand side of (58) is obtained from the left-hand side of (57) by setting $z = 1$. Hence,

$$(59) \qquad q_{it} \rightarrow -(b_0 - a)/b_1 \sum_j w^{ij}(1).$$

Since the only constant solution of the linearized policy equation (43) is the Cournot-Nash point given by

$$(60) \qquad q_i = -(b_0 - c)/[b_1(n+1)],$$

it follows that the limit on the right-hand side of (59) must satisfy the condition that

$$(61) \qquad \sum_j w^{ij}(1) = 1/(n+1), \qquad \text{for all } i = 1, ..., n.$$

We now summarize these results in the following theorem.

THEOREM 2. *Let the weights be admissible. Then the linearized policy equations (43) will have a solution convergent to the Cournot-Nash equilibrium point (60) if and only if the spectral radius of $M = W_0^{-1}W$ is less than one.*

Proof. The preceding argument has shown that the necessary and sufficient condition for the convergence of the solutions of the linearized policy equations is that the spectral radius of M is less than one. It only remains to check that admissibility of the weights of the expectations functions implies that the limit of the solutions is indeed the Cournot-Nash point. It is, therefore, enough to compute the inverse of $W(1)$ and show that (61) is satisfied. But

$$W(1) = \begin{bmatrix} 2 & 1 & 1 & \cdots & 1 \\ 1 & 2 & 1 & \cdots & 1 \\ & & \cdots & & \\ 1 & 1 & 1 & \cdots & 2 \end{bmatrix},$$

and it is easy to verify that

$$w^{ij}(1) = \begin{cases} n/(n+1), & \text{if } i = j, \\ -1/(n+1), & \text{otherwise.} \end{cases}$$

This implies that (61) is satisfied. QED.

Theorem 2 is not directly useful in checking whether a given expectations model results in a convergent sequence because of the difficulty in calculating the spectral radius of the matrix implied by a given expectations model. In particular cases there are better ways of checking convergence. We now study whether or not the three expectations models, the Cournot model, adaptive expectations, and the arithmetic mean, have convergent solutions of the linearized policy equation. We first treat the Cournot model, next adaptive expectations, and then arithmetic mean expectations.

In the first two cases $W(z)$ is as follows:

$$(62) \quad \begin{bmatrix} 2 & h(z) & \cdots & h(z) \\ h(z) & 2 & \cdots & h(z) \\ & & \cdots & \\ h(z) & h(z) & \cdots & 2 \end{bmatrix}$$

Since

$$4+2(n-2)h-(n-1)h^2 = (2-h)[2+(n-1)h],$$

$$W(z)^{-1} = 1/\{(2-h)[2+(n-1)h]\}$$

$$\times \begin{bmatrix} 2+(n-2)h & -h & \cdots & -h \\ -h & 2+(n-2)h & \cdots & -h \\ & & \cdots & \\ -h & -h & & 2+(n-2)h \end{bmatrix}.$$

Therefore,

$$q_i(z) = -b_i(z)[2+(n-2)h-(n-1)]/[2-h][2+(n-1)h]$$

$$(63) \qquad = -b_i(z)/[2+(n-1)h].$$

For there to be a convergent solution it is necessary and sufficient that $2+(n-1)h(z)$ be zero-free inside and on the closed unit circle in the complex plane. Now for the Cournot expectations model where $s_{it}^* = s_{i,t-1}$, $h(z) = z$. Hence,

$$(64) \qquad 2+(n-1)h(z) = 2+(n-1)z = 0,$$

which has a root given by

$$(65) \qquad z = -2/(n-1),$$

in the unit circle for $n \geqslant 3$. Therefore, the Cournot expectations model only converges if there are two firms, that is, duopoly. It fails to converge if there are three or more firms!

For the adaptive expectations model we have

$$(66) \qquad h(z) = \gamma z/[1-(1-\gamma)z].$$

For stability it is necessary and sufficient that

$$(67) \qquad 2+\frac{(n-1)\gamma z}{1-(1-\gamma)z} = 0$$

be zero-free on the closed unit disk. However, this will be true if and only if

$$(68) \qquad \gamma < 4/(n+1).$$

Assuming that $\gamma < 1$ in any case, the upper bound on γ imposed by (68) is no additional constraint for $n \leqslant 3$. But for $n > 3$, it forces γ toward zero. This proves theorem 3.

THEOREM 3. *The Cournot expectations model has a convergent solution of the linearized policy equation if there are not more than two firms. The adaptive expectations model with an adjustment coefficient γ identical for all firms has a convergent solution of the linearized policy equations if and only if γ satisfies (68).*

In view of the fact that arithmetic mean is a limiting case of adaptive expectations as γ approaches zero, the next result should cause no surprise.

THEOREM 4. *For arithmetic mean expectations and any finite number of firms, the linearized policy equations always have a convergent solution to the Cournot-Nash point.*

Proof. In this case

$$s_{it}^* = (1/t) \sum_{k=0}^{t-1} s_{ik}$$

gives the expectation, and the linearized policy equation is $2q_{it} + s_{it}^* = d$, where $d = -(b_0 - c)/b_1 > 0$. By the symmetry of the model with respect to the n firms it must be true that the solution $\{q_{it}\}$ must be the same for all i. Let $q_{it} = x_t$ so that the common policy equation becomes

$$(69) \qquad 2x_t + [(n-1)/t] \sum_0^t x_{t-1-j} = d.$$

Since $x_t = 0$ if $t < 0$, it follows that $x_0 = d/2$. Let σ_t denote the partial sum $\sum_0^t x_{t-j}$ so that (69) becomes

$$(70) \qquad 2tx_t + (n-1)\sigma_{t-1} = td.$$

Let

$$(71) \qquad x(z) = \sum_0^\infty x_t z^t$$

denote the generating function of the sequence $\{x_t\}$. Multiplying through (70) by z^{t-1} and summing over all t gives

$$(72) \qquad \sum_0^\infty tx_t z^{t-1} + (n-1)\sum_0^\infty \sigma_{t-1}z^{t-1} = d\sum_0^\infty tz^{t-1}$$

Now

$$x'(z) = \sum_0^\infty tx_t z^{t-1}$$

and

$$(1-z)\sum_0^\infty z^{t-1}\sigma_{t-1} = \sum_0^\infty x_{t-1}z^{t-1}.$$

With these substitutions we obtain

$$(73) \qquad 2x'(z) + [(n-1)/(1-z)]x(z) = d/(1-z)^2.$$

Let $e^t = 1 - z$, and (73) becomes a first order linear differential equation with constant coefficients as follows:

$$(74) \qquad -2x'(t) + (n-1)x(t) = de^{-t}.$$

The solution of (74) is given by

$$(75) \qquad x(t) = [d/(n+1)]\, e^{-t} + k_0\, e^{(n-1)t/2}$$

or, in terms of the original variable z, it is given by

$$(76) \qquad x(z) = \{d/[(n+1)(1-z)]\} + k_0(1-z)^{(n-1)/2}$$

Using the fact that the initial value of x_t is $x_0 = d/2$ so that

$$k_0 = [d(n-1)]/[2(n+1)],$$

the generating function of $x(z)$ becomes

$$(77) \qquad x(z) = [d/(n+1)]\, \{(1-z)^{-1} + [(n-1)/2]\, (1-z)^{(n-1)/2}\}.$$

The coefficients of the powers of z are x_t, and these coefficients obviously tend toward the Cournot-Nash equilibrium point. QED.

This result is notable in comparison with the two other expectations models because it gives a convergent solution for any number of firms.[9] For both the Cournot expectations model and the adaptive expectations model, convergence depends on the number of firms. In fact the Cournot model only converges for two firms while the adaptive expectations model converges if the expectations coefficient γ satisfies (68).

In the next section we continue our analysis of expectations models for a different situation, where the goods are not perfect substitutes and the firm's price may be its policy variable. As we shall see, in these cases the expectations models will have different properties.

3. Expectation Models for Price as the Policy Variable

So far it is assumed that the products of the n firms are perfect substitutes in a perfect market so that all firms must have the same price and can control only their quantity offered. Price can be a policy variable only if the products of the firms are not perfect substitutes. Thus, if products are imperfect

9. The arithmetic mean expectations are the same as the determinants of a strategy used to play zero-sum two-person games based on learning. Each player calculates the arithmetic mean choice of his opponent based on all previous moves and makes his choice to maximize his return given the value of this mean. It can be shown that with this method of play the actual choices converge to the mixed strategies of the saddle point solution of von Neumann. This method of play was first proposed by Brown, and the proof of convergence is due to Robinson. See Gale (1960, chap. 7, sec. 9) for an excellent exposition of this approach. The theorem in the text is more elementary because the "optimal" strategies are pure instead of mixed.

substitutes, then each firm can choose a different price without drastic effects on the quantity that it sells and the firms' equilibrium prices may differ. As a result every firm can have its own price policy. The policy equations relate the firm's price to its expectations of competing prices. Study of this situation brings out new results illuminating the general properties of expectation models.

In this case the demand function for good i has a linear approximation given by

$$
(1) \qquad\qquad q_i = b_i - \sum_{j=1}^{n} a_{ij} p_i \qquad\qquad (i = 1, ..., n).
$$

In matrix form the demand equations are as follows:

$$
(2) \qquad\qquad q = b - Ap,
$$

where p, q, and b are $n \times 1$ column vectors and A is an $n \times n$ matrix of the price coefficients. Now one must impose conditions on the demand equations that will imply the existence of positive quantities demanded corresponding to some positive range of prices. Assume that A is invertible so that

$$
(3) \qquad\qquad p = A^{-1}(b - q).
$$

Also, assume that $b > 0$, the own price slopes are negative, and the cross price slopes are positive. The latter assumption means that all goods of the n firms are substitutes. In addition assume that for any $q > 0$ such that $b - q > 0$, there is a positive price vector. It can be shown that these properties imply that the matrix A^{-1} has nonnegative elements (Varga 1962, thm. 3.11). Denote this by $A^{-1} \geqslant 0$.

To represent the negative own price slopes and the positive cross price slopes, let us write the matrix A as the sum of two matrixes as follows:

$$
(4) \qquad\qquad A = A_d - A_v,
$$

where A_d is a diagonal matrix whose diagonal elements are taken from A and A_v is a matrix with zero's on the diagonal whose off-diagonal elements coincide with the off-diagonal elements of A. Hence,

$$
(5) \qquad\qquad A_d > 0 \quad \text{and} \quad A_v \geqq 0.
$$

(The double equality sign allows $A_v = 0$.) These assumptions have important consequences as we shall see.

We now consider the cost conditions. Let each firm have a linear total cost function given by $c_i q_i$ so that firm i's net revenue, denoted by r_i, is defined by

$$
(6) \qquad\qquad r_i = (p_i - c_i) q_i.
$$

The noncooperative solution for this case is a set of prices $\{p_i^0\}$ that is the solution of the n equations as follows:

(7)
$$\frac{\partial r_i}{\partial p_i} = 0.$$

In addition these prices must have the property that

(8)
$$r_i^0 \{p_i^0\} \geqslant r_i \{p_1^0, \ldots, p_{i-1}^0, p_i, p_{i+1}^0, \ldots, p_n^0\}.$$

It is easy to verify that the assumptions about the cost and demand conditions imply that (8) is satisfied. Equation (7) gives the results as follows:

(9)
$$q_i - (p_i - c_i)a_{ii} = 0 \qquad\qquad (i = 1, \ldots, n).$$

Using the decomposition of the matrix A given in (4), (9) can be rewritten as

$$b - Ap - A_d(p-c) = 0, \quad \text{or}$$

(10)
$$b - 2A_d p + A_d c + A_v p = 0.$$

Hence the equilibrium set of prices for the noncooperative solution satisfies

(11)
$$(2A_d - A_v)p = A_d c + b.$$

We must show that the assumptions about A imply that (11) does in fact have a solution with $p \geqslant 0$. To this end we now prove lemma 3.

LEMMA 3. *If $A^{-1} > 0$ where $A = A_d - A_v$ with $A_d > 0$ and $A_v \geqslant 0$, then the spectral radius of $A_d^{-1}A_v$ is less than one and (11) has a solution with $p \geqslant 0$.*
Proof. Let $C = A^{-1}A_v \geqslant 0$ so that

$$A_d^{-1}A_v = (A+A_v)^{-1}A_v = (I+A^{-1}A_v)^{-1}A^{-1}A_v = (I+C)^{-1}C.$$

Let $\lambda > 0$ be the spectral radius corresponding to an extremal vector $x \geqslant 0$ of the nonnegative matrix C. Hence $Cx = \lambda x$ and

$$(I+C)^{-1}Cx = (I+C)^{-1}\lambda x = [\lambda/(1+\lambda)]x.$$

Denote the spectral radius of a matrix M by $\rho(M)$. With this notation

$$\rho(A_d^{-1}A_v) = \rho((I+C)^{-1}C) = [\lambda/(1+\lambda)] < 1.$$

This proves the first part of the lemma. A fortiori,

$$(1/2)\rho(A_d^{-1}A_v) < 1.$$

Therefore, since

(12)
$$(2A_d - A_v) = 2A_d[I - (1/2)A_d^{-1}A_v],$$

the solution of p satisfies

(13)
$$p = [I - (1/2)A_d^{-1}A_v]^{-1}(c + A_d^{-1}b)/2,$$

where the existence of the inverse of the matrix in (12) is guaranteed by the fact that the spectral radius of $(1/2)A_d^{-1}A_v$ is less than one, and the positivity of the solution is guaranteed by the positivity of the matrix $A_d^{-1}A_v$ together with the fact that $(1/2)(c+A_d^{-1}b) \geq 0$. QED.[10] (see also thm. VI.17).

For each firm, equation (9) can be regarded as a price policy equation that relates firm i's price to its own cost and demand parameters and the prices of competing products. In particular the firm must rest its decision on its forecast of the prices of competing products. Let $p_{ij,t}^* = $ expected value of firm i's price by firm j.

In general $p_{ij}^* \neq p_{ij'}^*, j \neq j'$ until the firms are in equilibrium. In addition $p_{ii,t}^* = p_{it}$ for all t because every firm knows its own price. The price policy equations as a function of the expected prices are given by

$$(14) \qquad b_i - 2a_{ii}p_{it} + a_{ii}c_i + \sum_j a_{ij}p_{ji,t}^* = 0, \qquad \text{for } i = 1, \dots, n.$$

In this analysis we shall consider the three important kinds of expectations models, the Cournot expectations in which the expected price is the price in the preceding period, adaptive expectations in which the expected price is a geometrically weighted average of past prices, and the arithmetic mean in which the expected price is an equally weighted average of all past prices. Consider first the Cournot hypothesis, where

$$(15) \qquad p_{ij,t}^* = p_{i,t-1}.$$

Substituting this into the price policy equations (14), and expressing the result in matrix form, we obtain a set of linear difference equations as follows:

$$(16) \qquad b - 2A_d p_t + A_d c + A_s p_{t-1} = 0.$$

It is convenient now to introduce the lag operator L which is defined by the condition that $Lp_t = p_{t-1}$. Using the lag operator permits a simplification of the difference equations in (16) as follows:

$$(17) \qquad (2A_d - A_v L)p_t = A_d c + b.$$

Let

$$B = (1/2)A_d^{-1}A_v \geq 0,$$

and

$$m = (1/2)(c + A_d^{-1}b) \geq 0.$$

In terms of B and m the price policy equations for the Cournot hypothesis of price expectations becomes

$$(18) \qquad (I - BL)p_t = m.$$

---

10. The proof that the spectral radius is less than one and that the dominant eigenvector is positive is adapted from the proof of theorem 3.10 of Varga (1962). This result is closely related to certain propositions familiar to economists about the stability of markets with n commodities that are gross substitutes. For a discussion of these, see Karlin (1959, vol. 1, chap. 9).

It is well known that the system of linear difference equations given in (18) has a convergent solution if and only if the proper values of the matrix B are inside the unit circle of the complex plane.[11] Since the absolute value of the proper values is bounded by the spectral radius of B, which is less than one by lemma 3, it follows that (18) has a convergent solution. Moreover, the solution is nonnegative because both B and m are nonnegative. This proves lemma 4.

LEMMA 4. *If A satisfies the same conditions as in lemma 3 and if expectations satisfy the Cournot hypothesis given in (15) then the price policy equations have a convergent and nonnegative solution.*

Observe that this result, which holds for any number of firms, differs from the conclusion of the preceding section. The preceding section analyzes the case in which the products of the firms are perfect substitutes so that there is a common price, and quantity must be the firm's policy variable. In that case the Cournot expectations hypothesis implies a convergent solution of the policy equations only if there are two firms. In the present case, in which the products are imperfect substitutes, so that price can be a policy variable, the Cournot expectations hypothesis implies a convergent solution of the policy equations for any number of firms. This raises the question of explaining why the results differ for the two cases.

The difference between the results is explained by the necessity of postulating a more specific structure for the system in which the goods are imperfect substitutes. Thus, for economically meaningful results, it is necessary to assume that positive quantities are demanded for some range of positive prices. Together with the postulate that the goods are substitutes so that the own price slopes are negative and the cross price slopes positive, we may conclude that $A^{-1} > 0$. Consequently, a special structure is imposed on the matrix of price coefficients. This structure implies that the spectral radius of B is less than one as shown in lemma 3. In fact, for the special case in which all own price slopes are equal and all cross price slopes are equal, it is easy to verify that $A^{-1} \geq 0$ implies that the sum of the rows of the matrix $2A_d - A_v$ must be negative, a condition known as diagonal dominance. However, with all goods perfect substitutes the basic demand equation is a simple linear equation that makes the common price a linear function of the total quantity offered. Hence it is sufficient to postulate a positive intercept for the demand relation to give economically meaningful results. In a sense this is a weak requirement though the assumption that the goods are perfect substitutes is a strong one. In the latter case convergence depends more delicately on the nature of the expectations.

---

11. The solution of (18) is given by $p_t = (I + B + B^2 + \ldots + B^t)m$. For convergence of the power series in B it is necessary and sufficient that the spectral radius of B be less than one. See Varga (1962, thm. 3.7) for the details.

3. *Price as the Policy Variable*

We now study the price policy equations for the other two expectations models.

For adaptive expectations

(19) $\quad p_{ji,t}^* = (1 - \delta_{ji})(1 - \delta_{ji}L)^{-1} p_{j,t-1}$ $\qquad (j \neq i$ and $0 \leqslant \delta_{ji} < 1)$,

where L is the lag operator defined above.[12] It follows that the Cournot expectations hypothesis is that special case of adaptive expectations in which $\delta_{ij} = 0$. With adaptive expectations we shall prove that the price policy equations (14) have a nonnegative and convergent solution.

THEOREM 5. *Let the expected price be given by (19) and let* $p_{ii,t}^* = p_{it}$. *Let the matrix of price coefficients A satisfy the hypotheses of lemma 3. Then the price policy equations (14) have a nonnegative and convergent solution.*

Proof. Define

$$\bar{p}_{it} = \sum_j a_{ij} p_{ji,t}^* \qquad (j \neq i),$$

as the ith component of the $n \times 1$ vector $\bar{p}_t$. Hence the price policy equations can be written as follows:

$$b - 2A_d p_t + A_d c + \bar{p}_t = 0,$$

where $\bar{p}_t = G(L)p_{t-1}$, and $G(L)$ is an $n \times n$ matrix in the lag operator L defined as follows:

$$\begin{bmatrix} 0 & a_{12}(1-\delta_{12})(1-\delta_{12}L)^{-1} & \cdots & a_{1n}(1-\delta_{1n})(1-\delta_{1n}L)^{-1} \\ a_{21}(1-\delta_{21})(1-\delta_{21}L)^{-1} & 0 & \cdots & a_{2n}(1-\delta_{2n})(1-\delta_{2n}L)^{-1} \\ a_{n1}(1-\delta_{n1})(1-\delta_{n1}L)^{-1} & a_{n2}(1-\delta_{n2})(1-\delta_{n2}L)^{-1} & \cdots & 0 \end{bmatrix}$$

Therefore, the price policy equations can be concisely written as follows:

(20) $\qquad [2A_d - G(L)L]p_t = A_d c + b.$

The characteristic matrix for the linear operator $[I - (1/2)A_d^{-1}G(L)L]$ is defined by

(21) $\qquad I - (1/2)A_d^{-1}G(z)z,$

where z is a complex scalar.

---

12. Strictly speaking one should write (19) as follows, $p_{ji,t}^* = \delta_{ji}[1 - (1-\delta_{ji})L]^{-1}p_{j,t-1}$, to conform with the definition of adaptive expectations given in (2.34). However, it simplifies the notation and should cause no confusion to use (19) in its present form.

Now for all $|z| \leqslant 1$ we have the inequality

$$|a_{ij}(1-\delta_{ij})/(1-\delta_{ij}z)| \leqslant |a_{ij}|^{13}$$

so that

(22) $$|(1/2)A_d^{-1}G(z)z| < |B| = B.$$

Hence given any value of z such that $|z| \leqslant 1$, the spectral radius of the matrix on the left-hand side of (22) is less than the spectral radius of B, which is less than one by lemma 3. Hence the characteristic matrix defined in (21) is nonsingular for all $|z| \leqslant 1$ which proves that the linear operator in (20) is stable, and the price policy equations have a convergent solution. That the solution is nonnegative follows at once from the fact that the elements of the matrix, $G(L)$, are all positive so that the elements of $A_d^{-1}G(L)$ are all positive. QED.

The last expectations model analyzed herein is the arithmetic mean. In this case we have

(23) $$p_{jt}^* = (1/t) \sum_{0}^{t-1} p_{j,s}.$$

Thus the expected price is an arithmetic mean of all past prices through period $t-1$. Let p_t^* denote the $n \times 1$ column vector whose jth element is p_{jt}^*. Then the price policy equations can be written as follows:

(24) $$b - 2A_d p_t + A_d c + A_s p_t^* = 0.$$

Just as in the preceding section it is most convenient to study this form of expectations with generating functions. Let

$$p_j(z) = \sum_{0}^{\infty} p_{jt} z^t$$

be the generating function for the price sequence $\{p_{jt}\}$ and let $p(z)$ be an $n \times 1$ column vector composed of $p_j(z)$. The generating functions of prices must satisfy the following relation as implied by the price policy equations:

(25) $$(1-z)^{-2}b - 2A_d p'(z) + A_d c(1-z)^{-2} + A_v p(z)(1-z)^{-1} = 0,$$

which reduces to

(26) $$p'(z) - (1-z)^{-1} Bp(z) = (1-z)^{-2} m,$$

a system of first order differential equations in the generating function $p(z)$. For arithmetic expectations there is a convergent nonnegative solution of

---

13. Here is a proof: First, it is easy to verify that the extreme points of $|1-\delta_{ij}z|$ are at $|z| = 1$. There are only two possible cases, a minimum corresponding to $z = -1$ and a maximum corresponding to $z = +1$ since $0 \leqslant \delta_{ij} \leqslant 1$. Moreover, $1-\delta_{ij} \leqslant 1$. Since $|1-\delta_{ij}z| \geqslant 1-\delta_{ij}$ the assertion in the text follows.

(24) if and only if the system of differential equations shown in (26) is stable and the coefficients of the generating functions are nonnegative. We now prove theorem 6.

THEOREM 6. *For arithmetic expectations the price policy equations have a convergent nonnegative solution if $B \geqslant 0$ and has a spectral radius less than one. That is, if A satisfies the hypothesis of lemma 3, then the arithmetic mean expectations yield a nonnegative and convergent solution of the price policy equations.*

Proof. We first convert the first order differential equations in (26) to a system of *linear* first order differential equations via the transformation $e^t = (1-z)$ (see the proof of theorem 4). This transformation implies that

$$-\frac{dp}{dz}(1-z) = \frac{dp}{dt}$$

so that (26) becomes

$$-p'(t) - Bp(t) = m\,e^{-t}$$

or

(27) $$p'(t) + Bp(t) = -m\,e^{-t},$$

where $m = c + (1/2)A_d^{-1}c \geqslant 0$ as above. Now the solution of (27) is of the form

(28) $$p(t) = e^{-Bt}x(t),$$

where

$$e^{Bs} = I + Bs + B^2(s^2/2) + B^3[s^3/(3!)] + \cdots$$

is the exponential power series in the scalar s and the matrix B. Since

$$p'(t) = -Be^{-Bt}x(t) + e^{-Bt}x'(t),$$

it follows that $x(t)$ is the solution of the differential equation

$$x'(t) = -e^{-t}e^{Bt}m,$$

which is given by

$$x(t) = -\left[\int_0^t e^{-s}e^{Bs}ds\right]m.$$

Integrating we obtain

$$x(t) = (I - B)^{-1} e^{Bt} e^{-t}m$$

so that the solution of (27) is

(29) $$p(t) = e^{-Bt}(I - B)^{-1} e^{Bt}m,$$

which reduces to

(30) $$p(t) = (I - B)^{-1} e^{-t} m.$$

Since $B \geqslant 0$ and has a spectral radius less than one, it follows that the arithmetic mean expectation implies a convergent and nonnegative solution of the price policy equations. QED.[14]

4. Summary

This chapter analyzes various expectations models under static demand conditions always assuming competition among the firms. There are several important findings. First, if the products of the n firms are perfect substitutes, then the Cournot expectations hypothesis implies a convergent solution only if there are two firms but not if there are three or more. With adaptive expectations there is a convergent solution for up to three firms but not for more than three unless the expectations coefficient falls rapidly enough as the number of firms rises. Arithmetic mean expectations give a convergent solution for any number of firms.

Second, if the n competing products are not perfect substitutes and the demand functions satisfy additional conditions consistent with economic theory, then there is a convergent solution of the system of price policy equations even under the Cournot expectations hypothesis for any number of firms. A fortiori there is a convergent solution for any number of firms under the other two expectations models.

---

14. The proof in the text omits some details herein supplied. To integrate the expression

(i) $$-\int_0^t e^{-s} e^{Bs} \, ds,$$

it is most convenient to use the power series expansion:

(ii) $$e^{-s} e^{Bs} = I e^{-s} + B s \, e^{-s} + B^2 s^2 \, e^{-s}/2 + B^3 s^3 \, e^{-s}/3! + \ldots$$

(iii) $$\int_0^t s^k e^{-s} \, ds = -t^k e^{-t} + k \int_0^t s^{k-1} e^{-s} \, ds$$

(iv) $$= -[t^k + k t^{k-1} + k(k-1) t^{k-2} + \ldots + k!] \, e^{-t}.$$

Hence the integral of the kth term in i above gives

(v) $$-\frac{B^k}{k!} [t^k + k t^{k-1} + k(k-1) t^{k-2} + \ldots + k!] \, e^{-t}.$$

Sum this expression over k from 0 to infinity. This gives

(vi) $$-e^{-t} \left[\sum_0^\infty \frac{B^k t^k}{k!} + B \sum_1^\infty \frac{B^{k-1} t^{k-1}}{(k-1)!} + B^2 \sum_2^\infty \frac{B^{k-2} t^{k-2}}{(k-2)!} + \ldots \right]$$

$$= -e^{-t} [e^{Bt} + B e^{Bt} + B^2 e^{Bt} + \ldots]$$

(vii) $$= -e^{-t} (I - B)^{-1} e^{Bt} = -e^{-t} e^{Bt} (I - B)^{-1}.$$

The latter result provides the link between (29) and (30).

Third, in every case the sequence of choices requires every firm to "know" only its own cost and demand conditions together with the past values of prices and quantities. The firms need not know about each others' cost and demand conditions.

Expectations models under dynamic demand conditions are more complicated, as we shall see in chapter VI. Dynamic demand arises under one or both of two circumstances; first, a changing level of demand and, second, feed-backs in demand. The latter means that the current demand depends on the past either because consumers have stocks of the good or because price information diffuses slowly among the buyers. One may also regard the latter situation as one in which the demand depends on expected prices as a function of past prices. A firm maximizing the present value of its net return when the demand has feed-back considers the future repercussions of its current actions. Consequently, the existence of an equilibrium is even more sensitively dependent on the properties of the price expectations when n competing firms all face dynamic demand conditions.[15]

It is now convenient to comment on some implications of constant unit costs in the simple dynamic case of geometrically growing demand. Suppose the following demand relations:

$$(1) \qquad\qquad q_t = b_t - Ap_t,$$

where the presence of the subscript denotes the possibility of change over time. The price policy equations under the Cournot-Nash equilibrium are given by

$$(2) \qquad\qquad (2A_d - A_v)p_t = A_d c_t + b_t.$$

If the market expands geometrically at the rate ρ, then $b_t = \rho^t b$, and even with constant unit costs, so that $c_t = c$, prices would rise at the rate ρ because with a linear demand and geometrically rising b_t, the demand for each firm's product is becoming less elastic. If market growth results from the addition of more customers just like the old ones, then it is better to postulate a demand relation of the following kind:

$$(3) \qquad\qquad q_t = \rho^t(b - Ap_t),$$

which preserves a constant price elasticity with geometric market growth. Therefore, for a constant number of firms the Cournot-Nash equilibrium would have constant prices and geometrically rising sales.

---

15. Many articles on the subject of expectations in an oligopoly model have appeared in the past ten years. For the most recent list giving fourteen references see Okuguchi (1970). Most of this work differs from mine in several important respects. First, average and marginal costs are rising and not constant; second, in studying adaptive expectations some stringent conditions are usually imposed on the coefficients δ_{ij}, typically, that all are the same; third, arithmetic mean expectations are not studied; fourth, the firms are assumed to make perfect substitutes with quantity and not price as the policy variable; and finally, the results in chapter VI below are considerably more general than anything in this literature.

Chapter VI continues the analysis of the equilibrium for dynamic demand relations and studies the existence of the monopoly and the Cournot-Nash equilibria for a general class of expectations models which embrace all three of the ones considered so far.

V

Competition or Collusion?

1. Introduction

Since it seems that firms can raise their net return by forming monopolies, why is there not a monopoly in every industry? This is especially hard to answer for a theory assuming that firms strive to maximize their net return, for then one must explain the state of competition in an industry as resulting from the fact that collusion is not always more lucrative than competition. That is, a group of colluding firms may encounter costs of collusion itself that would be absent under competition. One of the main tasks of this chapter is to study these costs and how they relate to industry structure.

An empirical survey of industry structure shows several important facts. First, average firm size by industry differs more than firm sizes within an industry. Equivalently, an industry tends to have firms clustered around a characteristic size. Second, several empirical studies have found that rates of return by industry tend to vary directly with the relative size of firms by industry. Chapter VIII studies some determinants of rates of return by industry and estimates how they depend on relative firm size by industry. The latter is measured by the share of sales of the four leading firms in the industry, also known as the four-firm concentration ratio. These empirical findings suggest the possibility of a link between the net return from collusion and the relative size of firms, a topic which will occupy our attention after investigating several other problems. First, we shall seek greater clarification of the two concepts, competition and collusion. This encounters the question of what determines the number of firms in an industry, leading to a study of the efficient organization of an industry under competition and under collusion. The project is straightforward for well-defined products, but there are some pitfalls if product quality is also allowed to be one of the firm's policy choices. If so, the concept of efficiency requires more careful handling. A paradigm for quality variation is the simple location problem where one must determine the output per firm, the number of firms, and the location of the firms on a road, assuming there is a uniform price-responsive demand at every point on the road and decreasing average production cost per plant. Section 4 treats the cost of maintaining collusion and shows how it relates the industry structure to the average rate of return. Finally, the last section studies various theories of how firms share the collusive net return. After rejecting several theories based on different kinds of game theoretic characteristic functions that were promising, a simple profit sharing formula is proposed which has two merits; first, under collusion the firms are led to maximize the joint net return of the group and, second, as a result, the durability of the collusion is enhanced.

2. The Nature of Competition and Collusion

Although perhaps not widely recognized as such, the Cournot-Nash approach provides a rigorous statement of the classical theory of competition. It differs from the classical theory in its attempt to derive precise implications about how the number and relative size of firms affect the equilibrium. To illustrate, let us consider an industry in which there are n firms who offer the same product to the buyers at a constant and common price per unit. For the present we defer the question of what determines the number of firms. Let

$$(1) \qquad\qquad p = f(q)$$

represent the aggregate demand for the product, where q_i is the quantity sold by firm i and

$$(2) \qquad\qquad q = \Sigma\, q_i.$$

Assume that total cost is a convex increasing function of q_i as follows:

$$(3) \qquad\qquad c_i = g_i(q_i)$$

such that $g_i(0) = 0$. Hence in contrast to our previous treatments, this allows for the possibility that firms may have different costs. The net revenue of firm i, denoted as usual by r_i, is

$$(4) \qquad\qquad r_i = pq_i - c_i.$$

Assume that the firm chooses q_i to maximize r_i. Frequently, perfect competition is defined as a state in which p is independent of q_i, but for a finite number of firms, this cannot be true. A better definition of competition is that the firms act independently. This is the notion given precision by the Cournot-Nash model. If a maximum r_i exists, then the corresponding q_i must satisfy

$$(5) \qquad\qquad p + q_i \frac{\partial f}{\partial q}\frac{\partial q}{\partial q_i} - \frac{\partial g}{\partial q_i} \leqslant 0.$$

There is equality if the optimal q_i is positive. Independence of action by the firms means that

(6)
$$\frac{\partial q_j}{\partial q_i} = \begin{cases} 1, & \text{if } i = j, \\ 0, & \text{if } i \neq j, \end{cases}$$

according to the Cournot-Nash model. It implies that

(7)
$$\frac{\partial q}{\partial q_i} = 1.$$

Therefore, (5) becomes

(8)
$$p(1 + m_i/\eta) - \frac{\partial g_i}{\partial q_i} \leq 0,$$

where the market share of firm i, given by $m_i = q_i/q$, and the price elasticity of the aggregate demand, given by

$$\eta = (p/q)\left(1 / \frac{\partial f}{\partial q}\right),$$

are both measured at the equilibrium point.

Collusion would be represented by a functional relation among the firm outputs, say, $\Psi(q_1, \ldots, q_n) = 0$. If so,

$$\frac{\partial q_i}{\partial q_j} = \frac{\partial \Psi}{\partial q_i} / \frac{\partial \Psi}{\partial q_j} \neq 0, \qquad \text{if } i \neq j,$$

in contrast to (6). In this case we would also have

$$\frac{\partial q}{\partial q_i} = \frac{\partial \Psi}{\partial q_i}$$

Returning to the Cournot-Nash model, observe that if there is one firm in the industry so that its market share must be one, then inequality (8) becomes the familiar necessary condition for the maximum net return of a monopoly. At the other extreme, when m_i/η is small, (8) is the exact condition for competition often approximated by ignoring the term m_i/η. Notice that for a given elasticity of the total demand, although it is true that the error of approximation from assuming that the firm's actions cannot affect the current price, falls with the market share; (8) also shows that in fact the error depends on the ratio m_i/η. For instance, if the aggregate price elasticity is -1, then the price would exceed marginal cost by 5 percent if the firm's market share were 5 percent, and the excess would be 10 percent for a firm market share of 10 percent. For the commonly observed range of price elasticities a firm's market share must be quite small and/or η quite large before one can safely ignore the term m_i/η.

Indeed a source of controversy about (8) is precisely over its continuous interpolation by means of market share between extreme competition and extreme monopoly. Where firms have market shares above 10 percent, many would argue that collusion is more likely than competition, the latter being represented by (8). I shall argue that collusion occurs whenever it is more profitable to all of the participants than their feasible alternatives. When this is so, collusion dominates competition. Otherwise, competition dominates collusion. For collusion to dominate competition not only must the joint profit maximum exceed the combined competitive return but also none of the parties to the collusion must be able to obtain a higher return by means of some feasible strategy given that all others adhere to the collusive agreement. Were it always true that collusion dominates competition, then there would be a world of monopolies. Since we do not observe this, there must be conditions under which competition dominates collusion, and we need a theory to find these conditions.

Such a theory should include two points. First, it should give conditions for dominance with a given number of firms; and, second, it should explain when entry will occur so that the theory should also determine the number of firms. To begin with, suppose that a given number of firms succeed in reducing their total output so that prices are high enough to yield a return above the competitive level. Assuming full employment, there must be more resources in the competitive sectors and less in the collusive or monopolized ones. The higher return in the latter sectors induces firms to enter these industries. If there are no insurmountable barriers to entry, eventually outside firms will enter and survive in the collusive sectors. This process continues until a competitive equilibrium prevails in all sectors. Under these hypothetical conditions there cannot be permanent monopoly, and the ease of entry limits the size and duration of the collusive return. This affects the expected present value of the net return to a given group of firms who contemplate collusion. Hence without loss of generality we may confine the analysis to a market with a given number of firms, which still raises the question of what determines the optimal number of firms in an industry. However, the hypothesis that firms seek a maximum net return and certain other postulates also determine the optimal number of firms, as we shall see.

Appropriately modified, the Cournot-Nash theory supplies a framework for determining when collusion dominates competition. Chapter III contains the elements of this approach. It covers the case in which two identical firms offer the same product to many small buyers. It is shown that a necessary condition for collusion to dominate competition is that trade in the good can last indefinitely. That analysis assumes that firms cannot make side payments and can only enforce collusion by temporarily returning to competition. The dominance of collusion over competition depends on the values of three parameters; the probability of trade continuing, the gain from collusion, and the gain from cheating. We shall see that several other pertinent factors make their presence felt via these three parameters, including the cost of policing the collusion, the rate of entry, and the stability of the market.

In considering where collusion is most likely, one should also recognize the effects of legal constraints; but it is naive to suppose that certain actions do not occur merely because they are illegal. Legal sanctions at best reduce the amount of illegal activity but hardly eliminate it altogether. Nevertheless, it now appears that antitrust laws do inhibit certain firms from some practices both because the firms are conspicuous and because the actions would be clearly visible. For instance, under present conditions the three leading U.S. automobile firms cannot merge. Still it does not follow that a competitive model can explain all of the automobile firms' behavior including such anomalies as their failure to raise prices after World War II despite the active black market in cars.

Although in practice it is necessary to recognize the effects of legal constraints on collusion, I shall not do so here. Instead, I shall assume that all forms of collusion are legal. This has several advantages. First, it does throw light on the effects of the law. This is because the problems of detecting and punishing violations of the cartel agreement are like those of detecting and punishing violations of the antitrust laws. Nor is this all. If we assume that the law does not prevent collusion and that collusion occurs only when it dominates competition, then we can deduce a variety of industry structures as a function of the factors which determine whether competition or collusion prevails. As a result one can compare the hypothetical with the real world in order to learn about the effectiveness and the enforcement of the antitrust laws.

A simple dichotomy facilitates the study of collusion. All varieties of collusion fit into one of the categories as follows.

1. cartel: confederation of independent firms;
   a. without side payments,
   b. with side payments;
2. merger: union of formerly competing firms.

By a cartel I mean an alliance of competing firms such that the members retain their separate identities and separate control over their policies subject to the terms of the cartel agreement. A cartel can dissolve and competition can resume. By a merger I mean a union of competing firms into a single enterprise under the control of one management. Although the merged firms can retain some of their original form so that the merger can also be undone and competition among the constituents restored, this is usually harder than dissolving a cartel. The possibility of making side payments means that the cartel members can freely transfer sums of money among themselves. These transfers can be independent of sales or outputs of the firms. Moreover, under some conditions the cartel can fine a member to punish his violation of a cartel rule.

What are the main tasks of collusion? A cartel agreement should provide for three factors. First, it must specify the areas of collusion. These range from simply staking out exclusive markets for the cartel members according to the nature of the product with respect to quality or other attributes to the whole spectrum of business policy. The former may result in only a loose confederation if it is based, say, on geographical territory. The latter includes the determination of what to make and sell and for how much, investment decisions including plant and equipment design, sales methods, research outlays, wage and labor policy, financial decisions, and so on. The wider the range of agreement the more a cartel approximates outright merger of its constituents. Thus a cartel may do more than merely set a common price and allocate output quotas to its members. Second, the cartel agreement must specify how the members will share the net return or the cartel cannot form. Merger has a similar task which is to determine how much the surviving firm will pay to the owners of the firms that it acquires, which is often equivalent to sharing the expected present value of the monopoly return among the merging firms. Third, the cartel must specify how to enforce the agreement, which includes the detection and punishment of violations of the cartel rules and especially how much to allocate for these purposes. The more costly is enforcement and maintenance of the cartel agreement, the smaller is the cartel net return and, therefore, the less is its advantage over competition. Although merger avoids these policing costs, it encounters problems of its own primarily because the entrepreneurial input, the locus of final authority, is smaller relative to the total size of the firm. As a result it is not a priori obvious whether a merger can obtain a larger net return than would its constituents operating with the greater autonomy of cartelists.

The problem of the efficient industry organization is now pertinent. For a given rate of total output the efficient number of firms is independent of the state of competition, but the rate of output is not independent of the state of competition, so for given demand conditions the optimal number of firms will depend on the competitive state of the industry. If the minimum average cost occurs at a positive scale, then there is a determinate finite number of firms necessary for efficiency. Therefore, we may assume there is the optimal number of firms already present in the above material dealing with the Cournot-Nash model for a given number of firms. We begin with the simpler case where a good of prescribed quality is to be produced by those firms best able to do so. Afterwards we shall allow for product variety.

The problem of efficiency divides itself into two parts, the optimal output per firm and the optimal number of firms. Only the former problem is relevant for the short run while both are relevant in the long run. Thus the problem of efficiency in the short run is to allocate a given total output q among the n given firms in order to minimize the total cost of production. If, as above, $c_i = g_i(q_i)$ gives the total cost of firm i as a function of its output rate q_i, then the allocation of the output among the n firms is efficient if, for given q with

$q = \Sigma \, q_i$ and $q_i \geq 0$, the q_i's are chosen in order to

$$\min_{\{q_i\}} \Sigma \, c_i.$$

The Lagrangian for this problem is given by

(9) $$L(q, \lambda) = \Sigma \, c_i + \lambda(q - \Sigma \, q_i).$$

For efficiency it is necessary that

(10) $$\frac{\partial c_i}{\partial q_i} - \lambda \geq 0, \qquad \text{where} \quad \frac{\partial c_i}{\partial q_i} \begin{Bmatrix} = \\ \geq \end{Bmatrix} \lambda \quad \text{according as} \quad q_i \begin{Bmatrix} > \\ = \end{Bmatrix} 0.$$

This result is familiar. It says that the efficient allocation of output among the active firms makes them all have the same marginal cost. The Cournot-Nash equilibrium satisfies this condition for any number of firms because for all active firms the marginal cost equals a common value given by the marginal revenue.

Next, we determine the optimal number of firms. The easier case is where all firms have the same total cost function; so, let us begin with it. The total cost is given by

(11) $$C = \Sigma \, c_i = nc,$$

where c denotes the common value of the total cost function for each firm, assuming that all produce the same output, as is necessary for short-run efficiency. Denote this common output by q^*. To give the essentials quickly, let us treat the integer n as if it were a continuous parameter thereby allowing the use of ordinary calculus. For efficiency it is necessary to choose n and q^* to minimize the total cost of producing a given total output. That is,

$$\min_{n, \, q^*} C, \qquad \text{subject to} \quad q = nq^* \quad \text{and} \quad n, q^* \geq 0.$$

The Lagrangian is given by

(12) $$L(n, q^*, \lambda) = nc + \lambda(q - nq^*).$$

If there is a minimum, it must satisfy the conditions as follows:

(13)
$$\frac{\partial L}{\partial q^*} = n\left(\frac{\partial c}{\partial q^*} - \lambda\right) \geq 0,$$

$$\frac{\partial L}{\partial n} = c - \lambda \, q^* \geq 0.$$

Assuming, as is reasonable, that the optimal n and q^* are positive, these conditions reduce to equalities, and we obtain

(14) $$\frac{\partial c}{\partial q^*} = c/q^*,$$

which is the familiar conclusion that the optimal number of firms and the output per firm is such that all produce where the average cost is a minimum.

Next, let us generalize the problem to include the case where firms have different cost functions. It is helpful to index the firms with a continuous parameter denoted by σ, which replaces the integer index i of the preceding work.

$$(15) \qquad\qquad c(\sigma) = g(\sigma, q(\sigma)) \qquad\qquad (0 \leqslant \sigma \leqslant 1),$$

gives the total cost function of firm i. The total cost function for the whole industry becomes

$$(16) \qquad\qquad C(s) = \int_0^s c(\sigma)F(\sigma)\, d\sigma \qquad\qquad (0 \leqslant s \leqslant 1),$$

in place of the sum $C = \Sigma\, c_i$. In (16), $F(\sigma) =$ number of firms of type σ, and the upper limit of the integral s denotes the value of the index of the "marginal" firm—that is, the firm that it just pays to operate, given the desired total output. The industry's total output is given by

$$(17) \qquad\qquad Q(s) = \int_0^s q(\sigma)F(\sigma)\, d\sigma.$$

In this approach one is free to order the firms with respect to the indexing parameter in any convenient way. One particular order commends itself for reasons we shall see. This is to order the firms with respect to their minimum average cost. Hence the minimum average cost becomes a rising function of σ.

The optimal s and q is the solution of the problem as follows:

$$\min_{s,\, q} C(s), \qquad\qquad \text{subject to } Q(s) \text{ a given constant.}$$

The Lagrangian is given by

$$(18) \qquad\qquad C(s) + \lambda\left(Q - \int_0^s qF\, d\sigma\right).$$

If there is a minimum, then the following relations must be satisfied:

$$(19) \qquad \frac{\partial C}{\partial q} - \lambda \int_0^s F\, d\sigma \geqslant 0,$$

$$\frac{\partial C}{\partial s} - \lambda q(s)F(s) \geqslant 0.$$

Since

$$\frac{\partial C}{\partial s} = c(s)F(s) \quad \text{and} \quad \frac{\partial C}{\partial q} = \int_0^s \frac{\partial c}{\partial q} F(\sigma)\, d\sigma,$$

it follows that

$$\int_0^s \left(\frac{\partial c}{\partial q} - \lambda \right) F(\sigma)\, d\sigma \geq 0,$$

(20)

$$c(s) - \lambda q(s) \geq 0.$$

By hypothesis the output of the marginal firm is positive or there would be no problem. Hence the necessary conditions given in (20) become equalities. Moreover, without loss of generality we may assume that $F(\sigma) > 0$ for all σ with $0 < \sigma < s$. These imply that

$$\frac{\partial c}{\partial q} = \lambda, \qquad \text{for all } \sigma \text{ with } 0 \leq \sigma \leq s,$$

(21)

$$\frac{\partial c(s)}{\partial q(s)} = \frac{c(s)}{q(s)}.$$

The first condition means that for all active firms there must be a common marginal cost. The second condition means that the scale of operation of the "marginal" firm is such that its average cost is a minimum. Moreover, for all extramarginal firms, that is, for all $\sigma > s$, $q(\sigma) = 0$. However, the intramarginal firms do not operate at rates of output that minimize their average cost, implying they obtain scarcity rents. These results are all familiar, at least since Viner's well-known article (1952, pp. 198–232).

Under monopoly or collusion a conscious mechanism secures efficiency because the cartelists or the monopolist actively seek a maximum net return. There results a choice of output and active firms giving the least total cost for that total output that maximizes the net return. Therefore, maximum net revenue and maximum efficiency are consistent under monopoly or collusion.

Under competition a less direct but no less powerful mechanism secures maximum efficiency. When there is competition, no inactive firm can obtain a larger net return than the least profitable of the active firms. This implies a selection of active firms such that the second condition of (21) is satisfied. Moreover, since all competitive firms obtain the same marginal revenue, the first condition of (21) is also satisfied. Therefore, all active firms have the same marginal cost. It follows that there is maximum efficiency under competition.

If the average cost functions are convex with respect to the rates of output, then the intramarginal firms operate at a scale above that which minimizes their average cost. Hence there is a finite number of active firms in the industry, each producing at a positive output rate. This number is determined by the conditions given in (21). As long as the firms' average cost functions are strictly convex, that is, not linear, the optimal number is determinate.

The next section continues the analysis of efficiency assuming there is product variety.

3. Equilibrium with Product Variety Illustrated for Spatial Competition

There are two reasons for studying the properties of an equilibrium when there is a variety of closely related products. First, the phenomenon is common, and we gain useful insights from a theory relating prices and quantities in this situation. Second, the frequent contention that firms compete or collude not only with respect to price but also with respect to product characteristics requires some discussion for a reasonably comprehensive theory of competition.

A full theory of product variety would use some advanced mathematical tools from functional analysis, a topic we shall study in the next chapter. For the present we can study the essentials of product variety in a special case, the problem of spatial competition, for which simple methods suffice. The next chapter treats the determination of the optimal price policy assuming intertemporal maximization. In order to concentrate on the salient aspects, the location example requires several simplifying assumptions whose implications and relevance to product variety will soon be apparent.

Assume there is a uniform distribution of demand on a straight road of length S. This means that the demand schedule is the same at every point on the road but that the quantities demanded may differ because of differences in the delivered price. Thus, the demand schedule relates the quantity demanded to the delivered price. The location of the firms on the road corresponds to the choice of what kind of product to make. The number of firms on the road corresponds to the number of product varieties. The rate of output of each firm corresponds to the output of each product type in the general case. We are interested in the properties of the equilibrium and whether there is social efficiency.[1]

Let the demand schedule at a given point s on the road be denoted as follows:

$$(1) \qquad q(s) = \phi(p^*(s)),$$

where $q(s)$ is the quantity demanded and $p^*(s)$ is the lowest per unit delivered price to s. Assume that price and quantity vary inversely, thereby ruling out absolute inelasticity. The delivered price is the price per unit of the good delivered to the customer's door which includes the transportation cost. We shall also assume that it is prohibitively costly for the firms to discriminate among their customers with respect to their location. Hence all customers

---

1. Much of this material follows Telser (1969a).

pay the same FOB price at the plant. It is also convenient, though hardly essential, to assume that the customers themselves bear the shipping cost. Let the transport cost per unit distance be a constant a that is independent of the amount shipped. Therefore, the delivered price to a point s units distant from a plant satisfies

$$(2) \qquad\qquad p^* = p + as.$$

In this case one may assume that the p^* appearing in (1) is the lowest delivered price of all those available. Hence one need not include a continuum of prices from plants other than the one offering the minimum FOB price. Generally, for product variation it would be necessary to postulate a more elaborate demand function relating the quantity demanded to the prices of all the available varieties. Since the same phenomenon presents itself in the determination of price policy for intertemporal optimization, we can better defer these complications until the next chapter.

Equation (1) gives the demand at one point on the road, but the pertinent demand schedule for the firm is an aggregate of all of these summed over its market territory, which is given as follows:

$$(3) \qquad\qquad Q = \int_0^{s_1} \phi(p^*)\, ds + \int_0^{s_2} \phi(p^*)\, ds.$$

The first integral gives the demand to the firm coming from its market on its left and the second from its market on its right. It is intuitively obvious that the firm maximizes the quantity sold at a given FOB price by locating at the midpoint of its market territory. Formally,

$$2\int_0^s \geqslant \int_0^{s_1} + \int_0^{s_2},$$

where $2s = s_1 + s_2$. This assertion is a consequence of the uniform distribution of demand on the road. Hence the firms have an incentive to gratify their customers' demand for a convenient location. In the general case of product variety, firms similarly want to gratify their customers' desire for particular varieties, but they typically offer only a finite number of varieties. This is equivalent to observing a finite number of firms on the road. Hence in the location example one must explain both the number of firms and their location.

Assume that all of the firms have the same cost function given by

$$(4) \qquad\qquad C = \begin{cases} k + cQ, & \text{if } Q > 0, \\[2mm] 0, & \text{if } Q = 0; \end{cases}$$

k denotes the "fixed" cost and c the marginal variable cost. The average cost of every firm is a decreasing function of its output. If fixed costs were zero, then there could be a plant at every point on the road, provided the demand at every point were large enough to cover the variable cost. However, if the fixed cost is so large relative to the rate of demand that a firm cannot cover its total cost from the sales forthcoming in its immediate neighborhood, then a single firm will serve a finite segment of the road, and there will be a finite number of firms. Generally, the presence of fixed cost and the limited demand for products of a given type such that total revenue is below total cost explains the offer of a finite product variety. In the location example the more plants there are, the smaller is the shipping cost and the larger is the average production cost. For efficient location one must strike a balance between these two costs.

At the outset suppose there are n firms to be located such that a given output can be supplied at the least total cost, the sum of the production and the transport costs. The solution is easy. Since both the total output and the number of firms are given, one must place the firms to minimize total shipping cost. This requires equal intervals between every pair of firms such that each firm occupies the midpoint of its market. Therefore, the efficient sites are

(5) $S/2n, \; 3S/2n, \; ..., \; (2n-1)S/2n.$

Thus every firm has a symmetric market of length S/n, and no customer is more than $S/2n$ units distant from a supplier.

Now, let us see whether attainment of efficiency in this sense depends on the competitive structure of the industry. If all the firms were under the control of a single party as would be true for a cartel or a monopolist, then clearly it would choose the efficient locations to maximize its net return. Would this also happen with competition? We need only consider the case where the fixed cost is too large relative to sales minus variable cost to support a firm at every point on the road. Let us assume at the outset that a single firm wants to start business on a road void of all competition. It would obviously choose the midpoint, the efficient site for one firm. Let a second firm desire entry given the occupation of the midpoint by the first firm. Some may claim that the second firm would locate midway between an end of the road and the site of the first firm which would set in motion a sequence of changes because the first firm would no longer wish to remain at the road's midpoint. If the firms continue to choose locations such that each is halfway between its competitor and the end of the road, then eventually the first firm would be at the point $S/3$ and the second at $2S/3$. Formally, let σ_1 denote the first firm's location and σ_2 the second firm's location. According to this argument, the final equilibrium would satisfy

$$\sigma_1 = (1/2)\sigma_2, \quad \sigma_2 = \sigma_1 + (1/2)(S - \sigma_1),$$

and the two firms would be closer together than would be prescribed by consideration of maximal efficiency. From such reasoning have come far-reaching conclusions that competition results in too much product similarity and a failure to satisfy adequately the diversity of taste in the population.

However, the reasoning is mistaken because it belies the firms' own goal of maximal net returns because it fails to minimize transport cost. Each firm would have twice as many customers toward the open end of the road as compared with those in between since the latter will be shared equally by the two firms. Instead of the process leading to the above two equations, we now derive the correct analysis on the hypothesis that the firms maximize their net return.

In accord with the Cournot-Nash theory of competition, which assumes that the firms do not cooperate, each firm chooses its site assuming a given site for the other. As above, let σ_i denote the location of firm i. Consider firm 1, which realizes that whatever be its exclusive market territory, the net return is largest at its midpoint. Hence, assuming for definiteness that firms are ordered from left to right, firm 1 will choose a location to satisfy $\sigma_1 = (1/2)(\sigma_2 - \sigma_1)$, and firm 2's location will satisfy $S - \sigma_2 = (1/2)(\sigma_2 - \sigma_1)$. This pair of equations has a solution $\sigma_1 = (1/4)\,S$ and $\sigma_2 = (3/4)\,S$. These are the most efficient sites for the two firms, and the same argument applies for n firms, as we now see.

With n firms in Cournot-Nash competition, the equations consistent with profit maximization are as follows:

$$\sigma_1 = (1/2)(\sigma_2 - \sigma_1), \quad \sigma_t - \sigma_{t-1} = \sigma_{t+1} - \sigma_t \quad (t = 2, \ldots, n-1),$$

$$S - \sigma_n = (1/2)(\sigma_n - \sigma_{n-1}).$$

The solution of these equations is the efficient locations given in (5).

We may conclude from these arguments that whether there is competition or collusion, the location of the plants would be efficient. Given a uniform distribution of the demand and the fact that all firms have the same cost functions, it follows that for maximum efficiency the firms should be equispaced along the road. For a nonuniform distribution of demand there would be an appropriate, but harder to describe, set of efficient locations. Nevertheless, whether there is competition or not, the locations would be efficient. Whether the number of plants and the output per plant is efficient is another problem to which we shall return.

This analysis implies that regardless of the nature of the relations among the firms, each firm has a demand schedule represented as follows:

(6)
$$Q = 2 \int_0^{S/2n} \phi(p^*)\, ds.$$

Hence the total revenue of a firm is pQ, and the total expenditure by its customers is $pQ + T$, where T, the transport cost is defined by

(7)
$$T = 2 \int_0^{S/2n} as \quad \phi(p^*)\, ds.$$

Hence the total outlay by the customers of all n firms is given by

(8) $$R = n\bar{R}, \quad \bar{R} = pQ + T.$$

The total cost for the n firms plus the total transport cost is given by

(9) $$G = n\bar{C}, \quad \bar{C} = k + cQ + T.$$

With collusion the firms would choose n and Q to maximize $R - C$, which equals the net return excluding the transport costs. Under competition each firm chooses a rate of output such that marginal revenue equals marginal cost. In either case there is a common FOB price at every plant, but under competition the number of firms is determined by the condition that the net return per firm is zero. Together these two conditions constitute Chamberlin's monopolistic competition equilibrium. Hence marginal revenue equals marginal cost and average revenue equals average cost. One can easily show that output per firm is smaller and the number of firms larger with competition than with collusion (Chamberlin 1933).

Thus far the analysis leads to similar conclusions as in the standard case with a single product. However, some pitfalls are present as we consider the question of efficiency with respect to the location of the firms and the output per firm, assuming a given total output denoted by X so that $X = nQ$. The solution of

$$\min_{n,\, Q} G, \qquad\qquad \text{subject to } nQ = X, \text{ a given total,}$$

determines the efficient combination of n and Q for producing a given total output X. The Lagrangian is as follows:

$$L(n, Q, \lambda) = G + \lambda(X - nQ).$$

If there is a minimum, it must satisfy the usual necessary conditions which imply

(10) $$\bar{C} = Q\frac{\partial\bar{C}}{\partial Q} - n\frac{\partial\bar{C}}{\partial n},$$

(11) $$\frac{\partial\bar{C}}{\partial Q} = \left(\bar{C} + n\frac{\partial\bar{C}}{\partial n}\right)\bigg/ Q = \lambda,$$

where the Lagrangian multiplier λ is the marginal cost. (In this treatment we make the harmless and useful simplification of treating the discrete variable n as if it were continuous.) Neither the competitive nor the collusive solutions satisfy these efficiency criteria.

The monopolistic and the collusive solution are inefficient because maximizing net revenue R with respect to n and Q is not equivalent to minimizing total cost for a given total output $X = nQ$, where the latter combination yields the maximum net revenue. Thus the necessary conditions for net revenue maximum under monopoly are as follows:

$$\frac{\partial R}{\partial Q} = \frac{\partial G}{\partial Q} \quad \text{and} \quad \frac{\partial R}{\partial n} = \frac{\partial G}{\partial n}.$$

For efficiency it is necessary that

$$(1/Q)\frac{\partial G}{\partial n} = (1/n)\frac{\partial G}{\partial Q} = \lambda.$$

Hence monopoly is efficient if and only if

$$\frac{\partial R}{\partial Q} = \lambda n \quad \text{and} \quad \frac{\partial R}{\partial n} = \lambda Q,$$

which is not true for the demand relations postulated in (6). Similarly, one can show that the competitive solution, which is Chamberlin's monopolistic competition equilibrium, would also be inefficient.

These conclusions are better understood by comparing the situation to the one where there is a single product treated in the preceding section. In the usual case the demand schedule relates the price per unit and the rate of purchase by a function as follows: $p = H(q)$. There is also a production function, which relates the output q to the inputs of, say, two factors of production, as follows: $q = F(y_1, y_2)$, where the y's denote the input rates of the two factors. The monopolist would choose q, y_1, and y_2 to maximize his net revenue given by $pq - (w_1 y_1 + w_2 y_2)$, subject to the constraint given by the production function, where w_i denotes the constant price per unit of factor input y_i. Let us prove that efficiency and maximum net revenue are compatible. Fix the total output at that rate which maximizes the net return and denote this output by q_m. This determines the price from the demand function which also determines total revenue. Consequently, the maximum net revenue for the given output leads to that choice of input rates, y_i, which minimize the total cost. Therefore, the maximum net revenue implies maximum efficiency. As is well-known, competition also results in efficiency in this case.

These conclusions are consequences of the properties of the demand function. In this case the demand for the good is independent of how it is made. The inputs are cost characteristics of the product that do not affect the demand. In contrast, the inputs do affect the demand in the location example. Thus in the location example the number of plants and the output per plant constitute the "inputs". Clearly, the total quantity demanded depends on these inputs. Hence fixing the total output $X \, (= nQ)$ does not fix the total revenue as in the case of a single product. Therefore, it does not follow that the maximum net return is equivalent to the minimum total cost if the total output is set at the optimal monopoly level. Nor does competition result in efficiency. Nevertheless, with either competition or collusion, the plants are located at the efficient points. To understand the distinction between efficiency for plant sites and efficiency for the total output, we should consider product variation more generally.

Generally a good is a combination of cost and demand characteristics. The former are the inputs of the factors of production necessary to make a given amount of the good. The quantity demanded at a given price does not depend on the combination of factor inputs used to make the good. That is, consumers are indifferent among goods identical in every respect save for differences in the methods of production. For example, the ratio of the inputs of labor to capital does not normally affect the demand for the product. However, there are product characteristics that do affect the demand and, therefore, these may be called the demand characteristics. For example, the physical specifications of a product constitute its demand characteristics. The horsepower of a car is a demand characteristic. But physical attributes are not the only demand characteristics. Services offered jointly with the good may also be demand characteristics. Examples are credit terms, delivery, advertising, and methods of sales promotion.

For a given set of demand characteristics the firms have an incentive to produce a given output at the lowest total cost regardless of the state of competition. This corresponds to the choice of plant location in the location example. To give another example, suppose that a Chevrolet car were a well-defined product. It follows that it would be in the interest of General Motors to produce a given number of Chevrolets at the least total cost. However, it is not in General Motors' interest to produce "cars" at the least total cost because producing a given number of "cars" at the least total cost is ambiguous until the commodity, a "car," is well defined in terms of its specifications. Once we have a well-defined "car" the question of efficient production methods becomes equivalent to the question of producing a given number of Chevrolets efficiently since by hypothesis Chevrolets are a well-defined product.

In the location example the definition of a product is ambiguous. The total output is not well defined until one specifies who is to obtain how much. The customers desire the good delivered to their doorstep in a usable form. A change in the spacing of the plants is equivalent to a change in the product variety. Implicitly, therefore, to determine the minimum total cost of producing a given total output is equivalent to determining the product variety. But the determination of product variety is not a question of efficiency; it is a question of consumer preference expressed by demand. Therefore, in the location example monopoly results in a different product variety than under competition, and in neither case is total output produced at the least total cost. Nevertheless, it is true that, given the product variety corresponding to monopoly and to competition, the plants are spaced efficiently. A more fundamental approach would determine the marginal cost, λ, of a given total output. To do so, one must decide the total output consistent with consumer welfare. This is equivalent to setting a value for λ that adequately represents consumer desires. In the location example neither competition nor monopoly does so.

The location example also illustrates another interesting point. This deals with the policing of the cartel agreement, the main topic of the next section. Suppose that the firms on the road collude such that output per firm and the number of firms are chosen to maximize the joint net return. The necessary conditions for a maximum given above imply that

(12) $\quad \bar{C} + n \dfrac{\delta T}{\delta n} = \bar{R} + n \left(Q \dfrac{\delta p}{\delta n} + \dfrac{\delta T}{\delta n} \right), \quad n \left(c + \dfrac{\delta T}{\delta Q} \right) = n \left(p + Q \dfrac{\delta p}{\delta Q} + \dfrac{\delta T}{\delta Q} \right),$

where, since n and Q are the control variables, p is a function of n and Q.[2] The term $(\delta T/\delta n)$ and $(\delta T/\delta Q)$ denote the partial derivatives of T with respect to n and Q, where p is permitted to change appropriately. It follows from (12) that the necessary conditions imply that

(13) $\qquad\qquad \bar{C} = \bar{R} + nQ \dfrac{\delta p}{\delta n}, \quad c = p + Q \dfrac{\delta p}{\delta Q},$

where Q, a function of p and n given by (6), can be written concisely as $Q = f(p, n)$. Consequently,

$$\dfrac{\delta p}{\delta Q} = \left(\dfrac{\partial f}{\partial p} \right)^{-1} \quad \text{and} \quad \dfrac{\delta p}{\delta n} = -\left(\dfrac{\partial f}{\partial n} \right) \left(\dfrac{\partial f}{\partial p} \right)^{-1}$$

The second equation in (13) seems to imply a harmony of interest between the cartel and its members because the output for both makes marginal cost equal marginal revenue. Hence once the cartel sets n and Q it would appear that an individual cartel member has no incentive to sell other than his assigned quota. In fact this is not so for perhaps subtle mathematical reasons that we shall now derive.

At the boundary of a market territory the delivered price is $p + aS/2n$, and

---

2. The necessary conditions for maximum monopoly net revenue are that n and Q equate their respective marginal costs and marginal revenues. Thus,

$$\dfrac{\partial G}{\partial n} = \dfrac{\partial R}{\partial n} = \bar{R} + n \dfrac{\partial \bar{R}}{\partial n}, \quad \dfrac{\partial G}{\partial Q} = \dfrac{\partial R}{\partial Q} = n \dfrac{\partial \bar{R}}{\partial Q}.$$

However,

$$\dfrac{\partial G}{\partial n} = \bar{G} + n \dfrac{\partial \bar{G}}{\partial n}, \quad \text{where} \quad \dfrac{\partial \bar{G}}{\partial n} = \dfrac{\delta T}{\delta n}, \quad \dfrac{\partial G}{\partial Q} = n \dfrac{\partial \bar{G}}{\partial Q} = n \left(c + \dfrac{\delta T}{\delta Q} \right);$$

$$\dfrac{\partial \bar{R}}{\partial n} = Q \dfrac{\partial p}{\partial n} + \dfrac{\delta T}{\delta n}, \quad \dfrac{\partial \bar{R}}{\partial Q} = p + Q \dfrac{\partial p}{\partial Q} + \dfrac{\delta T}{\delta Q}.$$

These relations imply (12) in the text.

the quantity sold is $q = \phi(p + aS/2n)$. However, the demand schedule is defined by the integral shown in (6). The value of an integral is unaffected by the value of the integrand at a single point such as at the boundary of a market. It follows that, except for the endpoints of the road, there is an indeterminacy at every boundary point of a firm's market territory because at these points there are two firms equally willing and able to supply the demand. Therefore, at these boundary points the cartel would have to police the agreement because here is where two firms directly compete. In practice the cartel must resolve this indeterminacy by stipulating who is to serve the customers at the boundary points. For n firms there are $n+1$ boundary points. Therefore, to give every firm the same share of the demand at these points implies that every firm receives $(n+1)/n$ of the quantity demanded at these points. For efficiency, however, it is necessary to give the two firms closest to the endpoints all of the sales there. From these conditions one can readily work through the implications of an equal share agreement.[3]

4. The Costs of Maintaining Collusion

Many issues relevant for this analysis emerge in the study of a cartel with only two members of equal size who make perfect substitutes. Chapter III began a discussion of this case. Let us now extend the analysis to cover the possibility of side payments and their implications. In addition we shall complete the discussion of the costs of collusion without side payments.

Consider figure 5.1. Firm 1's net revenue, r_1, is on the horizontal axis and firm 2's net revenue, r_2, on the vertical axis. The set of points OMN represents the outcomes feasible by means of price or output policies. The locus MN gives the cartel's efficient points since it is not possible to increase the net revenue of both firms by a movement along MN. The point J corresponding to the policy that maximizes the joint net return has an interesting geometric property.

The locus of points that satisfy

(1) $r = r_1 + r_2$

is tangent to the efficient set MN at the point J where r is a maximum. The slope of MN at J is -1 because at the maximum an increase of one firm's net return by one dollar entails a decrease of the other's net return by an equal amount. J gives both firms a larger net return than any of the points

---

3. If all of the firms receive equal shares of the revenue from the customers at the boundary points, then the shares are as follows:

Firm Number:	1	2	3	$\ldots$	n
Share:	$(1, 1/n)$	$((n-1)/n, 2/n)$	$((n-2)/n, 3/n)$	$\ldots$	$(1/n, 1)$

The left-hand value of each pair denotes the share of the firm from its left-hand boundary point, and the right-hand value from its right-hand boundary point.

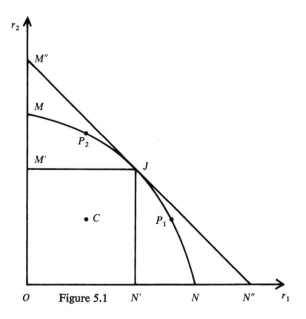

Figure 5.1

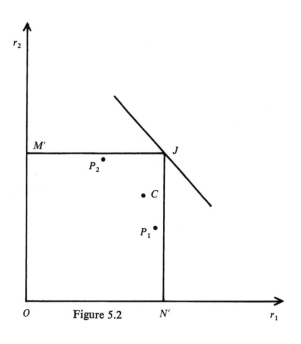

Figure 5.2

inside the rectangle $OM'JN'$. Hence, J directly dominates this rectangle, but J does not directly dominate the two sets $MM'J$ and $NN'J$, which are outside the rectangle. Firm 1 prefers any point in the latter to its imputation at J while firm 2 prefers any point in $MM'J$ to its imputation at J. Therefore, the point J does not directly dominate all feasible alternatives given our assumptions.

Presumably, the cartelists would confine their interest to the points on the segments MJ and NJ because these are on the frontiers of their preferred regions. Thus for any point in firm 1's preferred region, $NN'J$, there is at least one point on NJ at least as good for firm 2 and definitely better for firm 1. Therefore, the conflict between the cartelists over the division of the joint net return centers on MN, which dominates all of the points inside OMN, and none of whose points dominate each other. It is worth remarking that MN is a subset of the von Neumann and Morgenstern solution.[4]

Figure 5.1 can also represent the situation for two firms with an indefinite horizon without side payments. Let the point C represent the Cournot-Nash equilibrium which is inside the rectangle $OM'JN'$. P_2 represents the outcome if firm 1 adheres to the joint profit maximum strategy while firm 2 maximizes its own net return in violation of the cartel agreement with a corresponding interpretation for the point P_1. The possible outcomes are shown in the matrix as follows:

$$
(2) \qquad
\begin{bmatrix}
(u, u) & (u-\delta_1-\delta_3, u+\delta_2) \\
(u+\delta_2, u-\delta_1-\delta_3) & (u-\delta_1, u-\delta_1)
\end{bmatrix}.
$$

This same matrix is used to study the Cournot-Nash equilibrium for an indefinite horizon in section III.4. It is shown therein that competition dominates collusion if and only if

$$
(3) \qquad \alpha < \delta_2/(\delta_1+\delta_2),
$$

---

4. The definition of the von Neumann and Morgenstern solution is as follows:

DEFINITION. *Let there be n participants in a game with a characteristic function v(S) defined for all possible coalitions S. Let x_i denote the imputation of the ith participant and let $X = \{x_i | i = 1, \dots, n\}$ denote the set of all feasible imputations. The imputation of a coalition S is feasible if*

(i) $\qquad\qquad\qquad\qquad \Sigma_s x_i \leqslant v(S).$

A set of imputations $\{x_i\}$ dominates a set $\{x_i'\}$ with respect to a coalition T if

(ii) $\qquad\qquad\qquad\qquad x_i > x_i' \qquad$ *for all i in T.*

We say that an imputation x dominates an imputation y if there is a nonempty coalition S such that (i) *and* (ii) *are satisfied.*

The solution is the set of imputations Z such that if x and y are in Z, then they do not dominate each other, and, for any imputation w not in Z, there is at least one imputation in Z that dominates it.

The definition of domination in the text corresponds to this one. However, as we shall see in section 5, there is no useful definition of a characteristic function for oligopoly. Moreover, there are cogent objections to the von Neumann and Morgenstern solution concept, which reduce its usefulness. For a good summary, see Luce and Raiffa (1957, chap. 9). W. F. Lucas has recently shown that there are games for which a von Neumann and Morgenstern solution does not exist (1967 and 1968).

where, as above, α denotes the probability of trade continuing for one more period. The entry in row 1, column 1 gives the coordinates of J, the entry in row 2, column 2 gives the coordinates of C, and the location of the points P_j depends on the sense of the inequality shown in (3). If the probability of trade continuing is below the value on the right-hand side of (3), then J does not dominate P_j. Figure 5.1 shows this situation. Moreover, $P_{\cdot}$ is efficient for the cartel in this case. To illustrate, firm 2 maximizes its net return given that firm 1 offers the output corresponding to its assignment under J which gives the point P_2. However, if the probability of continuing trade exceeds the right-hand side of (3), then collusion dominates competition so that the points P_j are inside the rectangle $OM'JN'$ dominated by J. This situation is shown in figure 5.2.

According to the matrix, the joint profit maximum appears more lucrative to both parties than the competitive solution so they should both prefer it. However, the matrix represents the outcomes for only one period and does not show the present value of the outcome if there is an infinite horizon because of uncertainty about continuing trade. Moreover, given agreement on the strategy that secures J, each firm can calculate its maximum expected return assuming its rival abides by the agreement. Moreover, assume that the only penalty for one firm's violation of their agreement is a return to competition, the point C. Given these assumptions, we may calculate the expected net return to a potential violator of the cartel agreement. If this expected net return exceeds the firm's share of J, then C dominates J via the feasible intermediate points, P_j, so that once a firm chooses the strategy resulting in the Cournot-Nash outcome, its rival maximizes its net return by doing likewise. This property of the Cournot-Nash point is a consequence of the proposition that if all firms save one choose the Cournot-Nash outputs then the desire for a maximum net return by the remaining firm impels it to do likewise. Hence whether or not the competitive equilibrium can dominate the cartel solution depends on whether or not (3) is satisfied. This depends on the values of the three parameters, α, δ_1, and δ_2. Given δ_1 and δ_2, if the probability of continuing trade is high enough, so that the points P_j are inside the rectangle $OM'JN'$, and both sets $MM'J$ and $NN'J$ are empty, then collusion dominates competition, J dominates C. The figure which represents only the one period outcome cannot show whether this is so. One must compute the expected present value of the various strategies over an infinite horizon.

Let us consider the effects of policing costs while maintaining the assumption of no side payments. The larger the policing costs the smaller is δ_1, the difference between the net return under collusion and competition. There-fore, as shown in (3), given δ_2, the smaller is δ_1 the higher must be α in order that collusion dominates competition. Put differently, given α and δ_2, there is a lower bound for δ_1 as follows:

(4) $$\delta_1 > \delta_2(1-\alpha)/\alpha,$$

which determines whether or not collusion dominates competition. If δ_1 is below the right-hand side of (4) because of high policing costs, then competition dominates collusion. We shall return to policing costs below.

Ease of entry affects δ_1 like policing costs. The easier the entry, the lower is δ_1. For a given δ_1 in order that collusion dominate competition, the easier is entry, the higher must be the probability of continuing trade.

Let us now consider some of the determinants of δ_2, which gives the return in one period to that firm violating the cartel agreement while its rival offers its cartel-assigned quota. Consider figure 5.3, which shows the per firm cartel

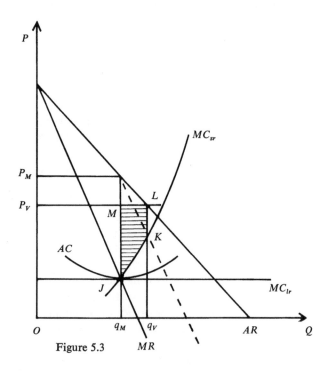

Figure 5.3

output, q_M, and the cartel price, p_M, assuming all cartel members have the same cost functions and make the same product. For this part of the analysis we assume that the cartel has the optimal number of active—that is, producing —members. As a result, assume also that the average cost of the cartel is constant and equal to the marginal cost. Let every individual cartel member have a U-shaped average cost function and operate at that scale at which the average cost is a minimum. The shaded area in figure 5.3 measures δ_2, the gain to a cartel member who violates the cartel agreement while the others remain loyal. Plainly, given the demand, δ_2 depends on the shape of the firm's marginal cost schedule. The flatter this schedule, the larger is δ_2.

Let us approximate the total cost function of the firm with a Taylor series in a neighborhood of that output rate which minimizes average cost.

(5) $$c \approx c_0 + c_1 q + c_2 q^2 + \text{error term.}$$

Denote the average cost minimizing output by $\bar{q}$. An approximation to $\bar{q}$ is given by

(6) $$\bar{q} \approx \sqrt{(c_0/c_2)},$$

where, by hypothesis, c_0 and c_2 are both positive. The slope of the marginal cost schedule is $2c_2$. Therefore, given the total quantity demanded and the overhead cost, c_0, the flatter the marginal cost schedule, the fewer is the number of active firms in the industry. This would imply that the gain to a violator of the cartel agreement would vary inversely with the optimal number of cartel members. However, it does not follow that competition is more likely to dominate collusion in industries where the optimal number of firms is small. This is because the parameters c_0 and c_2 may be related. For instance, if they vary inversely, then $\bar{q}$ would be constant, and the optimal number of firms would vary directly with the quantity demanded. Therefore, we could not deduce a relation between the gains from chiselling and the optimal number of cartel members.

The preceding analysis applies if there are no side payments. Suppose now that side payments are possible. Therefore, all the points on the triangle $OM''N''$ (fig. 5.1) become feasible since the area outside OMN, namely, $MM''J$ and JNN'', can be reached with transfer payments. Therefore, for any point on MN, there is a point on $M''N''$ preferred by one firm and acceptable to the other, with an obvious qualification for the point J. Side payments permit the cartel to separate the choice of a best strategy from the imputation of the net return. Moreover, a cartel can fine those members who violate the rules without disturbing the optimal strategy of the loyal members if there are side payments. These considerations suggest that collusion may dominate competition for a wider range of conditions.

Suppose a cartel member contemplates a violation of the cartel agreement, say, sales above his quota in period T after $T-1$ periods of faithful obedience. Suppose the cartel can detect violations after k periods, $k \geqslant 1$, where k depends on the cartel's expenditures on policing its members. The fine must be large enough to induce loyalty to the cartel. Therefore, the fine, denoted by F should satisfy the following inequality:

(7) $$F > \delta_2 \alpha^T (1 - \alpha^{k+1})/(1 - \alpha).$$

The value on the right-hand side gives the gain from the violation. Since k varies inversely with the policing outlay, the larger this outlay, the lower is δ_1. Also, the larger is the policing outlay, the smaller is the fine necessary to satisfy (7). Moreover, if the fine satisfies the following inequality

(8) $$F > \delta_2/(1 - \alpha),$$

then it surely satisfies (7) since

$$\delta_2/(1-\alpha) > \delta_2\alpha^T(1-\alpha^{k+1})/(1-\alpha).$$

These calculations suppose that violations are detected with certainty. If the probability of detection is β with $0 < \beta < 1$, then the expected fine is βF. Hence to make compliance more lucrative than noncompliance, the fine should satisfy the following inequality:

(9) $$\beta F > \delta_2\alpha^T(1-\alpha^{k+1})/(1-\alpha).$$

The condition corresponding to (8) is

(10) $$\beta F > \delta_2/(1-\alpha).$$

Thus, the larger is the probability that trade continues and the lower is the probability of detection, the larger is the necessary fine.

There is a complication as yet omitted. Loyal cartel members may count as part of their net return their share of the fines collected by the cartel, but this cannot be large. For if it were, then the cartel could not choose a β and an F capable of deterring violations of its rules. But then the cartel could not survive since it would always be more profitable for any one cartel-list to be disobedient. What is true for one is true for all. Therefore, a successful cartel does not collect much revenue from fines.

The amount that a cartel can spend on enforcing its rules is constrained by the inequality

(11) $$\delta_1 > 0,$$

which is a necessary condition for the very existence of the cartel. Since the probability of detecting violations of the cartel rules is an increasing function of the amount allocated to this purpose, (11) imposes an upper bound on β. We now take up the delicate question of whether the implied constraint on β thwarts the dominance of collusion by competition.

All along we assume that firms seek a maximum net return. Hence a cartel member would violate the agreement for a net positive expected return, that is, one that allows for the expected cost of the fine. Therefore, survival of the cartel depends on whether the members find this to be profitable. This can be true despite the upper bound on β implied by (11), if the cartel can impose and collect a large enough fine to ensure loyalty. One might think this is always true because a violator can always pay the fine from the proceeds of his transgression, suggesting that a cartel has only to make the fine exceed the gain and it can always secure loyalty.

However, the capability of imposing fines is not enough to ensure cartel survival and, moreover, there is a paradox. If violations cannot be detected with certainty, then as (9) shows the fine has to exceed the gain by a multiple of $1/\beta$ so that it may be uncollectable. An uncollectable fine is no deterrent. A rational violator would compute the expected value of the fine reckoning on the uncertainty surrounding the payment of the fine. Ultimately, the cartel has only two penalties. The first is expulsion from the cartel. In effect this implies the death of the cartel for if it pays for one to cheat then it pays for all. Second, the law may allow the cartel's fine as a legal liability of the offender. If so, a fine can force the firm into bankruptcy and the cartel can take it over. For a penalty of the first kind, the expected cost to a rule breaker is his share of the cartel's net return which is equivalent to having no side payments. In the second case, the penalty is the net worth of the firm, including the present value of its share of the monopoly return.

This analysis reveals some properties of successful collusion. The more capital intensive the industry, the larger is a firm's hostage to fortune and the easier is enforcement of collusion because, under these conditions, for a given gross return there is a larger cost imposed on the rule breaker. The less liquid forms of capital, such as plant and equipment specialized to the industry, are better hostages to good behavior than the more liquid forms such as inventories of raw or finished materials. A calculating offender can reduce his liquid capital while exceeding his assigned quota. Many forms of intangible capital also have a small transfer value such as the capital created by the firm's outlays on specific training of its labor force, the knowledge it acquires about the special requirements of its customers, possibly good will, and so forth. These factors increase the effectiveness of the cartel's penalties and raise the likelihood of the dominance of collusion over competition. Observe that capital intensive industries are more likely to be collusive for these reasons depending on the composition of their capital and not because in these industries there are greater barriers to entry due to "imperfect" capital markets. Thus the possibility of side payments is relevant both in facilitating compliance with the cartel's regulations by the penalties thereby made possible and because with side payments the assignment of production quotas can be separated from the imputation of the cartel's net return.

If fines sufficiently large to deter violations are infeasible, then the cartel must be prepared to take over the rule breaker when the law allows the cartel's fine as a legal liability. Therefore, the confederation of colluding firms approximates a merger, and the question of when a cartel with side payments dominates competition reduces to the problem of when a merger is the better device for extracting a monopoly return. The preceding analysis shows that collusion with side payments subject to constraints ultimately has two alternatives. Either the cartel operates as if there were no side payments so that penalties take the form of more or less long-lived suspensions of the cartel agreement or the cartel must coalesce into one firm under the control of one management.

Having said all this, there remains the possibility of a feasible combination of δ_1, β, and F such that collusion with side payments dominates competition so that compliance with the cartel rules maximizes a firm's net return. We have seen above that high intensity of certain capital components tend to have this result. Let us now consider some other conditions.

Collusion among sellers seems more likely if the buyers are numerous and relatively small. Suppose, on the contrary, there were a few large buyers with a fairly stable demand. These may be assigned to particular cartel members who can thereby share the market. However, the very fewness and relatively large size of the buyers work against a monopoly return altogether. Thus one of the buyers can threaten to withdraw his patronage from a seller who will not take lightly this threat to the bulk of his business, impelling the sellers to betrayal of the cartel. With few relatively large buyers the situation resembles bilateral monopoly, and it would be surprising to find that the sellers can obtain a monopoly return at the buyers' expense. Far more likely is a combination of buyers and sellers to exploit their common interest at the expense of third parties, their suppliers and their customers.

The history of cartels teaches us some of the methods used to maintain collusion in markets with relatively small individual buyers. The members may be given exclusive territories such as in the location example of the preceding section. Within each firm's market territory the buyers have only one supplier. Only the buyers at the market's boundaries would have a choice between two sellers. Geography is not the only basis of exclusive markets. Cartel members may specialize in certain product lines, quality ranges, or may deal with certain classes of customers, for example, households versus businesses, large versus small customers, and customers classified by industry. Sometimes the cartel allows its members to sell as much as they please within their exclusive markets. Depending on the substitutability among the products within different markets, this device serves to preserve the cartel. The most favorable situation is when exclusive markets are defined by national frontiers so that government trade restrictions can prolong the life of a cartel agreement.

Occasionally a cartel ensures compliance with its rules by transferring some of its activities to special organizations. For example, a cartel may have a single marketing agency through which it sells all of its members' products. The sales proceeds are shared according to the terms of the agreement. Sometimes the cartel itself undertakes research and development and assesses its members for the cost. Such cartels tend to approximate a single firm in so far as some activities are centralized while others are left to individual cartelist's discretion.

In some markets none of the above methods is suitable. For example, suppose there is a large number of similar customers whose individual demands for the product are small and subject to random change. If the cartel members have no exclusive markets and compete for the same population of buyers then preservation of the cartel requires statistical inspection methods to ascertain the extent of compliance with the cartel rules because it would be prohibitively costly to obtain figures for all transactions. The topic is worth some formal study.

Let the cartel consist of n firms who make the same product. The aggregate

4. *Costs of Maintaining Collusion*

demand given in 2(1) applies and is rewritten here for convenience as follows:

(12) $$p = f(q).$$

Let $q_i^* =$ firm i's cartel quota and let $q^* =$ optimal cartel output.

(13) $$q^* = \Sigma \, q_i^*.$$

Let the cartel choose a uniform price that corresponds to q^* and allow each member to satisfy the demand forthcoming to him at this price. Assume that the aggregate demand changes randomly so that the actual sales of every cartel member also changes randomly. It is helpful to invert the demand relation (12) and write

(14) $$q = h(p) + u,$$

where u denotes a random variable, $q^* = h(p^*)$, and $q = \Sigma \, q_i$. Hence, $u_i = q_i - q_i^*$, and $u = \Sigma \, u_i$. Equivalently, one may assume that the outputs are literally fixed and that the firms adjust their prices to equilibrate quantity in response to the randomly fluctuating demand schedule. It would then be necessary for the cartel to determine whether its members' prices are within limits consistent with the properties of the random component. Consider the case where the price is fixed so that the cartel's task is to collect sales data. Ultimately, the cartel needs both price and quantity data to see what happens during a violation of its rules.

Random fluctuations of demand come from two sources, the first is from the common factors affecting the sales of all firms in the same direction and the second from the specific factors due to the shifting of buyers among the sellers which cause their sales to move in opposite directions. The common factors impart positive correlations among the u_i's while the specific factors impart negative correlations. To measure the compliance with the rules, the members must distinguish the specific from the common factors and must see whether the former is actually random or results from deliberate actions by one or more cartelists.

The enforcement of the cartel agreement is more or less costly depending on how much an individual member can infer about market conditions from the data given in the records of his own transactions. Cartelists would often be willing to incur the cost of estimating total industry sales which would at least provide each of them with a way of estimating his market share. However, a cartelist who deliberately sells more than his assigned quota would not give accurate figures to the cartel. In that case firms must rely on an analysis of their own sales data to infer the extent of compliance with the cartel rules. Each firm will benefit from a study of the composition of its sales according to various criteria such as the size distribution, repeat purchase rates, and so on. Moreover, a firm can be expected to have some knowledge about general conditions in its trade. Allowing for these it can draw inferences about loyalty to the cartel from its own figures. The most favorable situation is where total sales are stable so that common factors are unimportant. Hence a well-established product free from cyclical fluctuations makes a good candidate for a cartel agreement.

Suppose there is the more favorable situation where the firms make frequent sales reports to each other. They can then compare their actual and assigned shares. Also assume that it is impractical for the cartel members to have such close contacts with their customers as would thwart secret price cuts. For example, if every firm had regular customers who buy at steady rates, it would be readily apparent when there are unusual market conditions. The more difficult situation arises when the cartelists must rely on samples and statistical analysis to determine compliance with the cartel agreement, the case we now consider. To bring out these aspects, assume a stable total rate of demand. Let $m_i^* = q_i^*/q^*$ denote the market share assigned to firm i and let $m_i = q_i/q$ denote its actual market share. If there is compliance with the cartel agreement, then the difference between m_i and m_i^* behaves like a random variable, denoted by w_i so that $w_i = m_i - m_i^*$. $\Sigma\, m_i = \Sigma\, m_i^* = 1$ implies that $\Sigma\, w_i = 0$.

To deduce more of the statistical properties of w_i, one must assume more knowledge of the nature of the demand. In the simplest case all of the buyers either buy the same amount or nothing. Hence the appropriate statistical model is the classical Bernoulli process with the probability of a sale by firm i equal to m_i^*.[5] Let sales reports be obtained from a random sample of N potential customers. It follows that

$$E\, w_i = 0,$$

$$\mathrm{Var}\, w_i = m_i^*(1 - m_i^*)/N,$$

$$\mathrm{Cov}\,(w_i, w_j) = -m_i^*, m_j^*/N.$$

Assuming there is compliance with the cartel agreement, plainly, the larger is the sample size, N, the smaller is the variance of w_i and the narrower the confidence limits for the true market share. Before going on to discuss the cost of the sample, let us pause to consider the scope of the Bernoulli process.

In general all customers do not buy goods at the same rate. Households have positive income elasticities so that wealthier ones tend to buy at higher rates. Larger business firms tend to buy larger quantities. Moreover, even within a given stratum of buyers classified by wealth or size, it is common to observe fluctuations in the purchase rates. Let it be true that customers are grouped into strata homogeneous with respect to mean purchase rates such that the rate of demand by a customer is an identically distributed

---

5. A Bernoulli process is a random process such that an event has a constant probability of occurrence. The distribution of the number of occurrences for a given number of trials is the binomial function (Wilks 1962, sec. 6.2).

random variable. The negative exponential, a plausible density for generating this random variable, is given by

(15) $$a e^{-ax} \qquad (x \geqslant 0),$$

where x denotes the random purchase rate. The mean and the standard deviation both equal $1/a$. This density is skewed to the left, and the probability of a zero rate of purchase can be said to be proportional to a. If follows that the distribution of the market share is given by a Beta density as follows:[6]

(16) $$[(N-1)!/(N_i-1)!(N-N_i-1)!]\, m_i^{N_i-1} \,(1-m_i)^{N-N_i-1},$$

where N is the sample size, and N_i is the number of firm i's customers in the sample. The mean of the Beta density is N_i/N which would equal m_i^* if there is obedience to the cartel. The variance is $N_i(N-N_i)/[N^2(N+1)]$. The equation $m_i^* = N_i/N$ implies that this variance, shown in (19) below, is of the same form as that for a Bernoulli process. Hence if the density of individual purchase rates is a negative exponential then the first two moments of the distribution of market share are unaffected by this. Also observe that the shape of the Beta density shown in (16) bears a strong family resemblance to the Bernoulli distribution.

The mode of the negative exponential is at zero. The class of Gamma densities has a positive mode and generates a Beta distribution of market share. The Gamma density is as follows:

(17) $$a^v x^{v-1}\, e^{-ax}/\Gamma(v), v > 0.$$

The mode is $(v-1)/a$, the mean is v/a, and the variance is v/a^2. The distribution of firm i's market share is given by

(18) $$[\Gamma(vN)/\Gamma(vN_i)\Gamma(vN)]\, m^{vN_i-1}(1-m)^{v(N-Ni)-1}$$

which is a Beta density with mean N_i/N and variance

$$N_i(N-N_i)/[N^2(vN+1)].$$

Both the Beta density and the Gamma from which it is derived have two parameters, v, and a. If $v = 1$, then the Gamma density reduces to the negative exponential. Moreover, the Gamma density is flexible enough to fit many observed densities of purchase rates.[7] If this is true, then one can approximate the true density of purchase rates with the still simpler Bernoulli process because the derived market share distributions have nearly the same means and variances as the Beta densities of market shares derived

---

6. See Wilks (1962, p. 174) for the details of the derivation and see also section 7.7 therein, which gives the moments of the Gamma and Beta distributions.

7. For example, Fourt and Woodlock (1960) report that the heavy buying third for typical consumer products account for 65 percent of the total volume, the middle third for 25 percent, and the light third for only 10 percent. A Gamma distribution would fit nicely these data.

from the Gamma distribution of purchase rates. Moreover, one can defend this approximation if one wishes to see how the sample size affects the accuracy of the estimates.

We wish to derive the relation between the relative size of the cartel members and the sample size necessary to estimate the market share with a given accuracy. Since the cost of the sample rises with its size, this will show how the cost of policing the cartel agreement depends on the relative size of the cartel members. Assume the cartel wants a large enough sample to estimate the market share with a P percent error at an R percent confidence level. Let a Bernoulli process generate the purchase rates so that the market share density is also Bernoullian. Suppose there is compliance with the cartel agreement. These assumptions provide enough information to calculate the necessary sample size.

These desiderata determine a relation as follows: $n_R \sigma_m / m^* \leqslant P$, where n_R is the value of the normal variate corresponding to the R percent confidence level using the normal approximation to the binomial. Since

$$(19) \qquad\qquad \sigma_m = \sqrt{[m^*(1 - m^*)/N]},$$

the sample size N must satisfy the condition as follows:

$$(20) \qquad\qquad N \geqslant [(1 - m^*)/m^*](n_R/P)^2.$$

For example, at a 95 percent confidence level and with the maximum error P not to exceed 5 percent, $n_R = 1.96$, and $P = .05$. Therefore, (20) becomes $N \geqslant 1600(1 - m^*)/m^*$. The important fact about (19) is that N varies inversely with m^*. This means that the smaller are the relative sizes of the cartel members the larger is the necessary sample size for inspection of the cartel's operations with a given accuracy. In addition the greater the desired accuracy and the higher the desired confidence level, the larger is the sample required. Thus, N varies directly with n_R and inversely with P. This simple argument explains why competition is more likely to dominate collusion, the smaller are the relative sizes of the firms in the industry.[8]

This analysis relies on the hypothesis of a single class of customers for whom there is a common density function generating their purchase rates. With many classes a more complicated sampling design is necessary so that the cost of policing the cartel agreement is all the greater. Even this paints too simple a picture of the difficulties before the cartelists. In practice it is as difficult to obtain the information for policing the cartel as it is to design and execute a sample to measure an economic statistic such as the Consumer Price Index. Indeed there are many common problems. This is a consequence of the complexity of real goods which are combinations of many parts. Not even the provision of invoices from a sample of the cartel's customers can

---

8. Stigler's statistical analysis of the effects of policing costs on the durability of collusion stresses the buyer characteristics (1968, chap. 5).

ensure the absence of secret price cutting in the form of selling higher quality goods in the guise of lower quality. Ultimately, a firm must rely on its own sales figures to assess its experience as a cartel member.

The nature of the good is, therefore, pertinent in determining its suitability for collusion. Simple products are better candidates than complex ones. Ready made goods are better than goods made to order. Moreover, the latter distinction also has relevance for explaining the size of firms. Firms that make goods to order must decide how much discretion to give their salesmen in quoting prices to their customers. Sometimes the salesmen must follow complicated written instructions designed to anticipate many possible contingencies. To the extent this procedure is successful it eases the task of management. For the same reasons it eases the task of collusion. However, sometimes mechanical instructions are unsuitable and the salesmen must have more discretion. If there are only a few employees to whom the managers can give such discretion or if the managers or owners must exercise close supervision over the transactions, then this limits the size of the firm. For example, steel is a product illustrating the first point. It is made to customer's specifications according to written instructions which the salesmen must obey. Building construction or carpentering illustrates the latter type of product. The entrepreneur himself quotes the price for the job. Hence according to these arguments steel would be a better candidate for a cartel agreement than construction.

Finally, let us consider merger as an alternative to a cartel agreement. We have seen that even the largest outlay on cartel enforcement that maximizes the probability of detecting violations together with a system of fines may fail to make collusion dominate competition, and only merger may be capable of achieving this. However, although merger is necessary for the dominance of collusion, it may not be sufficient because of scale diseconomies. The cost function of the firm formed by merger is not the envelope of the cost curves of its components. Before merger the individual firms were each under the control of their managers. All of these are replaced by a single control after merger. Hence the average cost curve of the merged firm is akin to the average cost curve of the most efficient firm out of the original number. If average cost is a rapidly rising function of output, then despite the prospect of a monopoly return, one firm may find it prohibitively costly to supply the market alone. Yet precisely under these conditions a cartel is also least likely to dominate competition. For if it is true that the firms have scale diseconomies, then even a cartel with side payments and fines encounters the largest cost of policing the cartel agreement. This is because the scale diseconomies imply an industry composed of firms of small relative size which raises cartel policing costs. Therefore, those conditions making it desirable for a group of colluding firms to coalesce into one by merger are frustrated by the scale diseconomies which also frustrate successful operation of a cartel. It would seem that intermediate firm size distributions, where the entrepreneurial input relative to industry output is larger, are the more promising candidates for a monopoly return by collusion instead of by merger.

According to these arguments collusion costs explain why competition dominates collusion in industries with relatively small firms. This provides a partial explanation for a positive association between rates of return and the relative size of firms by industry. Chapter VIII resumes this analysis with empirical evidence.

5. Sharing the Collusive Return

A common argument for explaining the dominance of competition over collusion is that the firms will be unable to agree on how to share the monopoly return. Moreover, this argument continues, any agreement is vulnerable to one firm's action. If the products are regarded as substitutes by the firms' customers, then, plainly, one firm can lower the net return of the others by increasing its rate of sales. One task of this section is to study limits for the imputations acceptable to the firms. This can be done with the help of characteristic functions which give the largest net return that one or more firms can assure themselves under the most adverse conditions. Assuming that the firms' imputations would satisfy conditions like the core constraints, this procedure would determine the limits of acceptable imputations, but whether the results would be of any use depend on the properties of the demand and the cost functions as well as the appropriate kind of characteristic function, as we shall see.

At the outset a theory of imputations derived from characteristic functions encounters the difficulty that there are many plausible characteristic functions for this purpose. Moreover, either these impose trivial limits on the firms' imputations or there may not be a set of output rates to enforce the values of the characteristic function. In the latter event the core is empty.

An alternative approach relates the firm's imputation to its contribution to the coalition's net return like the marginal productivity theory of wages. This seems more promising than the limits set by the core constraints. The latter part of this section develops this argument.

We first consider duopoly. Let

$$(1) \qquad\qquad p_i = f_i(q_1, q_2) \qquad\qquad (i = 1, 2),$$

denote the demand function for firm i's product and let $g_i(q_i)$ denote its total cost function. Consequently, a firm affects its competitors only via the demand relation. For substitutes and normal demand p_i varies inversely with both q_1 and q_2. The net revenue of firm i is defined by

$$(2) \qquad\qquad r_i = F_i(q_1, q_2) = q_i f_i(q_1, q_2) - g_i(q_i),$$

so that externalities to a firm arise from the demand function. A firm's net revenue is a monotone decreasing function of its rival's rate of sales. If r_i is a strongly concave function of q_i, (section VI.3 and theorem VI.8) then firm i can find a net revenue maximizing q_i for any given rate of sales $q_j, j \neq i$. For this reason it is often assumed that r_i is a strongly concave function of q_i. Having set the stage, we now consider various revenue sharing theories based on core constraints appropriate to various kinds of characteristic functions.

No one characteristic function commands widespread acceptance as being the best representative for duopoly or oligopoly. For each type of characteristic function there is a corresponding set of core constraints which the firms' imputations are supposed to satisfy. Hence there are as many such theories as there are plausible kinds of characteristic functions. Among these is the most pessimistic characteristic function leading to the α-core. The most pessimistic characteristic function is defined as follows:

$$(3) \qquad v(1) = \max_{q_1} \min_{q_2} F_1(q_1, q_2), \quad v(2) = \max_{q_2} \min_{q_1} F_2(q_1, q_2),$$

$$(4) \qquad v(1, 2) = \max_{q_1, q_2} (F_1 + F_2).$$

The latter is simply the maximum joint net return. The imputations, x_i, are said to be in the α-core if they satisfy

$$(5) \qquad x_i \geqslant v(i) \quad \text{and} \quad x_1 + x_2 \leqslant v(1, 2)$$

for some feasible set (q_1, q_2) (Aumann 1961). Since these characteristic functions give the least amount that a firm can assure itself under the most adverse conditions, the α-core gives the widest possible limits for the imputations and, as we shall see, is consequently not very helpful in narrowing the range of possible profit sharing schemes.

Another and less pessimistic definition of a characteristic function stems from the following inequalities:

$$\max_{q_1} \min_{q_2} F_1(q_1, q_2) \leqslant \min_{q_2} \max_{q_1} F_1(q_1, q_2),$$

$$\max_{q_2} \min_{q_1} F_2(q_1, q_2) \leqslant \min_{q_1} \max_{q_2} F_2(q_1, q_2).^{[9]}$$

Define the characteristic function

$$u(1) = \min_{q_2} \max_{q_1} F_1(q_1, q_2), \quad u(2) = \min_{q_1} \max_{q_2} F_2(q_1, q_2).$$

Therefore, $v(1) \leqslant u(1)$ and $v(2) \leqslant u(2)$. The characteristic function for the

---

9. For a proof of these inequalities, see von Neumann and Morgenstern (1947, sec. 13). There is equality in case the functions have a saddlepoint so that the function F_1 is concave in q_1 and convex in q_2. However, in the applications to oligopoly this will be the rare exception.

coalition of the two firms is the same as above, namely, $u(1, 2) = v(1, 2)$. The β-core consists of the imputations x_1 and x_2 satisfying the inequalities $x_i \geqslant u(i)$ and $x_1 + x_2 \leqslant u(1, 2)$ for some feasible rates of output, q_i.

The difference between these two kinds of characteristic functions can be interpreted in terms of which firm is thought to make the first move. With the most pessimistic characteristic function, v, and the α-core, a firm chooses a strategy that is effective for *any* strategy chosen by its rival. Pessimism is reflected in the property that the given firm must, therefore, assume the worst of its rival. For the β-core, the β-strategies of a firm depend on the competitor's choice so that the firm obtains a guaranteed return depending on what the rival does. It is as if the rival makes the first move to which the given firm responds. Following Aumann (1961, sec. 8) we may say that (1) a firm has an α-effective strategy if it can guarantee itself a payoff for *any* strategy chosen by its competitor; (2) a firm has a β-effective strategy if *given* the choice of a strategy by the competitor, the firm can guarantee itself a payoff depending on and good against the competitor's choice. Thus the α-effective strategy requires the given firm to behave as if it must make the first move while the β-effective strategy is as if the competitor must make the first move.

Another way of interpreting these concepts is to observe that the value of the α-characteristic function gives the amount that a firm can assure itself while the value of the β-characteristic function gives the amount that it can be held to by its competitors. There is the important result for our purposes that with side payments the α-core and the β-core are the same. This is a consequence of von Neumann's minimax theorem (Aumann 1962, pp. 91–93). However, there is the troublesome problem of finding whether there is a set of strategies capable of satisfying the core constraints. This is in contrast to the situations heretofore considered in which a coalition could isolate itself from the rest of the market, its members trade among themselves and thereby secure the value of the characteristic function. Now it may be impossible for both firms to choose rates of sale forcing them each to accept the least imputation given by the values of the characteristic function. Before examining an example, consider another candidate as a characteristic function to represent duopoly.

The preceding characteristic functions ignore the cost to a firm of lowering its competitor's net return. This can be remedied, at least formally, by defining a new characteristic function as a constrained function. Let

$$w(1) = \max_{q_1} \min_{q_2} \ F_1(q_1, q_2), \text{ subject to } F_2 \geqslant K_2,$$

with a similar expression for $w(2)$. Clearly, $w(i) \geqslant v(i)$ and, possibly, $w(i) \geqslant u(i)$ as well. However, this approach begs the question of what determines the values of the parameters, K_i. Thus, postulating a pair (K_1, K_2) is equivalent to settling the terms acceptable to a firm in exchange for imposing losses on its rival. This is because the pair (K_1, K_2) implies a pair of Lagrangian multipliers, the values of which give the size of the loss that one firm is willing to sustain so as to make its rival lose one dollar. What determines the values of the K's and the implied values of the Lagrangian multipliers? Answers to this question are designed to show how threats determine the firm's imputation. We shall not pursue this topic and refer the interested reader to the excellent discussion in Luce and Raiffa (1957).

We shall, however, explore some of the problems connected with the theory of imputations based on the α-cores and the β-cores for the simple case in which the demand functions are linear and there is constant average cost. The demand functions are given as follows:

$$p_1 = b_1 - a_{11} q_1 - a_{12} q_2,$$

(6) $\qquad\qquad (a_{ij} > 0).$

$$p_2 = b_2 - a_{21} q_1 - a_{22} q_2,$$

We shall impose two constraints on the problem, namely, the quantities and the prices must be nonnegative. The latter imposes upper bounds on the rates of sale because the firms cannot induce their customers to raise their purchase rates by offering them negative prices.

The net revenue functions are

(7) $\qquad\qquad r_i = (p_i - c_i) q_i,$

where c_i is the constant per unit cost of firm i. Assume that

(8) $\qquad\qquad b_i > c_i.$

This condition ensures an essential market.

The α-core, equivalent to the β-core because there are side payments, is nonempty and gives trivial limits. We may readily verify that there are $q_i > 0$ such that

(9) $\qquad\qquad p_i - c_i = 0 \quad \text{implying} \quad r_i = 0.$

Moreover, since the choice of $q_i = 0$ guarantees the firm at least a zero net return, (9) implies that its rival can prevent it from getting more. Hence $v(i) = u(i) = 0$. Of course, these are the classical competitive output rates but, unfortunately, these results reveal nothing about how the firms would share the joint maximum net return. The nonemptiness of the α-core and, therefore, the β-core, is an implication of a general result due to Scarf. In the context of oligopoly for n-firms, Scarf's theorem may be restated as follows:

SCARF'S α-CORE THEOREM. *There are n firms such that Q^i representing the set of possible rates of sale for firm i is closed, bounded and convex. Letting*

(10) $\qquad\qquad p_i = f_i(q_1, \ldots, q_n)$

denote the demand function for firm i's product, we define Q^i as follows:
$$Q^i = \{q_i / p_i \geqslant 0 \text{ and } q_i \geqslant 0\}.$$

The net revenue of firm i is defined by

(11) $r_i = F_i(q_1, ..., q_n) = p_i q_i - g_i(q_i),$

where $g_i(q_i)$ denotes the total cost function of firm i. It is assumed that F_i is a continuous quasi-concave function of its arguments, $q_1, ..., q_n$.[10]
For any coalition S, where I_n denotes the set of all n firms the characteristic function is defined as follows:

(12) $v(S) = \max_{S} \min_{I_n - S} \Sigma_{i \text{ in } S} r_i.$

Under these assumptions the α-core is not empty (Scarf 1970).

THEOREM 1. *If the goods sold by the n firms are substitutes so that*

$$\frac{\partial f_i}{\partial q_j} < 0, \qquad \text{for all } i, j = 1, ..., n,$$

there are constant unit costs given by c_i, the demand functions f_i are linear in prices and rates of output, $f_i(0) > c_i$, and $q_i \geqslant 0$ are restricted by $p_i \geqslant 0$; then $v(S) = 0$ for all $S \subset I_n$ and $S \neq I_n$, and there are positive rates of output to satisfy the α-core and, therefore, also the β-core constraints, such that $q_i > 0$ are the solutions of the equations $p_i = f_i(q_1, ..., q_n) = c_i$.

Proof. We can readily verify that F_i is in fact concave in this case and that the set of admissible output rates is bounded and convex. To show that $v(S) = 0$ as asserted form the two nonoverlapping coalitions S and $I_n - S$ so that there is in effect duopoly. We may then apply the simple argument in the preceding example to obtain the desired result. QED.

In general, the requirement that Q^i be bounded rules out, for example, constant elasticity demand functions. Moreover, the convexity of Q^i further limits the class of admissible nonlinear demand relations. Finally, the scope of Scarf's α-core theorem is restricted by its assumption that F_i is quasi-concave as he points out, noting, for example, that a Cournot-Nash point would exist if F_i were merely required to be concave in q_i alone.

The general case may be described graphically in terms of duopoly. Consider figure 5.4 where firm 1's output rate is plotted on the horizontal axis and firm 2's output rate on the vertical axis. The curve M_1M_1 depicts combinations of (q_1, q_2) such that firm 1 obtains a zero net return, that is, M_1M_1 is the locus of $r_1 = 0$ and N_1N_1 is the locus of $r_2 = 0$. If firm 1 chooses the output rate OA, then it can be held to a zero net return if firm 2 chooses an output rate above AB. Hence firm 1 can assure itself of a zero net return if

---

10. A function $\Phi(x)$ is called quasiconcave if for every x and y, $\Phi((x+y)/2) \geq \min [\Phi(x), \Phi(y)]$.

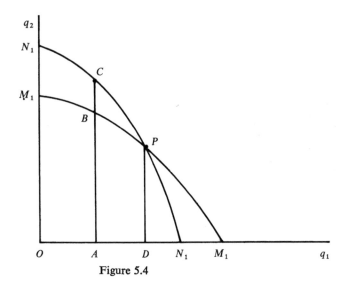

Figure 5.4

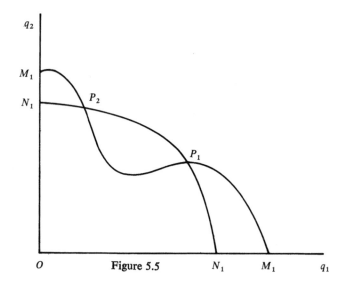

Figure 5.5

and only if it chooses the output rate OD. More generally, $v(i) = u(i) = 0$, if there is a nonnegative pair (q_1, q_2) that satisfies the two equations

(13) $$r_i = F_i(q_1, q_2) = 0 \qquad (i = 1, 2).$$

Figure 5.5 illustrates a situation in which there is no unique solution. Moreover, when no unique solution exists it does not appear possible to determine which will prevail. Of course, it is possible that no solution exists as is shown in figure 5.6. In this case firm 2 is in a dominant position because firm 1

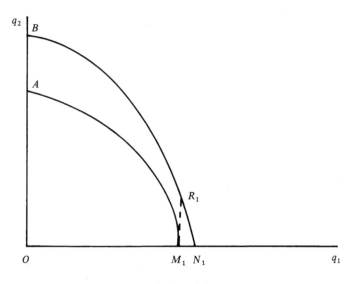

Figure 5.6

cannot hold it to a zero net return by choosing feasible output rates. For instance, firm 2 can only be held to the positive return implied by firm 1's choice of output rate OM_1 and, furthermore, firm 1 can be prevented from obtaining more than a zero net return if firm 2 chooses an output rate above OA. Thus figure 5.6 illustrates the possibility that

$$v(2) = u(2) > 0 \quad \text{and} \quad v(1) = u(1) = 0.$$

It should be noted that these figures assume that all of the points satisfying equations (13) also satisfy the demand conditions determined by

(14) $$p_i = f_i(q_1, q_2) \geqslant 0, \qquad \text{and } q_i \geqslant 0.$$

It is an easy calculation to verify that the situation as described by figure 5.6 can arise for linear demand relations and constant average cost only if

$$a_{22}/a_{12} > (b_2 - c_2)/(b_1 - c_1) \quad \text{and} \quad a_{21}/a_{11} > (b_2 - c_2)/(b_1 - c_1).$$

A general result is contained in the following lemma.

LEMMA 1. *Define the net revenue of firm i to be*

(15) $r_i = F_i(q_1, q_2) = q_i f_i(q_1, q_2) - g_i(q_i)$ $(i = 1, 2)$,

where f_i denotes the demand function and g_i the total cost function. If

(16) $F_i(0, q_j) = 0$ $(q_j \geqslant 0)$,

then $u(i) = v(i) \geqslant 0$. If there is at least one nonnegative pair (q_1, q_2) which satisfies

(17) $F_i(q_1, q_2) = 0$ $(i = 1, 2)$,

then $u(i) = v(i) = 0$. But if (17) has no nonnegative solution then for at least one firm the value of the characteristic function is strictly positive. It is understood that $g_i(q_i)$ is a strictly increasing function of q_i and that $f_i(q_1, q_2)$ is a strictly decreasing function of its arguments such that $p_i = 0$ for some nonnegative pair (q_1, q_2).

The proof is virtually self-evident in view of the preceding discussion in terms of figures 5.4–5.6.

These results readily extend to the problem of finding the values of the characteristic function for n-firm oligopoly. Define for all $S \subset I_n$, $S \neq I_n$ the following characteristic function:

(18) $u(S) = \displaystyle\min_{q_j \text{ in } I_n - S} \; \max_{q_i \text{ in } S} \; \sum_{i \text{ in } S} r_i,$

and

(19) $u(I_n) = \displaystyle\max_{q_i} \sum_{i \text{ in } I_n} r_i,$

subject to $p_i \geqslant 0$ and $q_i \geqslant 0$ for all i. If

(20) $r_i = F_i(q_1, q_2, \ldots, q_{i-1}, 0, q_{i+1}, \ldots, q_n) = 0$

for all $q_i \geqslant 0$, then the conditions of lemma 1 apply to the functions

(21) $r_S = \displaystyle\sum_S r_i \quad \text{and} \quad r_{I_n - S} = \sum_{I_n - S} r_j.$

The conditions implying a positive $u(S)$ depend on the properties of the demand and net revenue functions just as in lemma 1 with a coalition S in place of firm 1 and the complementary coalition $I_n - S$ in place of firm 2. The conclusion is conveniently summarized as follows:

THEOREM 2. *Let r_i be defined by (11), $u(S)$ by (18), and r_S and $r_{I_n - S}$ by (21). If (20) holds then $u(S) \geqslant 0$ for all $S \subset I_n$, $S \neq I_n$. Moreover, $u(S) = 0$ for all S if there is a set of nonnegative q_i to satisfy*

(22) $r_S = 0 \quad and \quad r_{I_n - S} = 0$

for any $S \subset I_n$. But if (22) has no nonnegative solution for some S, then for that coalition the value of the characteristic function is strictly positive. It is understood that $g_i(q_i)$ is a strictly rising function of q_i and that the demand functions are normal with all goods gross substitutes such that $p_i = 0$ for some nonnegative point $(q_1, ..., q_n)$.

The proof is exactly the same as for lemma 1, namely by appealing to the graphs. Observe that, as claimed at the outset, the core constraints would not narrow the range of admissible imputations if $u(S) = 0$ for all S. Moreover, there is the further condition that there must exist a set of q_i's that can implement the core constraints, which is not always true so that the members of S may be unable to secure $u(S)$ by the choice of feasible rates of sale, $\{q_i / i \text{ in } S, q_i \geq 0 \text{ and } p_i \geq 0\}$.

Oligopoly has some issues in common with public goods. Consider the bridge example (discussed in sec. I.8). Suppose that no individual can pay

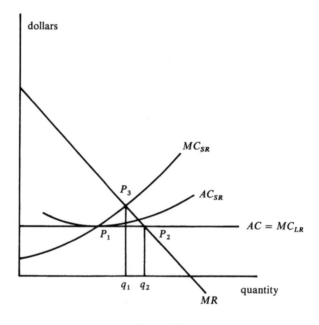

Figure 5.7

for the bridge alone and that the bridge is so costly that all must share in the expense if it is to be built at all. Consequently, $u(S) = 0$ for all $S \subset I_n$, $S \neq I_n$, $u(I_n) = $ cost of the bridge. Therefore, the core constraints would not narrow the range of admissible imputations although no individual damages the others as with oligopoly. The oligopoly in which $u(S) > 0$ for some coalitions is like a bridge the cost of which can be borne by some but not by all the members of the community so that all need not cooperate to have the bridge. If so, the core constraints do narrow the limits of admissible imputations. Now the analogy between public goods and oligopoly breaks down. In the bridge example the users can obtain the value of the characteristic function by themselves while in oligopoly a coalition of firms cannot necessarily secure the value of their characteristic function by their own efforts. This is why there seems to be no useful theory of imputations for a cartel which derives from the α-core or the β-core. Nor is this all. Since there is no canonical characteristic function to represent oligopoly, there is no agreement on which characteristic function best represents oligopoly.[11]

Now let us take a different approach to the problem of determining the firm's imputation in a cartel. This looks at the firm's contribution to the cartel instead of the largest return it can secure for itself under the most adverse conditions. Figure 5.7 illustrates the principal idea. Suppose that all of the firms have the same minimum average cost so that the long-run average cost is constant for all output rates. To find a firm's contribution to the cartel, we can calculate the total cost of a given output rate with n and with $n + 1$ firms, assuming in both cases an optimal output per firm. In figure 5.7, the cost difference is shown by the area under MC_{SR} less the area under MC_{LR} at a given output rate. The cartel maximizes its net return by correctly choosing the number of its members. For a given number, the cartel would offer, say, Oq_1, while for the optimal number it would offer Oq_2. In this way it would increase its net return by the area of $P_1P_2P_3$. It would seem that a cartel member can claim that this amount is his contribution to the cartel. However, although the cartel would not give a member more than this amount, it may well give him less depending on the available supply of potential cartel members.

The argument is akin to the marginal productivity theory of wages. Assume there is a supply of identical workers who are willing to offer their services at the terms described by a supply schedule. The firm's output depends on how many workers it hires and not on who they are. If the wage rate to the firm is independent of how many workers it hires, then the firm will hire that number at which the marginal product equals the real wage rate. The market equilibrium can be described in terms of an intersection of the supply of and demand for labor. Similarly, a cartel faces a supply of firms willing to join it. Though a firm contributes to the cartel's net return, for a large enough supply of candidates with no alternative better than a competitive return, a cartel need offer only a competitive return, and it can obtain as large a membership as it pleases. It is those who control entry into the cartel who can obtain the

11. Several bargaining theories combine threats and demands. In these the protagonists first present each other with their best threats and having previously agreed upon how to resolve the conflict, the threats determine their share of the net return. The best-known and cleverest of these theories is Nash (1950a and 1953). Shubik (1959b, chaps. 4 and 5) applies Nash's theory to duopoly. For a perceptive critique of these theories and related approaches consult (Luce and Raiffa 1957, chap. 5).

full monopoly return. Another way of stating this is to suppose that those who control entry sell licences for joining the cartel in a competitive market. The price of a licence will be such that a successful bidder can obtain only the competitive return from its use. The monopoly return goes to those who can sell the licences in the first place, and, even then, only if it is costless to control entry and to create a monopoly can the organizers of the cartel obtain a monopoly return.

Before taking up these matters, let us briefly study the more general case in which the cartel members differ. Either because of cost differences or because of product differences, cartel members may not all be equally lucrative to the cartel. Hence some potential cartel members can have more bargaining power than others. Competition for the more lucrative members by the given cartel or other groups bids up the price for their services above the competitive level just as workers of above average ability can command higher wage rates.

Returning to the main issues, one should not assume that the creation of monopoly is costless. A patent is an obvious example. Although it may be true that a lucky invention that is patentable may give its owner a rate of return far above the average, it does not follow that the class of inventors has more wealth per capita than any other class of equal talent and training. Similarly, although the rate of return on a winning lottery ticket may be astronomical, it does not follow that all those who buy lottery tickets become wealthy. If anyone can enter the inventing business, then it will yield a normal rate of return. In calculating the rate of return the failures are often overlooked, and this gives the erroneous conclusion that the rate of return is higher than in fact is true. Therefore, although a successful patent may yield a high return, partly as a result of the creation of monopoly, it does not follow that this road to monopoly is paved with riches.

Patents illustrate only one of the avenues to a monopoly return. Firms can obtain the fruits of unique advantages of varying durability in many other ways. One example is a successful advertising campaign that catches the public's fancy. New production methods such as the assembly line or new business methods such as the leasing of equipment can also be a source of temporarily high profits. Firms allocate resources to a systematic search for such methods of raising profits. The spectacular success of some must not blind us to the failures of many others. If nothing hinders firms from engaging in the quest, then the average return will not rise above the competitive level taking the failures with the successes.

These considerations also explain a positive association between the concentration ratio and the rate of return, the subject of the empirical research presented in chapter VIII. Let one or more firms secure an advantage over their competitors and they will obtain relatively high sales and profits. The effects need not last long nor need they be explained by collusion, although the temporary advantage can account for the common source of both the high concentration ratio and the high profit rate.

An often-discussed form of collusion refers to a group of firms who combine to reduce their output and raise prices without having some unique advantage that can deter entry. This begs two questions. First, how can they prevent or retard entry, and second, how will they share the combined net return? Without state protection which limits entry by the legal requirement of a license or the equivalent, it seems hardly possible for such a group of firms to hope for more than a temporary gain. High rates of return eventually attract new firms. Although the profits will not persist, they last for a while. It is hard to believe that rational businessmen will then fail to agree on how to share the gain.

A plausible profit-sharing formula is as follows. Firms of the same size who make the same product and are equally efficient would share the return equally. By the same logic, firms of different sizes would share the net return in proportion to their size. Hence in either case all parties to the agreement would receive the same rate of return. This formula increases the duration and stability of a cartel because it gives the firms less incentive to change their relative size in order to raise their net return. Moreover, if the return is proportional to the firm's capital, then there is the additional merit that the cartel will choose the policy that maximizes the joint net return even without side payments. Therefore, each firm's share of the maximum joint net return equals its contribution to the maximum. Consider for example the duopoly shown in figure 5.1. This argument implies that the firms choose those policies that give the maximum joint net return, the point J, firm 1 receives ON', and firm 2 receives OM', the coordinates of J.

VI

The Monopoly and Cournot-Nash
Equilibria under Dynamic Conditions

1. Introduction

We now resume the study of optimal policies and expectations for n competing firms begun in chapter IV. Our concern is with the dynamic properties of the competitive and monopoly equilibria. Despite the greater difficulties of the analysis, many conclusions applicable to the static model remain valid in a dynamic setting, but certain new problems appear, and we shall need more powerful mathematical tools to solve them.

Changing market conditions arise in two ways. First, the level of demand changes over time because of changes in wealth per household, the number of households, tastes, or the entry and exit of competing goods. Second, even with a market of constant size, the demand changes because customers change their expectations or their stocks. Both sets of forces introduce dynamic elements that complicate a firm's optimal policy.

Given dynamic demand functions, rational firms pursue appropriate goals. The basic one is the maximum present value of the expected net return over an infinite horizon instead of the one-period maximum net return. One might think it simpler to use a finite T-period horizon. This is not true. It is both mathematically simpler and economically more logical to use an infinite horizon. As to the latter, suppose that a firm has a finite T-period horizon instead of an infinite horizon. This literally means that the firm plans for T periods initially, $T-1$ periods in the next period, and so on, until in the next to the last period, it plans for the one remaining period. This has absurd implications that one may try to avoid by postulating a terminal market value of the firm. Thus, the new present value would be the sum of the present value of the net return of the T periods and the present value of the terminal market value of the firm. This approach obviously begs two questions; what determines the terminal market value, and what determines the length of the horizon? If the terminal market value does depend on events beyond period

218

T, then, implicitly, one has assumed an infinite horizon without incorporating the firm's future prospects in a satisfactory manner.

Typically, it is true that a firm does plan for a fixed interval of T periods into the future, and it does so in every period. This means that it implicitly uses an infinite horizon such that events beyond period T have a small effect on the present value. It is better to be explicit and to use an infinite horizon with an appropriate discount factor. More precisely, let u_t denote the t-period net return and let α denote the constant discount factor $\alpha = 1/(1 +$ discount rate), where $0 < \alpha \leqslant 1$. The present value of the net return is

$$PV = \sum_0^\infty \alpha^t u_t,$$

which is finite if and only if

$$|u_t| < K(\alpha + \delta)^{-t}, \qquad \text{for some } \delta > 0.$$

This means that the net revenue does not grow at a rate above the discount rate; or the present value would be unboundedly large. If the present value is finite, then for any $\varepsilon > 0$ there is a T such that

$$\sum_T^\infty \alpha^t |u_t| \leqslant \varepsilon,$$

which determines the length of the T-period planning horizon valid for every period in which action is to be taken.

This chapter treats the general class of problems as follows. Let $a_t > 0$ denote the t-period discount factor such that

$$\sum_0^\infty a_t < \infty.[1]$$

This implies that a constant net revenue stream would have a finite present value. The t-period net revenue of the firm depends on current and lagged sales, say, $u_t = f(q_t, q_{t-1})$. To represent market change over time, f would also have a subscript and the net revenue would be $u_t = f_t(q_t, q_{t-1})$. The firm's goal is to choose an infinite sequence $\{q_t\}$ in order to

$$\max_{\{q_t\}} \sum^\infty a_t f_t(q_t, q_{t-1})$$

subject to certain constraints. The sequence $\{q_t\}$ can be regarded as an

---

1. This notation deserves some explanation. It means that the series of partial sums $S_n = \sum_0^n a_t$ is bounded. However, in this case, since the a's are nonnegative, the boundedness of the partial sums implies the convergence of the series. More generally, an expression such as the following, $\sum |a_t| < \infty$, means that the series in question is convergent. It will always be clear in the context whether boundedness or convergence is meant and, moreover, almost always, it is the convergence of the series that concerns us. Perhaps an example would not be superfluous. The sequence $b_0 = 1$, $b_n = (-1)^n 2$, has bounded partial sums since $S_n = (-1)^n$, but the series $\sum b_n$ does not converge although it is bounded. The series $\sum |b_n|$ diverges.

element in an infinite dimensional sequence space. Part of our concern is to develop necessary and sufficient conditions for the solution of this problem and to study the properties of this solution.

The assumptions about the discount factor a_t deserve some attention. It is commonly observed that no one is willing to pay an infinite amount for a promise to receive a constant stream of receipts in perpetuity even if there is no risk of default, for example, British consols. Hence a dollar in period t is worth less than a dollar now. In fact, according to this formulation one would be willing to buy or to sell the following income stream for a_t dollars now: $0, 0, \ldots, 1, 0, \ldots$, where the 1 appears in period t. This price, a_t, is determined very much like any other. Opportunities present themselves which take time to develop and which yield a time profile of receipts that people find incongenial so that they wish to borrow or lend to achieve a more congenial pattern. Markets spring up enabling people to transact in income streams. The implied sequence of discount factors constitutes an infinite number of prices serving to clear this market. The nature of investment opportunities and the preferences of the traders interact to determine the equilibrating price sequence. A detailed study of this market forms a major part of the theory of capital. Our purpose is to derive some implications of the existence of the discount factors in one industry with different states of competition. Although the industry prospects had some part in determining the equilibrium discount factors, we shall ignore this fact. Nor is this all. We shall derive the industry equilibrium for *any* sequence of discount factors such that

$$(1) \qquad\qquad a_t > 0 \quad \text{and} \quad \sum_0^\infty a_t < \infty.$$

This is the positive orthant of the infinite dimensional linear space, denoted by the symbol l_1, the space of absolutely convergent series.[2] A subset of this space is the sequence of discount factors decreasing at a geometric rate given by

$$(2) \qquad\qquad \{\alpha^t\} \qquad\qquad (0 < \alpha < 1).$$

It is important to understand the difference between the problem of maximizing the present value for *any* positive element of l_1 and the problem of maximizing the present value for a *particular* element of l_1. This distinction appears below in the mathematical analysis.

The firm's net revenue at time t depends on how well it forecasts future events if the present depends on the past. This paradoxical assertion has a simple explanation. Suppose that a firm desires the maximum present value of its expected net return and that its current choices are constrained by past events which partly depend on its own past actions. Then the maximum present value requires a forecast of the future consequences of its current

---

2. The space of l_1 sequences $\{b_n\}$, where the b's are real numbers, consists of all sequences such that $\Sigma |b_n| < \infty$. Hence, it is the space of absolutely convergent series (see n.1).

choice. We shall concentrate on that particular set of factors making the present depend on the past via the demand relations.

To gain familiarity with this class of problems, the next section gives some properties of dynamic demand relations, and states some of the mathematical problems to be studied later. Sections 3 and 4 develop the mathematical equipment for solving the economic problems and analyze in detail the nature of infinite horizons. Section 5 uses these results to derive the necessary and sufficient conditions for the existence of an equilibrium taking expectations into account. There is also an analysis of the determinants of competition versus collusion under these conditions. Finally, there is a summary of the results.

2. Dynamic Demand Relations

Dynamic elements enter the demand functions primarily for two reasons. First, the current rate of demand may depend on the relation between the current and expected prices. If price expectations are derived from past prices then the current demand becomes a function of current and past prices. Second, the demand for durables is dynamic because a given stock of such goods yields services for more than one time period. In fact the durability of goods depends on the calendar length of the time period since virtually all goods are durable for sufficiently short periods. Both of these considerations, expectations and durability, introduce feedbacks even with markets of constant size. Obviously, changing market conditions due to changing affluence and number of customers also introduce dynamic demand elements.

We shall represent the demand conditions with two functions, an accounting identity relating purchase rates to stocks and a behavioral relation between desired stocks and current and expected prices. Assume there are n competing goods. Let

$s_t = n \times 1$ vector of customers' stocks at the end of period t,
$q_t = n \times 1$ vector of purchase rates during period t,
$p_t = n \times 1$ vector of t-period prices,
$c_t = n \times 1$ vector of t-period consumption rates.

The accounting identity is given by

(1) $$s_t = s_{t-1} + q_t - c_t.$$

It is convenient to assume that consumption rates are proportional to the available stock. Thus the ith component of the vectors satisfies

$$c_{it} = (1 - \gamma_i) s_{i, t-1}$$

In matrix form this becomes

(2) $$c_t = (I - G) s_{t-1},$$

where I is the $n \times n$ identity matrix, and G is an $n \times n$ diagonal matrix with diagonal entries γ_i. The size of the parameter γ depends both on the nature of the good and on the calendar length of a period. The longer the period, the smaller are the γ's. For a perfect perishable, $\gamma = 0$, and for a perfect durable, $\gamma = 1$. An advantage of this formulation is the ease with which it links continuous and discrete time demand functions. Thus, if h denotes the calendar length of the time period and ρ the instantaneous rate of consumption, then $\gamma_i = e^{-\rho_i h}$. For a perfect perishable ρ_i is infinite and for a perfect durable it is zero. In practice we should not expect to encounter either of these extremes.

With this specification of consumption rates instead of (1) we obtain

(3) $$s_t = Gs_{t-1} + q_t.$$

The price of the services of a good in period t is given by

(4) $$r_t = p_t - Gp_{t+1}.$$

This is like a rent, for if one buys a unit of the good in period t, γ units will remain in period $t+1$ that can be sold for p_{t+1} per unit. The desired stock is a function of these current and expected quasirents, r_t. A linear approximation to these functions is given by

(5) $$s_t = b_t - (A_0 r_t + A_{-1} r_{t+1} + \ldots + A_{-n} r_{t+n} + \ldots).$$

The price coefficients are represented by $n \times n$ matrices of constants, the A's. The term $\{b_t\}$ is a sequence of $n \times 1$ vectors to represent the effects of income, the number of customers, entry and exit of goods remotely related to those under study, and other variables that affect the desired stock. There are compelling reasons to postulate a demand relation of this form, as we shall see in the formal analysis in section 5. It is not implausible to assume that the desired stock depends on the current and expected quasirents although, so far as we know now, other formulations appear equally plausible.

It is cumbersome to write out the demand relations every time, so let us adopt a concise notation that reveals the analogies with the static system and facilitates a general analysis. Let L denote the lag operator, which shifts the sequence back one period, and E denote the lead operator, which shifts a sequence forward one period. $Lx_t = x_{t-1}$ and $Ex_t = x_{t+1}$. Define the power series in E by

(6) $$A(E) = A_0 + A_{-1}E + \ldots + A_{-j}E^j + \cdots,$$

which has matrix coefficients of real scalars given by A_{-j}. Using this notation the demand equation (5) becomes

(7) $$s_t = b_t - A(E)r_t.$$

$A(E)$, which is a linear operator in E, is analogous to an ordinary $n \times n$ matrix and does in fact reduce to a matrix of scalars if $A_{-j} = 0$ for $j \geqslant 1$.

Instead of (3) we now have

(8) $$(I - GL)s_t = q_t + d_t,$$

where

$$d_t \quad \begin{aligned} &= Gs_{-1}, &&\text{if } t = 0, \\ &= 0, &&\text{if } t > 0. \end{aligned}$$

To express purchase rates as a function of prices, multiply both sides of (7) by the linear operator $I - GL$ and obtain

$$(9) \qquad \begin{aligned} (I - GL)\, s_t &= (I - GL)b_t - (I - GL)A(E)r_t, \\ q_t &= (I - GL)b_t - d_t - (I - GL)A(E)r_t \\ &= (I - GL)b_t - d_t - (I - GL)A(E)(I - GE)p_t. \end{aligned}$$

In the last step we use (4) to replace r_t with an appropriate function of p_t. Therefore, given the calendar length of the time period such that some goods are durable, we see that purchase rates depend on both p_{t-1} as well as expected prices.

Other derivations of the demand functions are sometimes useful. Suppose the customers are households. It is traditional to endow them with utility functions and assume that the desired stocks maximize utility subject to a budget constraint. This procedure would impose certain conditions on the demand functions. Thus the demand functions consistent with utility maximization for one household would satisfy:

$$(10) \qquad s_t = b_t - [A'(L) + A(E)]\, r_t,$$

where

$$A'(L) = A'_0 + A'_{-1}L + \ldots + A'_{-j}L^j + \ldots,$$

and the primes denote the conjugate transpose of the matrices. $A'(L)$ is also called the *adjoint* operator of $A(E)$. According to (10), the current desired stock would have to depend on past and future prices such that the matrix of coefficients for r_{t+n} is the conjugate transpose of the matrix of coefficients for r_{t-n}. This is the dynamic generalization of the well-known Slutsky relation for the static case that requires symmetry of the price coefficient matrix. For dynamic demand the Slutsky condition is, therefore,

$$\frac{\partial s_{t+n}}{\partial r_{t-n}} = \frac{\partial s_{t-n}}{\partial r_{t+n}}.^3$$

Although it is interesting to derive the implications of these demand relations, it would be unwise to confine oneself to such demand functions for several reasons. First, the pertinent demand relations refer to aggregates of customers and not to individuals. It is by no means obvious that such aggregates obey the same laws as do the individuals. Equivalently, it is not necessarily true that there exists an individual who would make the same

---

3. For an elementary discussion of the Slutsky equation see Hicks (1946), and for an advanced treatment see Uzawa (1959, pp. 129–48). For an extended discussion in a dynamic setting see Telser and Graves (1971, chap. 3).

choices as the aggregate. Second, the theory should encompass nonhousehold demand as well as household demand. Although the nonhousehold demand ultimately comes from the household demand, their properties may differ. For example, the automobile manufacturers' demand for tires derives from the household demand for cars but depends on other factors as well and need not satisfy the same conditions as the household demand for tires. Finally, future prices do not stand on the same footing as present prices if the future is uncertain. There is no satisfactory theory of such uncertainty.

Nevertheless some hypotheses about the properties of the dynamic demand relations are plausible. To exhibit these, it is helpful to write $A(E)$ in another way. Let $A(E) = [a_{ij}(E)]$ be an $n \times n$ matrix whose elements are power series in E defined as follows:

$$(11) \qquad\qquad a_{ij}(E) = \sum_{k=0}^{\infty} a_{ij,\,k}\, E^k.$$

The scalar $a_{ij,\,k}$ is the ijth element of the matrix A_{-k} of the power series representation of $A(E)$ shown in (6). In the static case the elements of the matrix of price coefficients are real numbers; in the dynamic case they are power series in the lead operator, E. By analogy with the static case, write

$$(12) \qquad\qquad A(E) = A_d(E) - A_v(E),$$

where $A_d(E)$ is the diagonal matrix composed of the diagonal elements from $A(E)$, and $A_v(E)$ consists of the off-diagonal elements of $A(E)$ with zero's on the diagonal. Thus,

$$A_d(E) = \begin{bmatrix} a_{11}(E) & & & \\ & \cdot & & \mathbf{O} \\ & & \cdot & \\ \mathbf{O} & & \cdot & \\ & & & a_{nn}(E) \end{bmatrix},$$

and

$$A_v(E) = \begin{bmatrix} 0 & a_{12}(E) & \dots & a_{1n}(E) \\ a_{21}(E) & 0 & \dots & a_{2n}(E) \\ & & \dots & \\ a_{n1}(E) & & \dots & 0 \end{bmatrix}$$

[cf. IV.3 (4) and IV.3 (5)].

Equation (12) conveniently shows how competition affects the n goods. If these are gross substitutes, then all of the elements of $A_v(E)$ should be nonnegative. This means that the scalar coefficients of the power series $a_{ij}(E)$, $i \neq j$, should be nonnegative. We express this by writing

$$(13) \qquad\qquad A_v(E) \geqslant 0.^4$$

---

4. There is the possibility of ambiguity in referring to a positive or a nonnegative matrix, meaning a matrix with positive or nonnegative elements, respectively, and the matrix of a positive definite or semidefinite quadratic form. We can distinguish these cases by always using the word definite when it is ambiguous from the context which meaning is correct. See also n.13 below.

Consequently, s_{it} varies directly with $r_{i', t+n}$ for all n and $i \neq i'$. We also wish to assume that s_{it} varies directly with $r_{i, t+n}$ for $n > 0$ implying that a current stock of good i competes with past and future stocks of good i. Also s_{it} varies inversely with r_{it} so that the diagonal entries of A_0 are positive [recall the minus sign before $A(E)$ in the matrix representation of the demand equations]. Hence the constant terms of $a_{ii}(E)$ are all positive and all of the other coefficients are nonpositive. Therefore, a condition like (13) does not apply to $A_d(E)$; but, as we shall see in section 5, these assumptions about $a_{ii}(E)$ do imply that $A_d(E)^{-1} > 0$. At the corresponding stage in the static analysis given in chapter IV, we could prove theorems about the sign of A^{-1}. Similar results hold in the dynamic case, but before we can derive them, we must find sufficient conditions for the invertibility of linear operators such as $A(E)$, $A'(L)$, and $[A(E) + A'(L)]$.

To gain familiarity with the properties of these demand relations, consider the effects of hypothetical price changes. Assume that the process begins de novo in period 0 so that $p_{-t} = 0$ for $t > 0$.

First, let there be a once and for all change in r_0:

$$\Delta s_0 = -A_0 \, \Delta r_0, \quad \Delta s_t = 0, \qquad\qquad \text{if } t > 0.$$

From the stock-purchase relation (3), it follows that

$$\Delta q_0 = \Delta s_0, \quad \Delta q_t = \Delta s_t - G \, \Delta s_{t-1}, \qquad\qquad \text{if } t > 0.$$

Therefore,

$$\Delta q_1 = -G \, \Delta s_0, \quad \Delta q_t = 0, \qquad\qquad \text{if } t > 1.$$

Initially, purchases and stocks respond alike to the price change. Different responses appear in the period immediately following the time of price change. For a temporary rise in r_0, purchase rates initially fall, then rise, and finally return to their long-run equilibrium.

Next suppose there is a permanent change in r to a new level. Hence $\Delta r_t = \Delta r_0$, for all t. Therefore,

$$\Delta s_t = -\left(\sum_0^\infty A_{-j}\right) \Delta r_0, \qquad\qquad \text{for all } t,$$

$$\Delta q_0 = \Delta s_0,$$
$$\Delta q_t = (I - G) \, \Delta s_t, \qquad\qquad \text{if } t > 0.$$

The initial response of the purchase rate exceeds the long-run response.

Another interesting point emerges in this hypothetical experiment. The effect of the price change on the change in the desired stocks is given by a sum of matrices of price coefficients. For a finite response to a finite change, one must have

$$\sum_0^\infty A_{-j} < \infty,$$

which requires the elements of the matrices to form a convergent series thereby restricting the class of admissible demand functions. One can derive other restrictions by determining the effects of other patterns of hypothetical price changes, but it is more illuminating to discover appropriate restrictions on the demand relations from the study of a certain maximum problem.

Thus suppose a monopolist controls the output of the n goods. The present value of the net revenue is given by

$$(14) \qquad\qquad PV = \sum_0^\infty a_t(p_t'q_t - c_t),$$

where $p_t'q_t$ is the scalar product of gross receipts and c_t now denotes the t-period total cost. Assume the monopoly goal is maximum present value of net revenue by the choice of a sequence of price vectors subject to the constraint that quantity sold satisfies the inventory balance equations (3) and the customers' desired stocks satisfy their demand equations (7). Necessary and sufficient conditions for a solution of this problem impose restrictions on the linear operator $A(E)$ and on the matrix G as we shall see. Moreover, these restrictions explain why the desired stock should depend on the present value of the expected rents, r_t. With the help of these conditions we can derive the properties of the Cournot-Nash equilibrium.

The assumption of constant unit costs has several important consequences in this work. If, as is assumed in the demand relations, the n goods are imperfect substitutes, then it is demand and not cost that determines the optimal scale of the n firms. The Cournot-Nash equilibrium implies that these firms obtain a positive net return, begging the question of why entry does not occur until this return is driven down to zero. By hypothesis the product of every firm differs slightly from those of all of the others, and entry would result in the provision of more and different goods. Therefore, a satisfactory theory of entry would also determine the equilibrium array of product quality. If product quality is a continuum in the attributes of the good, then a rigorous analysis would require advanced mathematical treatment using techniques from functional analysis, which would be beyond the scope of this study. However, the determination of the equilibrium under dynamic conditions also requires some rudiments of functional analysis, which is the task of the next two sections.

3. Some Fundamentals on Optimal Policies

The present value of a net revenue stream exemplifies a mathematical object known as a functional. Let X denote an element in an infinite dimensional space of sequences, $X = \{x_n\}$, where x_n denotes the nth coordinate of X. Most of our work concerns functionals of the type given by

$$(1) \qquad\qquad \phi(X) = \sum_0^\infty a_n f_n(x_n, x_{n-1}),$$

where $f_n(x_n, x_{n-1})$ corresponds to the n-period net revenue as a function of the scalar control variables, x_n and x_{n-1}, and a_n is the positive n-period discount factor. Another illuminating way of writing this expression is as follows:

(2)
$$(A, Y) = \sum_0^\infty a_n y_n,$$

where

(3)
$$y_n = f_n(x_n, x_{n-1}).$$

According to (2), the functional is a scalar product of the two sequences $A = \{a_n\}$ and $Y = \{y_n\}$, which are elements of certain infinite dimensional sequence spaces. (A, Y) can also be regarded as a bilinear functional of A and Y since it is linear in one element for given values of the other. The functional $\phi(X)$ is not necessarily linear in X unless y_n in (3) is linear in its arguments.

We seek necessary and sufficient conditions for an extremum of the functional $\phi(X)$, where A may be *any* element in the positive orthant of the space of convergent series, that is,

(4)
$$\sum_0^\infty a_n < \infty, \qquad\qquad a_n > 0.$$

From now on, where the indexes of summation are omitted, they are understood to run from $n = 0$ to $n = \infty$. Also, it is slightly more convenient to develop this material for a minimum. The methods of ordinary calculus are helpful up to a point in solving this problem, but we must expect to encounter new difficulties in view of the infinite horizon. In fact a continuous version of this problem is related to those forming the subject matter of the calculus of variations. Unfortunately, this discipline gives us little guidance because rarely does it treat problems with an infinite horizon. Hence we shall have to develop for ourselves the necessary mathematical tools for solving the economic problems.

Before tackling the minimum problem, we face three basic questions. First, when is a functional $\phi(X)$ given in (1) finite for a sequence of finite x_n? Second, when is $\phi(X)$ a continuous functional? Third, when is $\phi(X)$ differentiable? The answers to all three questions take us beyond the ordinary calculus.

We begin with the first question. The most elementary examples show that something stronger than the finiteness of every y_n is needed to ensure the convergence of the infinite series $\sum a_n y_n$. An important result is contained in the following theorem.

HADAMARD'S THEOREM. *In order that* $\sum a_n y_n$ *should be convergent whenever* $\sum a_n$ *is convergent, it is necessary and sufficient that*

(5)
$$\sum |\Delta y_n| = \sum |y_n - y_{n+1}| < \infty.$$

For a proof, see Hardy (1949, thm. 7, 3.5).[5] For our purposes, it is sufficient
to remark that this theorem derives from a line of reasoning that begins
with the following:
Let

$$s_n = \sum_0^n a_\sigma \quad \text{and} \quad t_n = \sum_0^n a_\sigma y_\sigma.$$

Hence

$$t_n = \sum_0^{n-1} (y_\sigma - y_{\sigma+1})s_\sigma + s_n y_n,$$

since $s_{-1} = 0$ by definition. This procedure is akin to integration by parts. A
sequence $\{y_n\}$ satisfying (5) is said to be of bounded variation. Hence the
functional $\phi(X) = (A, Y)$ is finite for an element X with finite coordinates if
and only if the corresponding sequence $\{y_n\}$ is of bounded variation. This
has an important implication.

THEOREM 1. *If $\{y_n\}$ is of bounded variation, then y_n must approach a limit,
denoted by y.*
Proof.

(6) $$y_0 - y_n = \sum_0^{n-1} (y_\sigma - y_{\sigma+1}) \leqslant \sum_0^{n-1} |y_\sigma - y_{\sigma+1}|.$$

Since $\Sigma |y_\sigma - y_{\sigma+1}|$ is convergent, it follows that $y_0 - y_n$ is convergent, and
we must have $y_n \to y$. QED.[6]

It should be noted that Hadamard's theorem applies to a broader class of
discount factors than those discussed in section 1. Hadamard's theorem
applies to any sequence of discount factors $\{a_n\}$ for which the infinite series
Σa_n is convergent. This class of discount factors includes the space of
absolutely convergent series, namely, $\Sigma |a_n| < \infty$; but in both cases there is
no requirement that the discount factors a_n must be nonnegative. Presumably
it is the class of nonnegative or even strictly positive discount factors which is
of greatest relevance in economics. Hadamard's theorem gives necessary and
sufficient conditions for the broadest class of discount factors but only
sufficient conditions for the narrower class of absolutely convergent series.

---

5. There is another interesting result given in Hardy (1949, thm. 8, 3.5). In order that $\Sigma a_n y_n$
be convergent whenever $S_n = a_0 + \ldots + a_n$ is bounded, it is necessary and sufficient that (5) be
satisfied and that y_n should tend to zero.
   Observe that in both the theorem quoted in the text and in this theorem, the convergence of
$\Sigma a_n y_n$ is required to hold for *any* sequence $\{a_n\}$ with the properties given in the hypothesis of
the theorem. Generally, there are fewer restrictions imposed on y_n if more restrictions are
imposed on the sequence $\{a_n\}$. In fact we shall encounter an example of this in n.7 below.
6. This proof uses a familiar fact from the theory of infinite series, namely, a series Σx_n is
certainly convergent if the series $\Sigma |x_n|$ is convergent (Knopp 1951, chap. 4, secs. 16, 85).

For the latter we have the following result:

COROLLARY.

i. *The series* $\Sigma\, a_n y_n$ *is absolutely convergent if* $\{|y_n|\}$ *is bounded and the series* $\Sigma\, a_n$ *is absolutely convergent.*

ii. $\Sigma\, a_n y_n$ *is absolutely convergent for all absolutely convergent series* $\Sigma\, a_n$ *only if* $\{|y_n|\}$ *is bounded.*

iii. $\Sigma\, a_n y_n$ *is absolutely convergent for all bounded sequences* $\{y_n\}$ *only if* $\Sigma\, a_n$ *is absolutely convergent.*

Proof. i. $\Sigma\, a_n y_n \leqslant |\Sigma\, a_n y_n| \leqslant \Sigma\, |a_n y_n| \leqslant \max\,\{|y_n|\}\,\Sigma\,|a_n|.$

ii. By contradiction, suppose $\{|y_n|\}$ is not bounded so that it must contain a monotonically increasing subsequence $\{\,|\,y_{n_j}\,|\,\}$. A theorem of Knopp (1951, sec. 41, chap. 9) asserts that for an arbitrary monotone descending sequence ε_n with limit zero, a convergent series $\Sigma\, c_n$ and a divergent series $\Sigma\, d_n\, c_n$, $d_n > 0$, can always be specified such that $c_n = \varepsilon_n d_n$. Therefore, we may choose $a_{n_j} = (1/y_{n_j})\, d_{n_j}$ giving a contradiction of the hypotheses that $\Sigma\, a_n y_n$ is absolutely convergent.

iii. By contradiction, suppose $\Sigma\, |a_n|$ is divergent. Choose $y_n = \text{sign } a_n$ so that $\Sigma\, a_n y_n = \Sigma\, |a_n|$ giving a contradiction of the hypothesis that $\Sigma\, a_n y_n$ is absolutely convergent. QED.

Next we study the continuity of the functional by considering the behavior of $\phi(X)$ in a neighborhood of the given element X. A variety of neighborhoods with different properties and implications is at our disposal. For the present we shall use the neighborhood defined as follows:

DEFINITION. *The element $X+H$ is in a δ-neighborhood of X if*

(7) $$\max\,\{|h_n|\} \leqslant \delta.$$

Observe that this does not require the coordinates of X itself to be bounded but only the boundedness of the coordinates of H. The set of sequences with absolutely bounded coordinates forms a linear subspace of the space of all sequences which is denoted by l_∞. We see that a particular neighborhood is associated with a particular way of measuring the distance between elements of the space. A measure of distance is called a *norm*, and it is denoted by vertical bars. For l_∞ the norm is given by $\|H\|_\infty = \max\,\{|h_n|\}$. We shall have occasion to use other norms and other kinds of neighborhoods in the course of our work.

DEFINITION. $\phi(X)$ *is continuous at the point X, where the functional is defined and finite, if for an arbitrary $\varepsilon > 0$ there is a $\delta > 0$ depending on ε and, possibly, X, such that for all $\|H\|_\infty \leqslant \delta$, $|\phi(X+H) - \phi(X)| \leqslant \varepsilon$.*

This definition closely resembles the one given for functions in a finite dimensional space.

LEMMA 1. *Suppose $\Sigma\,|\,f_n - f_{n+1}\,|$ converges for some X and that every $f_n(x_n, x_{n-1})$ is continuous. Then for all H in a δ-neighborhood of X, we have convergence of $\Sigma\,|\,f_n - f_{n+1}\,|$.*

Proof.

$$\Sigma \,|\, f_n(x_n + h_n,\, x_{n-1} + h_{n-1}) - f_{n+1}(x_{n+1} + h_{n+1},\, x_n + h_n)\,|$$
$$= \Sigma \,|\, f_n(x_n + h_n,\, x_{n-1} + h_{n-1}) - f_n(x_n,\, x_{n-1}) + f_n(x_n,\, x_{n-1})$$
$$- f_{n+1}(x_{n+1} + h_{n+1},\, x_n + h_n) + f_{n+1}(x_{n+1},\, x_n) - f_{n+1}(x_{n+1},\, x_n)\,|\,.$$

By continuity of the *f*'s,

$$\Sigma \,|\, f_n(x_n + h_n,\, x_{n-1} + h_{n-1}) - f_n(x_n,\, x_{n-1})\,| \leqslant \varepsilon/2,$$

and

$$\Sigma \,|\, f_{n+1}(x_{n+1} + h_{n+1},\, x_n + h_n) - f_{n+1}(x_{n+1},\, x_n)\,| \leqslant \varepsilon/2.$$

Also, since the *f*'s are of bounded variation,

$$\Sigma \,|\, f_n(x_n,\, x_{n-1}) - f_{n+1}(x_{n+1},\, x_n)\,| < \infty.$$

Therefore,

$$\Sigma \,|\, f_n(x_n + h_n,\, x_{n-1} + h_{n-1}) - f_{n+1}(x_{n+1} + h_{n+1},\, x_n + h_n)|$$
$$\leqslant \varepsilon + \Sigma \,|\, f_n(x_n,\, x_{n-1}) - f_{n+1}(x_{n+1},\, x_n)\,|\,. \text{ QED.}$$

Observe that the hypotheses of the lemma are too weak to show that $\phi(X)$ is continuous; they are only strong enough to show that if $\phi(X)$ is finite then $\phi(X + H)$ is also finite for all H in a δ-neighborhood of X.

THEOREM 2. *If f_n is continuous and if for an arbitrary $\varepsilon > 0$ there is a bounded sequence $\{\delta_n\}$ such that*

(8) $$|\, f_n(x_n + h_n,\, x_{n-1} + h_{n-1}) - f_n(x_n,\, x_{n-1})\,| \leqslant \varepsilon/\Sigma\, a_n$$

for all $|\, h_n\,| \leqslant \delta_n$, then $\phi(X)$ defined in (1) is continuous at X.

Proof. By lemma 1 the hypotheses ensure the finiteness of $\phi(X + H)$ for all $\|\, H\,\|_\infty \leqslant \delta$.

$$\phi(X + H) - \phi(X) = \Sigma\, a_n[\, f_n(x_n + h_n,\, x_{n-1} + h_{n-1}) - f_n(x_n,\, x_{n-1})]$$

so that

$$|\, \phi(X + H) - \phi(X)\,| \leqslant \Sigma\, a_n\,|\, f_n(x_n + h_n,\, x_{n-1} + h_{n-1}) - f_n(x_n,\, x_{n-1})\,|\,,$$

and, appealing to (8), it follows that

$$|\, \phi(X + H) - \phi(X)\,| \leqslant \Sigma\, \varepsilon\, a_n/\Sigma\, a_n = \varepsilon$$

for all sequences $\{h_n\}$ satisfying $|\, h_n\,| \leqslant \delta_n$, where $\{\delta_n\}$ is bounded. QED.

It is instructive to apply these results to an example. Let

(9) $$\phi(X) = \Sigma\, a_n(x_n^2 - b_n x_n) = \Sigma\, a_n(x_n - b_n/2)^2 - (1/4)\,\Sigma\, a_n b_n^2.$$

Consequently,

$$\phi(X) \geqslant -(1/4)\,\Sigma\, a_n b_n^2.$$
$$\Sigma \,|\, f_n - f_{n+1}\,| = (1/4)\,\Sigma\, |\, b_{n+1}^2 - b_n^2\,|$$

if $x_n = b_n/2$. By theorem 1, it is necessary that b_n be convergent, but it is not necessary that $\Sigma\, b_n^2$ be finite. Moreover, if $|\, b_n\,|$ is bounded then the functional has a finite minimum. For this functional

$$\phi(X + H) - \phi(X) = \Sigma\, a_n[2x_n - b_n + h_n]h_n.$$

Hence it is possible to satisfy (8) and have continuity if, for an arbitrary $\varepsilon > 0$, there is a bounded sequence $\{\delta_n\}$ such that for all $|h_n| \leqslant \delta_n$

$$|[2x_n - b_n + h_n]h_n| \leqslant \varepsilon / \Sigma \, a_n,$$

and this can be satisfied if a bounded $\delta_n > 0$ gives

(10) $\qquad\qquad |2x_n - b_n + \delta_n| \, |\delta_n| \leqslant \varepsilon / \Sigma \, a_n.$

Define the function $g(\delta)$ as follows:

$$g(\delta) = \delta^2 + (2x_n - b_n)\delta - \sigma^2,$$

where, for the sake of brevity, $\sigma = \varepsilon / \Sigma \, a_n$. Plainly, $g(0) = -\sigma^2$ and $g''(\delta) = 2$. Hence $g(\delta) = 0$ has a positive root, $\hat{\delta}$, given by the larger of the two values in the expression as follows:

$$(1/2) \left\{ -(2x_n - b_n) \pm \sqrt{[(2x_n - b_n)^2 + 4\sigma^2]} \right\}.$$

Hence (10) can be satisfied by any $\delta_n > 0$ if $\delta_n < \hat{\delta}$. Moreover, $\hat{\delta}$ is of the same order of magnitude as $|2x_n - b_n| + \sigma$. Hence the given functional is continuous at a point X if $|2x_n - b_n|$ is bounded. This brings to light an interesting point, namely, that the class of admissible x_n's depends on the properties of the functions f_n since for continuity we require the x's to remain close to b_n.[7]

---

7. In example (9), the choice of $x_n = b_n/2$ gives the functional the value of its lower bound, which may be infinite. However, if $\{|b_n|\}$ is bounded and if $\Sigma|a_n| < \infty$, then the value of the minimum is finite. Hence, the choice of $x_n = b_n/2$ gives the minimum value of the functional. We can now illustrate the point referred to in n.5 above that convergence of y_n is not necessary for the existence of a finite value of the functional, provided additional restrictions are imposed on the series $\Sigma \, a_n$. This can be illustrated by an example which has some interest of its own.
Let

$$f(x_n, x_{n-1}) = b_0 x_n^2 + b_1 x_{n-1}^2 + c_0 x_n + c_1 x_{n-1}$$

for which

$$\phi(X) = \Sigma \, a_n[b_0 x_n^2 + b_1 x_{n-1}^2 + c_0 x_n + c_1 x_{n-1}]$$
$$= \Sigma (a_n b_0 + a_{n+1} b_1)x_n^2 + \Sigma (a_n c_0 + a_{n+1} c_1)x_n,$$

and it is assumed that $x_n = 0$, if $n < 0$.
This functional is strictly convex (see the definition in sec. VI.3) if and only if $a_n b_0 + a_{n+1} b_1 > 0$.
Also

$$\phi(X) \geqslant -(1/4) \Sigma (c_0 a_n + c_1 a_{n+1})^2 / (b_0 a_n + b_1 a_{n+1}).$$

Hence,

(i) $\qquad\qquad \phi(X) \geqslant -(1/4) \Sigma \, a_n[c_0 + c_1(a_{n+1}/a_n)]^2 / [b_0 + b_1(a_{n+1}/a_n)].$

According to Hadamard's theorem, this functional has a finite lower bound for any convergent series $\Sigma \, a_n$ if and only if a_{n+1}/a_n is convergent. However, there are certainly some series for which the lower bound is finite although the sequence $\{a_{n+1}/a_n\}$ is not convergent. Thus, suppose the sequence $\{a_n\}$ is given as follows:

$$1, 1, \alpha, \alpha, \alpha^2, \alpha^2, \ldots,$$

where $0 < \alpha < 1$. Then,

$$a_{n+1}/a_n = \begin{cases} 1, & \text{if } n \text{ is odd,} \\ \alpha, & \text{if } n \text{ is even.} \end{cases}$$

The choice of

(ii) $\qquad\qquad x_n = -(a_n c_0 + a_{n+1} c_1)/[2(a_n b_0 + a_{n+1} b_n)]$

secures the value given on the right-hand side of (i), which is, moreover, finite, although x_n given by (ii) is only bounded but is not convergent. This, therefore, illustrates the point that with further restrictions on the discount factors, one may secure a finite value of the functional $\Sigma \, a_n y_n$ although the sequence $\{y_n\}$ is not of bounded variation. The point is that Hadamard's theorem gives necessary and sufficient conditions when $\Sigma \, a_n$ is *any* convergent series, but it does not apply if $\Sigma \, a_n$ is *any absolutely* convergent series.

The assertions of theorems 1 and 2 together yield more information about the family of functions $\{f_n\}$. Theorem 1 asserts that the convergence of f_n to f is necessary for the finiteness of $\phi(X) = (A, Y)$ for arbitrary convergent series Σa_n. Define the functional

$$(11) \qquad F_N(X) = \sum_0^N a_n f_n + \sum_{N+1}^\infty a_n f.$$

For any point where $\phi(X)$ is finite and for an arbitrary $\varepsilon > 0$, there is an N such that

$$|\phi(X) - F_N(X)| \leqslant \sum_{N+1}^\infty a_n | f_n - f | \leqslant \varepsilon.$$

If in addition the f_n's are continuous then also $\phi(X + H)$ is finite for all $\| H \| \leqslant \delta$ according to lemma 1. Therefore, in this neighborhood it must also be true that

$$|\phi(X + H) - F_N(X + H)| \leqslant \varepsilon.$$

It is, therefore, natural to raise the question of whether $F_N(X)$ is itself continuous, and for this to be true it is obviously sufficient that f be continuous, leading us to ask when f is continuous.

We require some additional knowledge of functional analysis. First, there is an extension of the concept of uniform continuity to an infinite family of functions which is called equicontinuity.

DEFINITION. *Let a family of continuous functions be defined on the closed segment* $[c, d]$, *that is,* $c \leqslant x_n \leqslant d$ *so that a δ-neighborhood satisfies* $\delta \leqslant |d - c|$. *If to every $\varepsilon > 0$ there corresponds a $\delta > 0$ such that for* $|x'' - x'| \leqslant \delta$, *where x'' and x' lie in $[c, d]$ for an arbitrary function f_n in the family, we have*

$$| f_n(x'') - f_n(x') | \leqslant \varepsilon,$$

then the functions of the family are called equicontinuous.

DEFINITION. *An infinite set of elements is called compact if every infinite subset contains a convergent sequence.*

Finally, there is the well-known result of functional analysis based on the Arzelá-Ascoli theorem.

THEOREM 3. *A necessary and a sufficient condition that an infinite set of continuous functions defined on the closed interval $[c, d]$ be compact is that the functions are uniformly bounded and equicontinuous.*

For a proof, see Natanson (1961, vol. 2, chap. 17, sec. 3, thm. 4). In our case it is, therefore, necessary to require the family of functions to be equicontinuous, since, if every f_n is continuous on the closed interval $[c, d]$, the functions will automatically be uniformly bounded. This completes the proof of the following corollary.

COROLLARY. *For any convergent series $\Sigma\, a_n$, a necessary condition for the finiteness of $\phi(X+H)$ for all $\| H \|_\infty \leq \delta$, given that all f_n are continuous on the set $X+H$, is that $f_n \to f$, a continuous function, and that $F_N(X)$ defined in (11) be continuous for every N.*

This result is not applicable when the discount factors may be any *absolutely* convergent series.

The next step in our program is to determine when there is a linear approximation to the given functional. The best way to pose this question is to ask when the functional has a derivative. We begin by defining a linear transformation in an infinite dimensional space.

DEFINITION. *P is called a linear transformation from the normed linear space E into the normed linear space E_1 if the following conditions are satisfied:*
  i. for any X in E, $PX = Y$ is in E_1;
  ii. $P(X_1 + X_2) = PX_1 + PX_2$ for any X_1 and X_2 in E;
  iii. $\| PX \| \leq |P| \cdot \| X \|$ where $|P| = \sup \| PX \|$ for $\| X \| = 1$.

The second property is called additivity and the third is called boundedness. The derivative of a functional is a linear transformation as is shown in the following definition.

DEFINITION. *$\phi(X)$ has a derivative at X, denoted by $\phi'(X)$, if*

$$(12) \qquad \phi(X+H) - \phi(X) = \phi'(X) \cdot H + \alpha(X, H),$$

where $\phi'(X)$ is a linear transformation from the linear space containing H into the space of real numbers and

$$(13) \qquad \lim |\alpha(X, H)|/\| H \| = 0 \text{ as } \| H \| \to 0.$$

Observe that in this definition the linear space on which $\phi'(X)$ is defined conforms to the norm of H that defines the neighborhood of X. It is also true in this case that the expression $\phi'(X) \cdot H$ is a bilinear functional. Condition (13) means that given an arbitrary $\varepsilon > 0$, there is a $\delta > 0$ depending on ε and, possibly, X such that for all $\| H \| \leq \delta$

$$(13') \qquad | \phi(X+H) - \phi(X) - \phi'(X) \cdot H \| \leq \varepsilon \| H \|.$$

This is a more convenient analytical form for ascertaining whether a functional is differentiable.

Before giving a general result, let us consider the example (9). In this case, a natural try for $\phi'(X) \cdot H$ is given by

$$\phi'(X) \cdot H = \Sigma\, a_n(2x_n - b_n) \cdot h_n,$$

which implies that

$$\alpha(X, H) = \Sigma\, a_n h_n^2.$$

The derivative exists only if

$$\Sigma\, a_n h_n^2 \leq \varepsilon \cdot \| H \|_\infty, \qquad\qquad \text{for all } \| H \|_\infty \leq \delta.$$

Choose $\delta \leqslant \varepsilon / \Sigma a_n$ so that $\|H\|_\infty \leqslant \delta$ implies

(14) $$\|H\|_\infty \leqslant \varepsilon / \Sigma a_n.$$

Since

$$\Sigma a_n h_n^2 \leqslant \|H\|_\infty^2 \, \Sigma a_n,$$

together with (14) we have established some of the conditions necessary for the existence of a derivative.

In addition we must show that $\phi'(X)$ is a linear transformation on the space containing H. In particular it must be bounded so there must be a finite solution of the problem as follows:

$$\sup |\Sigma a_n(2x_n - b_n)h_n|, \quad \text{subject to } \max \{|h_n|\} = 1.$$

Clearly,

(15) $$|\phi'(X) \cdot H| = |\Sigma a_n(2x_n - b_n)h_n| \leqslant \Sigma a_n |(2x_n - b_n)|$$

for $\|H\| = 1$, and, by choosing $h_n = \pm 1$ according to the sign of $(2x_n - b_n)$, it is possible to attain the maximum. Hence the infinite series

(16) $$\Sigma a_n(2x_n - b_n)$$

must converge absolutely or the derivative does not exist. By the corollary of theorem 1, the series in (16) is convergent for any absolutely convergent series Σa_n if and only if $\{|2x_n - b_n|\}$ is a bounded sequence. Consequently, admissible sequences $\{x_n\}$ must not stray too far from b_n. Thus we encounter again the phenomenon that conditions on the sequence of functions $\{f_n\}$ impose conditions on the sequence of admissible controls $\{x_n\}$.

This example illustrates virtually all of the relevant conditions for the following theorem:

THEOREM 4. $\phi(X)$ *defined in (1) is differentiable at X for any positive element A in l_1 if the following conditions are satisfied:*

   i. *every f_n is continuously differentiable;*

   ii. $\Sigma \left| a_n \dfrac{\partial f_n}{\partial x_n} + a_{n+1} \dfrac{\partial f_{n+1}}{\partial x_n} \right| < \infty;$

   iii. *for any $\varepsilon > 0$ there is a bounded positive sequence $\{\delta_n\}$ such that for all $|h_n| \leqslant \delta_n$, we have*

$$|f_n(x+h) - f_n(x) - df_n(x; h)| \leqslant \varepsilon \max \{|h_n|\},$$

where $df_n(x; h)$ denotes the differential as follows:

(17) $$df_n(x; h) = \frac{\partial f_n}{\partial x_n} h_n + \frac{\partial f_n}{\partial x_{n-1}} h_{n-1}.$$

Proof.
Let

$$\phi'(X) \cdot H = \Sigma \, a_n \left[\frac{\partial f_n}{\partial x_n} h_n + \frac{\partial f_n}{\partial x_{n-1}} h_{n-1} \right]$$

(18)
$$= \Sigma \left[a_n \frac{\partial f_n}{\partial x_n} + a_{n+1} \frac{\partial f_{n+1}}{\partial x_n} \right] h_n.$$

For $\|H\|_\infty = 1$, we may choose $h_n = \pm 1$ according to the sign of the term within the square brackets and thereby conclude that

(19)
$$|\phi'(X)| = \Sigma \left| a_n \frac{\partial f_n}{\partial x_n} + a_{n+1} \frac{\partial f_{n+1}}{\partial x_n} \right|,$$

which is finite by hypothesis. It remains only to show that

$$|\alpha(X, H)| = |\phi(X+H) - \phi(X) - \phi'(X) \cdot H| \leqslant \varepsilon \|H\|_\infty,$$

if $\|H\|_\infty \leqslant \delta$. This follows immediately from iii since

$$\alpha(X, H) = \Sigma \, a_n [f_n(x+h) - f_n(x) - df_n(x; h)]. \text{ QED.}$$

Just as the existence of the first derivative implies the validity of a linear approximation to the given functional in a neighborhood of X, so too the existence of a second derivative can be used to approximate the functional up to quadratic terms. The second derivative is also a linear transformation denoted by $\phi''(X)$ and the second differential of a functional is a bilinear functional denoted by $(\phi''(X) \cdot H, H)$.

To illustrate, for the functional defined by

$$\phi(X) = \Sigma \, a_n f_n(x_n),$$

where every f_n is twice continuously differentiable, the formal expression for the second differential is given by

$$(\phi''(X) \cdot H, H) = \Sigma \, a_n f_n''(x_n) h_n^2.$$

For the functional defined by (1) the linear transformation $\phi''(X)$ can be represented as a matrix with an infinite number of row and columns given as follows:

$$\phi''(X) = \begin{bmatrix} a_0 \frac{\partial^2 f_0}{\partial x_0^2} + a_1 \frac{\partial^2 f_1}{\partial x_0^2} & a_1 \frac{\partial^2 f_1}{\partial x_0 \partial x_1} & 0 \cdots \\[2ex] a_1 \frac{\partial^2 f_1}{\partial x_1 \partial x_0} & a_1 \frac{\partial^2 f_1}{\partial x_1^2} + a_2 \frac{\partial^2 f_2}{\partial x_1^2} & a_2 \frac{\partial^2 f_2}{\partial x_1 \partial x_2} \\[2ex] 0 & a_2 \frac{\partial^2 f_2}{\partial x_2 \partial x_1} & a_2 \frac{\partial^2 f_2}{\partial x_2^2} + a_3 \frac{\partial^2 f_3}{\partial x_2^2} \\[2ex] 0 & 0 & a_3 \frac{\partial^2 f_3}{\partial x_3 \partial x_2} \\ & \cdots & \end{bmatrix}$$

DEFINITION. *The second derivative of the functional $\phi(X)$, denoted by $\phi''(X)$, exists at a point X if*

$$lim \; |\alpha(X, H)|/\|H\|^2 = 0 \; as \; \|H\| \to 0,$$

where

$$\alpha(X, H) = \phi(X+H) - \phi(X) - \phi'(X) \cdot H - (1/2)(\phi''(X) \cdot H, H),$$

and if $(\phi''(X) \cdot H, H)$ is a bounded bilinear functional on the space containing H.

Sufficient conditions for the existence of a second derivative of the functional $\phi(X)$ closely resemble those given in theorem 4 for the first derivative.

THEOREM 5. *$\phi(X)$ defined in (1) has a second derivative at X for any positive A in l_1 if the following conditions are satisfied:*

    i. *every f_n is twice continuously differentiable with respect to its arguments;*

    ii. $\Sigma \left| a_n \dfrac{\partial^2 f_n}{\partial x_n \partial x_{n-1}} + a_n \dfrac{\partial^2 f_n}{\partial x_n^2} + a_{n+1} \dfrac{\partial^2 f_{n+1}}{\partial x_n^2} \right| < \infty;$

    iii. *for any $\varepsilon > 0$ there is a bounded sequence $\{\delta_n\}$ such that for all $|h_n| \leqslant \delta_n$ we have*

$$| f_n(x+h) - f_n(x) - df_n(x; h) - d^2 f_n(x; h) | \leqslant \varepsilon \cdot max \; \{h_n^2\}.$$

Since the proof is only slightly more tedious than the one given for theorem 4 and presents no difficulties, it is omitted.

Theorem 5 completes the analysis of regularity conditions for the functional. Next on our program is the study of necessary and sufficient conditions for the existence of a minimum. We begin with some definitions, proceed to a necessary condition, and finally give some sufficient conditions.

DEFINITION. *A closed set is convex if it contains the segment $\alpha X + (1-\alpha)Y$, $0 \leqslant \alpha \leqslant 1$, whenever it contains the points X and Y.*

DEFINITION. *$\phi(X)$ defined on a closed convex set is said to be convex if*

$$(20) \qquad \phi(\alpha X + (1-\alpha)Y) \leqslant \alpha \phi(X) + (1-\alpha)\phi(Y) \qquad (0 \leqslant \alpha \leqslant 1),$$

for all X and Y in the set. The functional is called strictly convex if there is strict inequality in (20) for all α with $0 < \alpha < 1$.

DEFINITION. *$\phi(X)$ has a greatest lower bound, called the infimum (inf), on a closed set S if for all X in S, there is an m such that $\phi(X) \geqslant m$, and if for an arbitrary $\varepsilon > 0$, there is an element X_0 in S such that $m \leqslant \phi(X_0) \leqslant m+\varepsilon$.*

DEFINITION. *$\phi(X)$ has a minimum on a closed set S if it has an infimum on S that it attains for some X in S.*

Thus, for $\phi(X)$ to have a minimum on S, there must be an X in S such that $\phi(X) = m$, where m is the infimum. Strict convexity is not enough to ensure the existence of an infimum, let alone a minimum. For example, consider the strictly convex functional as follows: $f(x) = -x + (1/x)$ on the closed convex set, $S = \{x/x \geqslant 0\}$. Obviously, this functional has no lower bound on S and, a fortiori, no infimum. The strictly convex functional $g(x) = 1/x$ has an infimum on the given set S, namely zero, but has no minimum. The simple quadratic functional (9) does have a minimum for any absolutely convergent series $\Sigma\, a_n$ provided the sequence $\{|b_n|\}$ is bounded. Before giving a strengthened sufficient condition for the existence of a minimum of a convex functional, let us prove a necessary condition akin to the Euler equation in the calculus of variations.

THEOREM 6. *If the differentiable functional*

$$(21) \qquad \phi(X) = \Sigma\, a_n f_n(x_n, x_{n-1}),$$

defined on the space of sequences, has a minimum at the point X, then the coordinates of $X = \{x_n\}$ must satisfy

$$(22) \qquad a_n \frac{\partial f_n}{\partial x_n} + a_{n+1} \frac{\partial f_{n+1}}{\partial x_n} = 0.$$

Proof. Consider the sequence $\{x_n + \sigma h_n\}$, where σ is a scalar and $X = \{x_n\}$ gives the minimum. Hence, $\phi(X + \sigma H)$ is an ordinary function of the scalar σ which assumes the minimum at $\sigma = 0$. Therefore,

$$\frac{d}{d\sigma} \phi(X + \sigma H) \bigg|_{\sigma = 0} = \Sigma \left(a_n \frac{\partial f_n}{\partial x_n} + a_{n+1} \frac{\partial f_{n+1}}{\partial x_n} \right) h_n = 0$$

for all sequences $\{h_n\}$. In particular this must be true for the sequence $\{h_n\}$ with zero's everywhere except for the nth place where $h_n = 1$. This implies (22). QED.

In general, as in finite dimensional analysis, a necessary condition for a minimum of the differentiable functional $\phi(X)$ is that there exists an X such that

$$(23) \qquad \phi'(X) \cdot H = 0$$

for all H. The typical situation giving rise to an infimum instead of a minimum occurs when no X can satisfy the necessary condition. For example, the function $1/x$, which has an infimum equal to zero on the positive reals, has a necessary condition given by $-1/x^2 = 0$, which cannot be satisfied by any finite x so that $1/x$ has no minimum.

It is natural to wonder whether a sufficient condition like ordinary calculus will work as well as the necessary condition, namely, is

$$(24) \qquad (\phi''(X) \cdot H, H) > 0$$

together with (23) sufficient to guarantee the existence of a minimum provided $\phi(X)$ is convex? Although one may answer this question affirmatively, it is nevertheless possible for strange things to happen in an infinite dimensional space as is shown by the following example. Let $\phi(X) = \Sigma\, a_n f_n(x_n)$ and assume that

$$f_n''(x) > 0 \quad \text{for all} \quad x, \quad \text{and} \quad \Sigma\, a_n f_n''(x_n) < \infty.$$

The necessary condition for a minimum requires x_n to satisfy $f_n'(x_n) = 0$. Assuming X satisfies this condition, it follows that

(25) $$\phi(X+H)-\phi(X) = \Sigma\, a_n f_n''(x_n)h_n^2 + \alpha(X, H),$$

where

$$|\alpha(X, H)| \leqslant \varepsilon\,\|H\|^2 \quad \text{for all} \quad \|H\| \leqslant \delta.$$

Let

$$H_m = \{h_{m.n}\}, \text{ where } h_{m,n} = \begin{cases} \delta, & \text{if } m = n, \\ 0, & \text{if } m \neq n. \end{cases}$$

Hence,

$$\phi(X+H_m)-\phi(X) = a_m f_m''(x_m)\delta^2 + \alpha(X, H_m),$$

and $\|H_m\| = \delta$. Moreover, as $m \to \infty$, $a_m f_m''(x_m) \to 0$ because of the convergence of $\Sigma\, a_n f_n''(x_n)$, which is required by the hypothesis that the functional have a second derivative. Hence although $\|H_m\|$ does not approach zero, it is true that $\phi(X+H_m)$ does approach the minimum $\phi(X)$ as closely as is desired. This is strange.

This reasoning has an interesting economic interpretation. Suppose that $f_n(x_n)$ represents the n-period average cost as a function of the n-period output rate, x_n. To minimize the present value of the average cost, the outputs should satisfy $f_n'(x_n) = 0$. Although it is true that $\{x_n\}$ does minimize the average cost in every period, consideration of neighboring sequences $\{x_n+h_n\}$, constrained only by the condition that the h_n's be bounded, generates alternative solutions giving present values that come arbitrarily close to the minimum, although we measure the distance between optimal policies in such a way that there is a constant difference between the policies. Put differently, although the policies differ by a constant amount δ, the outcomes measured in terms of present values approach arbitrarily close to the minimum. This makes little economic sense, but, fortunately, there are several mathematical devices for bringing the problem closer to the underlying economic considerations, which we now consider.

Suppose that $\phi(X)$ is twice continuously differentiable so that it can be expressed as follows:

(26) $$\phi(X+H) = \phi(X)+\phi'(X)\cdot H+(1/2)(\phi''(X)\cdot H, H)+\alpha(X, H),$$

where for an arbitrary $\varepsilon > 0$, there is a δ depending on ε and possible X such that

$$|\alpha(X, H)| \leqslant \varepsilon\cdot\|H\|^2, \qquad \text{for all} \quad \|H\| \leqslant \delta.$$

Denote the quadratic part by $Q(H; X)$ so that

(27) $$Q(H; X) = \phi(X)+\phi'(X)\cdot H+(1/2)(\phi''(X)\cdot H, H),$$

and

$$\phi(X+H) = Q(H;X) + \alpha(X,H).$$

In the δ-neighborhood of X

$$\phi(X+H) \leqslant Q(H;X) + \varepsilon \cdot \|H\|^2 \leqslant Q(H;X) + \varepsilon \cdot \delta^2.$$

Hence,

$$\min_H \phi(X+H) \leqslant \min_H Q(H;X) + \varepsilon \cdot \delta^2.$$

It follows that the existence of a minimum for the quadratic approximation is necessary for the existence of a minimum of the functional. Let us, therefore, study the quadratic.

Assume that the quadratic does have a minimum denoted by H_0. Then

(28)

$$Q(H_0+\Delta H) - Q(H_0) = (\phi'(X) + \phi''(X) \cdot H_0)\Delta H - (1/2)(\phi''(X)\Delta H, \Delta H) \geqslant 0,$$

for all $\Delta H \neq 0$. But then it is necessarily true that

(29)
$$\phi'(X) + \phi''(X) \cdot H_0 = 0,$$

and, for (28) to hold, it is sufficient to require that

(30)
$$(\phi''(X) \cdot \Delta H, \Delta H) > 0, \qquad \text{for all} \quad \Delta H \neq 0.$$

We summarize these results in the following theorem.

THEOREM 7. *If $\phi(X)$ is a twice continuously differentiable functional having a minimum, then, necessarily, in a δ-neighborhood of an arbitrary X, the quadratic functional $Q(H;X)$ given in (27) must be bounded from below.*

Q will satisfy this condition if it has a minimum in the δ-neighborhood of an arbitrary X. At this minimum, denoted by H_0, it is necessary that H_0 satisfy (29) and that (30) is satisfied. In other words it must be possible to solve (29) for H_0.

This result does give an important clue toward the sought-for sufficient conditions for a minimum since it directs our attention toward the question of when the necessary condition for the minimum of a quadratic functional is solvable. In addition, we require sufficient conditions for the existence of a minimum of a quadratic functional. It turns out that both requirements are met by a single condition, namely, that

(31)
$$(\phi''(X) \cdot H, H) \geqslant m \|H\|^2$$

for some positive constant m. However, we can no longer use the neighborhoods defined by the space of bounded sequences because these would be too large. It is necessary to work with smaller neighborhoods given by the space of square summable sequences.

At this point we can no longer remain in the familiar terrain so much like the finite dimensional space, and our search for sufficient conditions takes a definite turn appropriate to the situation in infinite dimensional spaces. We have seen that strictly convex functionals are not even necessarily bounded from below so that they can hardly have a minimum on every closed convex set. Hence additional restrictions in terms of convexity are required to yield the sought-for sufficient conditions. Intuitively, these restrictions require the gradient of a convex functional (or its support) to rise rapidly enough so that the functional itself cannot fall without limit. A class of convex functionals with the desired property is described in the following definition.

DEFINITION. *A continuous convex functional defined on a closed convex set S in a Banach space is called strongly convex if there is an α inside the closed interval* [0, 1], $0 < \alpha < 1$, *such that*

$$(32) \qquad (1-\alpha)\phi(X) + \alpha\phi(X+H) - \phi(X+\alpha H) \geqslant m \parallel H \parallel^k$$

for some $m > 0$ *and* $k > 0$ *and every* X *and* $X+H$ *in S.*

The following result is proved in Telser and Graves (1971, chap. 4, thm. 3.1).

THEOREM 8. *Every strongly convex functional has a minimum on any closed convex set.*

A twice continuously differentiable convex functional is strongly convex if and only if condition (31) is satisfied. Finally, Polyak (1965) has shown that a twice continuously differentiable strongly convex functional can only be defined in a Hilbert space (for our purposes the space of square summable sequences). (See also Telser and Graves 1971, chap. 4, thm. 3.3.) Hence to apply these results to twice continuously differentiable functionals, we can no longer use neighborhoods defined by bounded sequences, and we must use smaller neighborhoods, namely, those that are defined by the space of square summable sequences.

First, let us see why neighborhoods defined by the space of bounded sequences will not do. We need some preliminary facts about linear spaces.

A square summable sequence is a sequence $\{h_n\}$ such that Σh_n^2 is convergent. The norm for square summable sequences is defined by

$$(33) \qquad \parallel H \parallel_2 = \sqrt{(\Sigma h_n^2)} < \infty.$$

The space of square summable sequences is denoted by l_2. Previously, we have worked with two other linear spaces, the space of absolutely convergent series, $\Sigma |a_n| < \infty$, denoted by l_1, and the space of bounded sequences denoted by l_∞. In fact one can show that[8]

$$l_1 \subset l_2 \subset l_\infty.$$

---

8. Obviously, if $\Sigma|h_n|$ is convergent, then $\{|h_n|\}$ must be bounded and, similarly, if Σh_n^2 is convergent, then h_n^2 must be bounded. Hence both l_1 and l_2 are included in l_∞. It remains to show that $l_1 \subset l_2$. Assume that $\Sigma|h_n|$ is convergent. Then,

$$\Sigma h_n^2 \leqslant \max \{|h_n|\} \Sigma |h_n|.$$

Hence the convergence of $\Sigma | h_n |$ implies the convergence of Σh_n^2, but not conversely, for example, take $h_n = 1/n$.

For example, the bounded sequence $h_n = 1/\sqrt{(n)}$, $n = 1, 2, \ldots$ is not square summable. Every square summable sequence is obviously bounded. Similarly, one can show that every summable sequence must be square summable. However, the square summable sequence $n^{-(1/2+\sigma)}$ is not summable for $0 < \sigma < 1/2$. The theorems on continuity and differentiability remain valid if H is confined to l_2-neighborhoods of a given point X instead of l_∞-neighborhoods.

Now consider the quadratic functional as follows:

$$(34) \qquad \phi(X) = \Sigma \, a_n(x_n^2 - b_n x_n).$$

For this functional to be strongly convex (it is obviously strictly convex), it must be true that

$$(35) \qquad \alpha(1-\alpha) \, \Sigma \, a_n h_n^2 \geqslant m \, \|H\|^k, \quad \text{for some index } k > 0,$$

as is readily verified by applying the definition of strong convexity to (34). The value $\alpha = 1/2$ makes the left-hand side of (35) as large as possible. In the l_∞-neighborhood where $\|H\|_\infty = \max \{|h_n|\}$, one cannot satisfy (35) given that $\Sigma \, a_n$ is any convergent series since we can choose the largest h_n for a large n, where a_n is small and easily violate the required inequality in (35). Moreover, even for l_2-neighborhoods, the functional $\phi(X)$ is not strongly convex as it stands because in such neighborhoods we would require that

$$(36) \qquad \Sigma \, a_n h_n^2 \geqslant 4m \, \|H\|_2^k = 4m \, (\Sigma \, h_n^2)^{k/2}.$$

For (36) to hold a_n must satisfy $a_n > 4m$, where $m > 0$, contradicting the hypothesis that $\Sigma \, |a_n|$ is convergent. This can be verified by noting that (36) must hold for the sequence $\{h_{j,n}\}$ with $h_{j,n} = 1$ if $j = n$ and $h_{j,n} = 0$ if $j \neq n$. However, the quadratic functional would be strongly convex for square summable sequences under the following transformation:

$$(37) \qquad y_n = \sqrt{(a_n)} x_n \quad \text{and} \quad c_n = \sqrt{(a_n)} b_n$$

so that

$$(38) \qquad \phi(Y) = \Sigma \, (y_n^2 - c_n y_n).$$

The functional $\phi(Y)$ is strongly convex for square summable $\{y_n\}$ if and only if

$$\Sigma \, h_n^2 \geqslant 4m \, (\Sigma \, h_n^2)^{k/2}, \qquad\qquad k > 0,$$

which can be satisfied by any m such that $0 < m < 1/4$. The same transformation would not give strong convexity if $\{|h_n|\}$ were merely required to be bounded.

This example illustrates a general principle applying to twice continuously differentiable strongly convex functionals. For such functionals one can always transform the original variables so that the transformed variables become square summable as was done in (37). Observe that (37) imposes no conditions on x_n itself beyond the requirement that the sequence of functions

$\{f_n(x_n)\}$ must be of bounded variation if Σa_n is any convergent series, or, if $\Sigma |a_n|$ is any absolutely convergent series, then $\{|f_n(x_n)|\}$ must be bounded. In the former case $\{x_n\}$ would have to be convergent while in the latter case it would have to be bounded.

If the element H must be in the space of square summable sequences, and if $\phi''(X)$ is a linear transformation defined on this space, then (31) implies that $\phi''(X)$ is invertible. Consequently, not only is it true that (31) is equivalent to strong convexity for twice continuously differentiable convex functionals, but it is also sufficient for ensuring the solvability of the necessary condition for the minimum of the quadratic functional $Q(X, H)$ defined in (27). The point still in need of further clarification is the meaning of solvability for an equation such as (28) and finding necessary and sufficient conditions for solvability.

4. The Solvability of Certain Linear Equations

The demand equation 2(7) is an instance of a linear equation of the type

$$(1) \qquad\qquad A(E)\, x_t = y_t.$$

A related equation is

$$(2) \qquad\qquad A'(L)\, x_t = y_t,$$

where $A'(L)$ denotes the adjoint operator of $A(E)$ so that if

$$(3) \qquad\qquad A(E) = A_0 + A_{-1} E + A_{-2} E^2 + \ldots,$$

then

$$(4) \qquad \begin{aligned} A'(L) &= A_0' + A_{-1}' L + A_{-2}' L^2 + \ldots \\ &= A_0 + A_1 L + A_2 L^2 + \ldots, \end{aligned}$$

where $A_j = A_{-j}$, the transpose of A_{-j}, $A_0 = A_0'$. [The reason for calling $A'(L)$ the adjoint operator will appear shortly.] Finally, the necessary condition for the minimum of a quadratic functional requires the solvability of an equation such as

$$(5) \qquad\qquad [A(E) + A'(L)]\, x_t = y_t.$$

In all these cases the x's and y's are $n \times 1$ column vectors and the operators are $n \times n$ matrixes whose elements are power series in L or E.

Roughly speaking, to solve an equation means to find a sequence of vectors $\{x_t\}$ that satisfies the equation for a given sequence of vectors $\{y_t\}$. Each sequence is regarded as a point in an infinite dimensional space. We shall confine our attention to the space of square summable vectors, which is a Hilbert space.

DEFINITION. *A sequence of vectors $\{x_t\}$ is called square summable if the infinite series*

(6) $$\Sigma\, x_t' x_t$$

is convergent.

The norm of this space is defined by

(7) $$\|X\|_2 = \sqrt{(\Sigma\, x_t' x_t)}.$$

To bring out the analogies with finite dimensional spaces it is desirable to adopt a suggestive notation. Thus an expression such as (6) is denoted by (X, X), where $X = \{x_t\}$ is a point in the Hilbert space so that

(8) $$(X, X) = \Sigma\, x_t' x_t,$$

and $\|X\|_2^2 = (X, X)$. The quadratic part of a quadratic functional is given by $\Sigma\, x_t' A(E)\, x_t$, and there is the identity

(9) $$\Sigma\, x_t'\, A(E)\, x_t \equiv \Sigma\, x_t'\, A'(L) x_t.$$

Let $y_t = A(E)x_t$, and we may write
$$(X, Y) = \Sigma\, x_t' y_t = \Sigma\, x_t'\, A(E) x_t,$$

(10)
$$(Y, X) = \sum y_t' x_t = \Sigma\, x_t'\, A'(L) x_t.$$

More generally,

(11) $$Y = AX,$$

where A denotes a linear transformation, and we may write

(12) $$(X, Y) = (X, AX), \quad (Y, X) = (A'X, X).^9$$

To motivate the study of the solvability of equations such as (3), we now consider in more detail the quadratic approximation of a given twice continuously differentiable functional. Generally,

(13) $$Q(X, H) = \phi(X) + \phi'(X)\cdot H + (1/2)(\phi''(X)\cdot H, H)$$

of which an example is afforded by

(14) $$\Sigma\, a_t b_t' h_t + \Sigma\, a_t h_t' A(E) h_t,$$

where a_t = a scalar t-period discount factor, $a_t > 0$. There is another way to write (14) showing the correspondences with (13).

However, before this can be done, there are compelling reasons for restricting the class of discount factors to those for which the discount rate is the same in every period. Thus for the remainder of this chapter we shall assume that

(15) $$a_t = \alpha^t,$$

---

9. This shows the reason for calling A' the adjoint transformation. In a finite dimensional space the analogous relations are as follows: $(x,y) = x'Ax$, and $(y,x) = x'A'x$.

where $0 < \alpha < 1$ so that $\alpha = 1/(1+\text{discount rate})$. This restriction is necessary for a combination of economic and mathematical requirements.[10] We have seen that reasonably general sufficient conditions for a minimum of a twice continuously differentiable functional require the use of neighborhoods defined for square summable sequences. The economics of the problem, on the other hand, demands consideration of situations in which there may be geometric growth of the control variables. This is met by discounting. Thus the control variables can grow while the discounted control variables can be square summable. With the discount factor of (15), (14) becomes

$$(16) \qquad \Sigma \alpha^t b_t' h_t + \Sigma \alpha^t h_t' A(E) h_t.$$

Now change the variables as in the preceding section. Let $g_t = \alpha^{t/2} h_t$, and $c_t = \alpha^{t/2} b_t$. The sequences $\{g_t\}$ and $\{c_t\}$ are square summable provided their elements do not grow at a rate in excess of the discount rate, that is, $|h_t| < (\alpha + \sigma)^{-t/2}$ for some $\sigma > 0$. The linear operators also undergo changes because of the use of this discount factor. Thus the following identities are readily verified:

$$(17) \qquad \Sigma \alpha^t h_t' A(E) h_t \equiv \Sigma \alpha^t h_t' A'(\alpha^{-1} L) h_t,$$

$$(18) \qquad \Sigma \alpha^t h_t' A(E) h_t \equiv \Sigma \alpha^{t/2} h_t' A(\alpha^{-1/2} E) \alpha^{t/2} h_t,$$

$$(19) \qquad \Sigma \alpha^t h_t' A'(\alpha^{-1} L) h_t \equiv \Sigma \alpha^{t/2} h_t' A'(\alpha^{-1/2} L) h_t.$$

The matrix representing the linear operator $A(E)$ is upper triangular with a countably infinite number of rows and columns as follows:

$$\begin{bmatrix} A_0 & A_{-1} & A_{-2} & \cdots \\ 0 & A_0 & A_{-1} & \cdot\cdot \\ 0 & 0 & A_0 & \cdots \\ & & \cdots & \end{bmatrix}.$$

Similarly, the matrix of the linear operator $A'(L)$ is given by

$$\begin{bmatrix} A_0 & 0 & 0 & \cdots \\ A_1 & A_0 & 0 & \cdot\cdot \\ A_2 & A_1 & A_0 & \cdots \\ & & \cdots & \end{bmatrix}.$$

Also,

$$E(\alpha^{1/2})^t = (\alpha^{1/2})^{t+1} = \alpha^{1/2} \alpha^{t/2},$$
$$L(\alpha^{1/2})^t = (\alpha^{1/2})^{t-1} = \alpha^{-1/2} \alpha^{t/2}.$$

---

10. The compelling mathematical reason for working with constant t-period discount factors is that otherwise we could not ensure the existence of solutions for the necessary conditions of an extremum problem. This is because discount factors $a_n > 0$ such that Σa_n is convergent imply that a_n tends to zero as n approaches infinity. The use of these discount factors is equivalent to transforming the original variable X by a diagonal matrix D whose nonzero diagonal elements are a_n. However, D^{-1} does not exist in this case so that it would not be possible to discover the solution in terms of the original variables, given a unique solution in terms of the transformed (discounted) variables. The same difficulty does not arise with the use of constant discount rates.

With the help of the identities (17) through (19), the quadratic functional (16) becomes the following:

(20) $\qquad \Sigma\, c_t'g_t+(1/2)\,\Sigma\, g_t'[A(\alpha^{-1/2}E)+A'(\alpha^{-1/2}L)]g_t.$

In this form the correspondences with (13) are apparent. Thus,

$\phi'(X)$ corresponds to $\{\alpha^{t/2}b_t\}=\{c_t\},$
$\phi''(X)$ corresponds to $[A(\alpha^{-1/2}E)+A'(\alpha^{-1/2}L)].$

The necessary conditions for a minimum of the quadratic functional Q with respect to H requires the solvability of

$$\phi'(X)+\phi''(X)\cdot H=0.$$

Therefore, in terms of (20) the solution $\{g_t\}$ should be square summable and should satisfy

(21) $\qquad c_t+[A(\alpha^{-1/2}E)+A'(\alpha^{-1/2}L)]g_t=0.$

Moreover, for the quadratic functional to be twice continuously differentiable, it is necessary that the linear operator

$$A(\alpha^{-1/2}E)+A'(\alpha^{-1/2}L)$$

should be bounded. Since α satisfies $0<\alpha<1$ and since the operator is a power series with matrix coefficients, for boundedness it is necessary that

$$A_{-j}\alpha^{-|j/2|}\to0,\qquad\qquad \text{as }|j|\to\infty.$$

Thus the elements of A_{-j} must approach zero at a geometric rate above the discount rate.

Since the concept of square summable solutions of a system of difference equations is unfamiliar in economics, let us consider two examples which are both *scalar* difference equations.

In the first example,

$$A(E)x_t=(1-\beta E)x_t=x_t-\beta x_{t+1}=y_t.$$

Its formal solution is given by

$$\begin{aligned}x_t&=(1-\beta E)^{-1}y_t\\&=(1+\beta E+\beta^2E^2+\ldots)y_t\\&=y_t+\beta y_{t+1}+\beta^2y_{t+2}+\ldots.\end{aligned}$$

Given that $\{y_t\}$ is square summable, we seek necessary and sufficient conditions in terms of β so that $\{x_t\}$ will be square summable.

$$\begin{aligned}\sum_0^T x_t^2&=\sum_0^T(\sum\beta^jy_{t+j})^2=\sum_0^T\sum_{j,\,j'}\beta^j\beta^{j'}y_{t+j}y_{t+j'}\\&=\sum_{j,\,j'}\beta^j\beta^{j'}\sum_0^Ty_{t+j}y_{t+j'}\leqslant\sum_{j,\,j'}\beta^j\beta^{j'}\sum y_t^2\\&\leqslant\sum\beta^{2j}\sum y_t^2,\qquad\qquad\text{for all }T.\end{aligned}$$

The Cauchy-Schwartz inequality is used twice in the last two lines.[11] By hypothesis $\Sigma\, y_t^2$ is convergent. Hence, $\Sigma\, x_t^2$ is convergent if and only if $\Sigma\, \beta^{2j}$ is convergent. This is true if and only if $\beta^2 < 1$. Hence the solution is square summable if and only if the difference equation is stable. So far there are no surprises.

The second example is

$$[A(E)+A(L)]x_t = [(1-\beta E)+(1-\beta L)]x_t = y_t.$$

In this case the characteristic equation of the linear operator is given by

$$1-\beta z+1-\beta z^{-1} = 0.$$

The complex scalar z replaces E and z^{-1} replaces L. This equation is clearly unstable since if r is a root then also r^{-1} is a root, and either both roots are of modulus one or the modulus of at least one of them exceeds one. To see if there is a square summable solution we proceed as follows:

$$\begin{aligned}
(1-\beta z)+(1-\beta z^{-1}) &= -\beta z^{-1}[1-(2/\beta)z+z^2]\\
&= -\beta z^{-1}(r-z)(r^{-1}-z)\\
&= \beta r^{-1}(1-rz^{-1})(1-rz).
\end{aligned}$$

Since

$$[(r-z)(r^{-1}-z)]^{-1} = (r^{-1}-r)^{-1}\{[1/(r-z)]-[1/(r^{-1}-z)]\},$$

after some algebraic reductions we find that

$$\begin{aligned}
\beta^{-1}[(1-\beta z)+(1-\beta z^{-1})]^{-1} &= r[(1-rz^{-1})(1-rz)]^{-1}\\
&= (r^{-1}-r)^{-1}[(1-rz^{-1})^{-1}+rz(1-rz)^{-1}]\\
&= (r^{-1}-r)^{-1}[\ldots+rz^{-1}+1+rz+\ldots].
\end{aligned}$$

We may write the formal inverse of $A(E)+A'(L)$ in a power series as follows:

$$[A(E)+A'(L)]^{-1} = \sum_1^\infty \gamma_j L^j + \gamma_0 + \sum_1^\infty \gamma_{-j} E^j.$$

Therefore, the formal solution may be expressed as follows:

$$x_t = \sum_{-\infty}^\infty \gamma_{-j} y_{t+j},$$

$$x_t^2 = \sum_{j,\,j'} \gamma_{-j}\gamma_{-j'} y_{t+j} y_{t+j'},$$

$$\sum_0^T x_t^2 = \sum_0^T \sum_{j,\,j'} \gamma_{-j}\gamma_{-j'} y_{t+j} y_{t+j'} = \left(\sum_{j,\,j'} \gamma_{-j}\gamma_{-j'}\right)\left(\sum_0^T y_{t+j} y_{t+j'}\right).$$

---

11. The Cauchy-Schwartz inequality asserts that $\Sigma|a_n b_n| \leqslant \sqrt{(\Sigma\, a_n^2)}\sqrt{(\Sigma\, b_n^2)}$. In the norm notation this becomes $|(X,Y)| \leqslant \|X\|\cdot\|Y\|$ for any square summable X and Y (Halmos 1957, sec. 5 thm. 2; Natanson 1961, chap. 7, sec. 1).

Therefore,

$$\sum_0^T x_t^2 \leqslant \sum \gamma_{-j}^2 \sum y_t^2,$$

where in the last step we appeal to the Cauchy-Schwartz inequality. Since $\sum y_t^2$ is convergent by hypothesis, $\sum \gamma_{-j}^2$ is convergent if and only if the series given by

$$+\ldots+r^4+r^2+1+r^2+r^4+\ldots$$

is convergent because

$$\gamma_j = (r^{-1}-r)^{-1} r^{|j|}.$$

Therefore, the solution of the given equation is square summable if and only if $r^2 < 1$. This is equivalent to saying that the roots of the characteristic equation should not have modulus one.

In general, for equations like (1), (2), and (3), the solvability also depends on the location of the roots of an appropriate generalization of the characteristic function for the scalar case. Before proving the pertinent theorem, let us continue our inquiry into the question of invertibility for linear operators in a Hilbert space.

In ordinary arithmetic the number 0 is not invertible because the equation

(22) $$0x = 0$$

has no unique solution. Thus any real number x can satisfy (22). Similarly, in n-dimensional space the homogeneous equation

(23) $$Ax = 0,$$

where A is an $n \times n$ matrix of real scalars, is not solvable if A is singular. A is said to be singular if and only if there is an $x \neq 0$ that can satisfy (23). When this is true, like ordinary arithmetic, (23) cannot have a unique solution because any vector of the form σx, σ a nonzero scalar, can satisfy (23) if x satisfies (23). Although there is a wider range of possibilities in a Hilbert space, the same basic idea holds. This is shown in theorem 9.

THEOREM. 9. *A linear operator A defined on the space of square summable vectors is invertible if there is an m > 0 such that for all X in the space,*

(24) $$\|AX\| \geqslant m \|X\|.$$

Proof. This assertion follows immediately from the results given in Halmos (1957, sec. 21; see also thm. 12 below). QED.

(From now on only square summable sequences appear so that the subscript "2" on the norm is understood).

This result aids our intuitive understanding of the condition $(\phi''(X) \cdot H, H) \geqslant m \|H\|^2$, for some $m > 0$. Thus let B be a self-adjoint operator corresponding to $\phi''(X)$, where $B = A + A'$. By the Cauchy-Schwartz inequality,

$$|(BH, H)| \leqslant \|BH\| \|H\|.$$

Hence,

$$(BH, H) \geqslant m \|H\|^2 \quad \text{implies} \quad \|BH\| \geqslant m \|H\|.$$

Therefore, appealing once more to the results in Halmos (1957, sec. 21), we conclude that B is invertible. There is also the slight additional complication of showing that the range of B is dense in the space, which can be taken for granted with operators on the space of square summable sequences of vectors, such as are considered here. This proves theorem 10.

THEOREM 10. *Let B be a self-adjoint operator defined on the space of square summable vectors $H = \{h_t\}$. If there is an $m > 0$ such that $(BH, H) \geqslant m$ $\|H\|^2$, then B is invertible.*

We need to translate these invertibility results into concrete terms applicable to our problems. This can be done with the help of generating functions. Let

$$(25) \qquad\qquad Y(z) = \sum_{-\infty}^{\infty} z^t y_t$$

denote the generating function of the sequence of vectors $\{y_t\}$, where, as usual, z is a complex scalar. Similarly, the generating function of the sequence $\{x_t\}$ is defined by

$$(26) \qquad\qquad X(z) = \sum_{-\infty}^{+\infty} z^t x_t.$$

If the sequences are nonzero for only the nonnegative subscripts then we may write

$$X^+(z) = \sum_{0}^{\infty} z^t x_t, \quad Y^+(z) = \sum_{0}^{\infty} z^t y_t.$$

Let

$$A(z) = \sum_{0}^{\infty} A_{-j} z^j.$$

Hence,

$$A'(z) = \sum_{0}^{\infty} A'_{-j} z^j = \sum A_j z^j,$$

recalling that $A'_{-j} = A_j$.

The equation

$$(27) \qquad\qquad y_t = A(E)x_t = \sum_{0}^{\infty} A_{-j} x_{t+j}$$

is a special kind of multiplication known as a convolution for which generating functions are especially helpful. Similarly,

$$y_t = [A(E) + A'(L)]x_t$$

$$= \sum_{0}^{\infty} A_{-j} x_{t+j} + \sum_{0}^{\infty} A_j x_{t-j}$$

$$(28) \qquad\qquad = \sum_{-\infty}^{\infty} A_j x_{t-j}.$$

We shall derive the invertibility conditions for the operators in terms of the location of the roots of their generating functions. The results are given in theorem 11.

THEOREM 11. *The operator $A(E)+A'(L)$ is invertible if and only if the matrix $A(z)+A'(z^{-1})$ is nonsingular for all $|z|=1$. The operator $A(E)$ is invertible if and only if the matrix $A(z)$ is nonsingular for all $|z| \leqslant 1$.*

Proof. We outline only the main ideas.

We may rewrite (28) so that

$$y_t = \sum_{-\infty}^{\infty} A_{t-j} x_j.$$

(29)
$$\sum_t z^t y_t = \sum_t \sum_j A_{t-j} z^{t-j} z^j x_j$$

$$= \sum_t A_{t-j} z^{t-j} \sum_j z^j x_j$$

$$= [A(z)+A'(z^{-1})]X(z),$$

or

(30)
$$Y(z) = [A(z)+A'(z^{-1})]X(z).$$

Some deeper knowledge of complex analysis would be needed to complete a rigorous proof. Thus it can be shown that $\{x_t\}$ is square summable for a given square summable $\{y_t\}$, where the relation (30) holds, if and only if the Hermitian matrix $A(z)+A'(\bar{z})$ is nonsingular on the unit circle, where $|z|=1$ and $z^{-1} = \bar{z}$ (the complex conjugate of z). For further details, see Widom (1957, secs. 3a and 3b).

A similar line of reasoning leads to a relation like (30) in proving the second assertion of the theorem, where it is convenient to adopt the convention that the generating function for $A(E)$ is extended to the positive integers j by setting $A_j = 0$, if $j > 0$. If this is done, we obtain $Y(z) = A(z)X(z)$, and, more exactly, $X^+(z) = A(z)^{-1}Y^+(z)$. For the x's to be square summable, $A(z)$ must be nonsingular for all $|z| \geqslant 1$. QED.[12]

This completes the analysis necessary to study the properties of the collusive and noncooperative equilibria under dynamic market conditions, the topic of the next section.

---

12. Starting from (30) in the text, it is shown that

$$\int_{|z|=1} X'(z) X(z) \, dz < \infty \quad \text{if and only if}$$

$$\int |A(z)+A'(z)|^{-1} |Y(z)| \, dz < \infty$$

given that

$$|Y(z)| = \sqrt{[\int_{|z|=1} Y(z)' \, Y(z) \, dz]} < \infty$$

5. Properties of the Cooperative and the Noncooperative Dynamic Equilibria

We can now resume our study of the nature of the dynamic equilibrium begun in section 2. A linear approximation of the demand for the n competing goods is given in 2(7), which becomes in terms of p_t the following:

(1) $$s_t = b_t - A(E)(I - GE)p_t.$$

The stocks and purchase rates are related by

(2) $$(I - GL)s_t = q_t + d_t,$$

where

$$d_t = \begin{cases} Gs_{-1}, & \text{if } t = 0, \\ 0, & \text{if } t > 0. \end{cases}$$

G is a diagonal matrix whose diagonal entries γ_i satisfy $0 \leqslant \gamma_i \leqslant 1$. If h denotes the calendar length of a period, then also

(3) $$\gamma_i = e^{-\rho_i h}$$

so that ρ_i denotes the corresponding instantaneous rate of consumption.

The firms determine the present value of their net revenue by their choice of the sequence of prices. If they cooperate, then the sequence $\{p_t\}$ is chosen to maximize

(4) $$\Sigma \, \alpha^t (p_t' q_t - c_t' q_t),$$

which is the present value of their joint net return and $c_t = n \times 1$ vector denoting the t-period unit costs. Hence (4) assumes that total cost is proportional to output so that the t-period average and marginal costs are equal. The subscript of c_t allows for the possibility that costs may change over time. If the firms compete (do not cooperate), then each chooses its price sequence to maximize the present value of its net return given its expectations about its rivals' price policies.

At the outset assume that $G = 0$. Not only does this simplify the analysis, but also it brings out an important economic fact. The choice of $G = 0$ can represent the effect of choosing periods of long calendar length. We shall see that if the conditions for a maximum are satisfied for $G = 0$ then they are also satisfied if $0 \leqslant \gamma_i \leqslant 1$. Therefore, the existence of an equilibrium for long time periods implies the same is true for all shorter periods.

We now wish to alter the form of (4) and apply the results of the preceding two sections. Let

(5) $$x_t = \alpha^{t/2} p_t.$$

It is assumed that the discount factor α is high enough to make the sequence

$\{x_t\}$ square summable. Starting with (4) and assuming $G = 0$ so that $q_t = s_t$ we obtain

$$
\begin{aligned}
PV &= \Sigma \, \alpha^t \{p'_t b_t - p'_t A(E)p_t - c'_t [b_t - A(E)p_t]\} \\
&= \Sigma \, \alpha^t [p'_t b_t - p'_t A(E)p_t + c'_t A(E)p_t] + \Sigma \, \alpha^t c'_t b_t \\
&= \Sigma \, \alpha^t [p'_t b_t - p'_t A(E)p_t + p'_t A'(\alpha^{-1}L)c_t] + \Sigma \, \alpha^t c'_t b_t \\
&= \Sigma \, \alpha^t [p'_t k_t - p'_t A(E)p_t] + \Sigma \, \alpha^t c'_t b_t,
\end{aligned}
$$

(6)

where

(7) $$k_t = b_t + A'(\alpha^{-1}L)c_t.$$

To obtain (6), we also use 4(17). Therefore, dropping the constant term $\Sigma \, \alpha^t c'_t b_t$,

$$
\begin{aligned}
PV &= \Sigma \, \alpha^{t/2} p'_t [\alpha^{t/2} k_t - A(\alpha^{-1/2}E)\alpha^{t/2}p_t] \\
&= \Sigma \, x'_t \{g_t - (1/2)[A(\alpha^{-1/2}E) + A'(\alpha^{-1/2}L)]x_t\}.
\end{aligned}
$$

(8)

This result relies on 4(17) to 4(20). Also,

(9) $$g_t = \alpha^{t/2} k_t$$

is assumed to be square summable.

For a maximum to exist, it is necessary that a square summable $\{x_t\}$ satisfies the following equation, as implied by the necessary condition, $\phi'(X) = 0$:

(10) $$g_t - [A(\alpha^{-1/2}E) + A'(\alpha^{-1/2}L)]x_t = 0.$$

A sufficient condition for a maximum is that for all square summable sequences $\{x_t\}$, there is an $m > 0$ such that

(11) $$\Sigma \, x'_t [A(\alpha^{-1/2}E) + A'(\alpha^{-1/2}L)]x_t \geq m \, \Sigma \, x'_t x_t.$$

Hence the sequence of equilibrium prices for monopoly (collusion) must satisfy (10). Moreover, by theorem 10, (11) is both sufficient for the existence of a maximum and it implies the solvability of the necessary condition. This is true because the self-adjoint linear operator B as follows,

(12) $$B = A(\alpha^{-1/2}E) + A'(\alpha^{-1/2}L)$$

satisfies the hypotheses of theorem 10 so that (11) is necessary and sufficient for a global maximum. We now derive sufficient conditions for the existence of the Cournot-Nash equilibrium. We do this in two steps. First, we shall assume that every firm knows all of the current and future prices but still ignores the repercussions of its actions on its rivals. Second, while retaining the latter part of this assumption, we assume that instead of knowing current and future prices the firms choose their current actions on the basis of expectations derived from past prices.

Consider firm 1 for whom the demand equation is given by

(13) $$q_{1t} = b_{1t} - [a_{11}(E)p_{1t} - \sum_j a_{ij}(E)p_{jt}].$$

This uses the representation of $A(E)$ shown in 2(12), namely,
$$A(E) = A_d(E) - A_v(E).$$

The present value of firm 1's net revenue is defined by

(14) $$PV_1 = \Sigma \, \alpha^t (p_{1t} - c_{1t}) q_{1t},$$

in which all of the components are now scalars. Substituting from (13) we obtain

$$PV_1 = \Sigma \, \alpha^t [p_{1t} - c_{1t}][b_{1t} - a_{11}(E)p_{1t} + \sum_j a_{1j}(E)p_{jt}]$$

(15) $$= \Sigma \, \alpha^t p_{1t}[b_{1t} - a_{11}(E)p_{1t} + a_{11}(\alpha^{-1}L)c_{1t} + \sum_j a_{1j}(E)p_{jt}]$$

$$-\Sigma \, \alpha^t [c_{1t} \sum_j a_{1j}(E)p_{jt} - c_{1t}b_{1t}].$$

Since firm 1's control variable is p_{1t} and under the Cournot-Nash equilibrium the prices of all other firms are regarded as given, in deriving the necessary conditions for a maximum PV_1, we may ignore all terms not involving p_{1t}. Hence the necessary condition for maximum PV_1 is given as follows:

(16) $$g_{1t} - [a_{11}(\alpha^{-1/2}E) + a_{11}(\alpha^{-1/2}L)]x_{1t} + \sum_j a_{1j}(\alpha^{-1/2}E)x_{jt} = 0,$$

using the same substitutions as in the development of (6) through (10). The equilibrium path of the solution for all firms in matrix form is given by

(17) $$g_t - [A_d(\alpha^{-1/2}E) + A_d'(\alpha^{-1/2}L)]x_t + A_v(\alpha^{-1/2}E)x_t = 0,$$

or, equivalently,

(18) $$g_t - [A(\alpha^{-1/2}E) + A_d'(\alpha^{-1/2}L)]x_t = 0.$$

[It is instructive to compare this expression with the corresponding one for the static case given in IV.3(11).] The existence of a Cournot-Nash equilibrium depends on whether (18) is solvable. That is, is the linear operator
$$A(\alpha^{-1/2}E) + A_d'(\alpha^{-1/2}L)$$
invertible? To answer this question and others concerning the properties of the demand equation as promised in section 2, we shall study in more detail the monopoly equilibrium.

Let A denote $A(\alpha^{-1/2}E)$. Hence the self-adjoint linear operator B can be concisely expressed as follows:

(19) $$B = A + A', \quad B_d = A_d' + A_d, \quad B_v = A_v' + A_v, \quad B = B_d - B_v.$$

We can take it for granted in all of this work that the range of the operators is dense in the space of square summable sequences. Hence, according to Halmos (1957, thm. 3, sec. 21) an operator C in a Hilbert space is invertible if and only if there exists a positive real number δ such that

(20) $$\|CX\| \geqslant \delta \|X\|$$

for every point X in the Hilbert space.

The first result asserts that if there is a monopoly equilibrium then the demand operator A must be invertible.[13]

THEOREM 12. *If the self-adjoint linear operator B is positive definite, then*

$$(21) \qquad \|AX\| \geqslant \delta \|X\|$$

for some $\delta > 0$ and for every square summable sequence $X = \{x_t\}$.

Proof. By hypothesis the operator B satisfies $(X, BX) \geqslant \beta (X, X)$ for some $\beta > 0$ and for all X. By theorem 10 this shows that B is invertible. We proceed by showing that the hypothesis that A is not invertible gives a contradiction. Suppose A is not invertible. Then there is a sequence $\{\varepsilon_n\}$, $\varepsilon_n > 0$ and $\varepsilon_n \to 0$ such that for each ε_n, there is an $X_n \neq 0$ which satisfies

$$\|AX_n\| < \varepsilon_n \|X_n\|.$$

$$(22) \quad \|(A+A')X_n - AX_n\| \geqslant \|(A+A')X_n\| - \|AX_n\| > \|(A+A')X_n\| - \varepsilon_n \|X_n\|,$$

but

$$\|(A+A')X_n - AX_n\| = \|A'X_n\| = \|AX_n\|.$$

Therefore, (22) implies that

$$\|AX_n\| > \|(A+A')X_n\| - \varepsilon_n \|X_n\|$$

so that

$$(23) \qquad \beta \|X_n\| \leqslant \|(A+A')X_n\| < 2\varepsilon_n \|X_n\|.$$

Since β is a fixed positive number, $X_n \neq 0$ so that $\|X_n\| \neq 0$, and $B = A + A'$, (23) gives the desired contradiction. QED.

The next series of results establishes the fact that the existence of a monopoly equilibrium implies the existence of a Cournot-Nash equilibrium.

LEMMA 2. *If B is positive definite, then B_d is positive definite.*

Proof. The argument in this case is the same as in the finite dimensional case provided we choose a basis with basis vectors of the form

$$X_j = \{x_{jt}\}, \quad \text{and} \quad x_{jt} = \begin{cases} 1, & \text{if } j = t, \\ 0, & \text{if } j \neq t. \end{cases}$$

The details, which are straightforward, may well be omitted. QED.

LEMMA 3. *If B is positive definite, then $B + B_d$ is positive definite.*

Proof. By the preceding result the positive definiteness of B implies that of B_d. But the sum of two positive definite linear operators is positive definite. QED.

---

13. The invertibility of A, which is equivalent to the stability of the linear operator $A(E)$, is a necessary but not a sufficient condition for the existence of a monopoly equilibrium. Thus $\|AX\| \geqslant m\|X\|$ for some $m > 0$ and for all square summable X does not imply that $A' + A$ is invertible, and a fortiori it does not imply that $A' + A$ is positive definite. An example of this is given by a second order scalar difference equation in Telser and Graves (1971, chap. 1).

We can now show that the positive definiteness of B, which is necessary and sufficient for the existence of the monopoly equilibrium, implies the existence of a Cournot-Nash equilibrium.

THEOREM 13. *If B is positive definite, then $A + A'_d$ is invertible.*

Proof. By lemma 3 it follows that $B + B_d$ is positive definite. Let $C = A + A'_d$ and $C' = A' + A_d$. Therefore, $C + C' = B + B_d$ so that $C + C'$ is a positive definite operator and, consequently, invertible. Theorem 12 gives the desired conclusion. QED.

These results give a few of the many implications following from the existence of a monopoly equilibrium for a set of n substitutes. We now introduce expectations. To prepare the way, let us rewrite the equilibrium conditions for the monopoly (10) and the Cournot-Nash model (18) more concisely. The policy variables in the monopoly equilibrium satisfy

$$(24) \qquad\qquad (A + A')X = G,$$

where $G = \{g_t\}$ is a sequence of square summable vectors. The Cournot-Nash equilibrium satisfies

$$(25) \qquad\qquad (A_d + A'_d - A_v)Y = G,$$

where Y is the square summable element consisting of the discounted prices. Observe that both X and Y are unique because the linear operators $(A + A')$ and $(A_d + A'_d - A_v)$ are both invertible.

These relations show that future values of g_t determine the solution because the inverses of the operators $(A + A')$ and $(A_d + A'_d - A_v)$ involve powers of the lead operator E. Nor is that all. In the Cournot-Nash equilibrium the rivals do not know future values of each others' prices. Hence they chose their current prices using forecasts of future competing prices. It is, therefore, necessary to describe the nature of these forecasts, which means to have a model of expectations. As in chapter IV, we shall assume that expected values of unknown variables depend on the past values of the same variable. To illustrate, consider expected prices. Let expected prices depend on past prices so that

$$p_t^* = w_1 p_{t-1} + w_2 p_{t-2} + w_3 p_{t-3} + \dots$$
$$(26) \qquad\qquad\quad = w(L)p_{t-1},$$

where p_t^* denotes the expected price and $w(L)$ is a power series in L defined by

$$(27) \qquad\qquad w(L) = w_1 + w_2 L + \dots$$

(see sec. IV.3). Since the demand equations contain $A(E)$, which means that future prices enter, we shall assume that the firms replace any unknown future price with a forecast depending on past prices given by (26). This procedure generates a complete sequence of forecasts. In addition it requires the invertibility of the operator $w(L)$. Hence, assuming invertibility, (26) implies

$$(28) \qquad w(L)^{-1} p_t^* = p_{t-1}, \quad w(L)^{-1} p_{t+1}^* = p_t^* = w(L) p_{t-1}.$$

Therefore,

$$p_{t+1}^* = w(L)^2 p_{t-1},$$

and

$$w(L)^{-1} p_{t+2}^* = p_{t+1}^* = w(L)^2 p_{t-1}$$

implies

$$p_{t+2}^* = w(L)^3 p_{t-1}.$$

The general formula is given by

(29) $$p_{t+k}^* = w(L)^{k+1} p_{t-1} \qquad (k = 0, 1, \ldots).$$

A similar argument applies to the components of $G = \{g_t\}$.

Let $p_{ij,\,t+k}^* =$ firm i's forecast of firm j's price at time t for time $t+k$, $k > 0$.

(30) $$p_{ij,\,t}^* = w_{ij}(L) p_{j,\,t-1}.$$

According to this formulation, although every firm uses the same information from the past in making its forecast, it does not necessarily give the same weight to a past price, $p_{j,t-k}$. Hence every firm has its individual forecast of competing prices. Since the corollary of theorem 3 asserts that the convergence of g_t is necessary for an equilibrium, we postulate rational expectations in the sense that if

$$p_{jt} \to p_j \quad \text{as} \quad t \to \infty,$$

then

$$p_{ij,\,t}^* \to p_j \quad \text{as} \quad t \to \infty.$$

This condition will be satisfied if

(31) $$w_{ij}(1) = 1, \qquad \text{for all } i, j = 1, \ldots, n.$$

So as to embrace the three expectations models in chapter IV, let the coefficients of the power series $w_{ij}(L)$ be nonnegative. We denote this by

(32) $$w_{ij}(L) \geqslant 0.^{14}$$

In the Cournot model of expectations, $w_1 = 1$ and $w_j = 0$ if $j > 1$ so that $w_{ij}(L) \equiv 1$. In adaptive expectations

$$w_{ij}(L) = (1 - \delta_{ij})/(1 - \delta_{ij} L) \qquad (0 \leqslant \delta_{ij} < 1).$$

In arithmetic mean expectations, all past prices receive the same positive weight; arithmetic mean expectations is the limiting case of adaptive expec-

---

14. A possible ambiguity arises because an operator may have a positive definite quadratic form such as, for example, the self-adjoint operator $A(E) + A'(L)$ so that

$$\Sigma x_t'[A(E) + A'(L)] x_t \geqslant m \Sigma x_t x_t \qquad (m > 0),$$

or the operator itself may be positive, meaning that it maps any positive element X into a positive element Y. For the latter property, the coefficients of the operator must be nonnegative (see n.4). Thus, (32) refers to the latter interpretation.

tations as $\delta_{ij} \to 1$. Also, observe that Cournot expectations are included in adaptive expectations by setting $\delta_{ij} = 0$.

In terms of price expectations there is a perceived demand equation for each firm's product. The term "perceived" indicates that a firm responds not to its rivals' current prices but to its forecasts of these prices. Since each firm chooses its own price, the diagonal terms of the operator in the perceived demand equations remain as before, namely, $A_d(E)$; but in place of the off-diagonal matrix $A_v(E)$ we have

(33) $$C_v = [a_{ij}(w_{ij}(L))w_{ij}(L)L].$$

Thus, in the perceived demand equations the linear operators $w_{ij}(L)$ replace E. We may write the system of perceived demand equations as follows:

(34) $$q = b - (A_d - C_v)\, p.$$

As in 2(13) we assume that the n goods are gross substitutes so that $A_v(E) \geqslant 0$. Together with (32) this implies $C_v \geqslant 0$.

The link between expectations and the equilibrium conditions relies on a certain fact about positive definite self-adjoint linear operators of the type $A(E) + A'(L)$ which is given as follows:

LEMMA 4. *A positive definite self-adjoint linear transformation defined on the space of square summable sequences of n-vectors $\{x_t\}$ can be represented as the product of two invertible linear transformations $M\,M'$ such that $B = M\,M'$. In particular*

(35) $$A(E) + A'(L) = M(E)M'(L),$$

where $M(E)$, and, therefore, $M'(L)$ are invertible.

A proof would be beyond the scope of this study. See Helson (1964, lecture 11). A constructive proof for polynomial matrixes is given in Telser and Graves (1971, chap. 1, thm. 8.5). The essential idea of a proof is contained in the example above on pp. 246–247.

Using lemma 4 we can rewrite the equilibrium condition for the monopoly solution (24) as follows: $M(E)M'(L)x_t = g_t$, so that

(36) $$M'(L)x_t = M(E)^{-1}g_t = M(w(L))^{-1}w(L)g_{t-1},$$

where $w(L)$ denotes the forecasting operators pertinent to the sequence $\{g_t\}$. If the firms collude, then they do not forecast each others' prices, presumably because these are mutually agreed upon, but future values of g_t are unknown and must be forecast as is shown in (36). Hence the monopoly or collusive solution becomes a function of current and past observable variables. A monopoly equilibrium exists if the sequence on the right-hand side of (36) is square summable. This follows from the invertibility of $M(E)$ and the postulated properties of the forecasting operators.

To find sufficient conditions for the existence of a Cournot-Nash equilibrium when firms respond to expected prices we shall proceed with an argument analogous to the one given in section IV.3. The pertinent operator for the Cournot-Nash equilibrium is given by

$$B_d - C_v = B_d(I - B_d^{-1}C_v)$$

[see (19)]. Since B is positive definite, so is B_d, which is, therefore, invertible. Hence $B_d - C_v$ will be invertible if the following inverse

(37) $$(I - B_d^{-1}C_v)^{-1}$$

exists. This suggests that we should seek conditions that would permit the expansion of the operator shown in (37) as a power series in powers of $B_d^{-1}C_v$.

First there is a general result called the Neumann series theorem.

NEUMANN SERIES THEOREM. *If T is a linear operator with $\|T\| < 1$, then $(I - T)^{-1}$ exists and $\|(I - T)^{-1}\| \leq 1/(1 - \|T\|)$. Further, if T is defined on the space of square summable sequences, then for any square summable Y satisfying $(I - T) \cdot X = Y$, there is a unique square summable X given by $X = (I + T + T^2 + ...)Y$ (Liusternik and Sobolev 1961, sec. 20, thm. 2).*

Recall that the norm of the linear transformation T is given by $\|T\| = \max \sqrt{(TX, TX)}$ subject to $(X, X) = 1$. Hence it is necessary to determine the value of $\|T\|$ for the linear operators relevant to our work. Fortunately, this can be done by studying the proper values of certain $n \times n$ Hermitian matrices as we shall see.

We can study whether the inverse shown in (37) exists by investigating the properties of the associated generating function which is an $n \times n$ matrix. With perfect foresight the generating function of the Cournot-Nash linear operator is given by

$$A_d(z) + A_d'(z^{-1}) - A_v(z),$$

which is derived from the linear operator by substituting z for E and z^{-1} for L. Similarly, the generating function with expectations is given by

$$A_d(z) + A_d'(z^{-1}) - A_v(w(z^{-1}))w(z^{-1}).$$

It turns out that we only have to consider the properties of the generating function on the unit circle, where $|z| = 1$. Moreover, for these values of z we can determine the bound of the operator from the largest proper value of the matrix generating function on the unit circle. These assertions are consequences of the following lemma, which is a special case of the Herglotz-Bochner theorem:

LEMMA 5. *If the $n \times n$ Hermitian matrix as follows:*

   i. $A(z) + A'(z) - \beta I$, $\beta > 0$ is positive definite for all $|z| = 1$, then
   ii. $\Sigma x_t'[A(E) + A'(L)]x_t \geq \beta \Sigma x_t'x_t$

for all square summable real n-vectors, x_t. For complex n-vectors, x_t, conditions i and ii are equivalent.

For a proof when $A(E)$ is a polynomial so that it has only a finite number of powers of E, see Telser and Graves (1971, chap. 1, thm. 8.2). The Herglotz-Bochner theorem is elegently proved by Loève (1955, pp. 207–10).

It is simple to apply this result to an $n \times n$ matrix operator whose elements are power series in E and L. Although the given linear operator T is not necessarily self-adjoint, the norm of T can be found from the norm of the self-adjoint operator $T'T$ since

$$\|TX\|^2 = (TX, TX) = (X, T'TX),$$

and $T'T$ is a self-adjoint linear operator. Therefore, setting $(X, X) = 1$, the norm of T is given by the square root of the largest σ such that

$$[\sigma - (X, T'TX)] \geq 0.$$

But this is equivalent to finding the largest proper value of the Hermitian matrix, $T'(\bar{z})T(z)$, on the unit circle where $|z| = 1$. Consequently,

$$\| T \| = \max \sqrt{[\sigma(z)]} \quad \text{subject to} \quad |z| = 1,$$

where $\sigma(z)$, a nonnegative real valued function of the complex scalar z, $|z| = 1$, gives the proper values of $T'(\bar{z})T(z)$. Therefore, $\| T \| < 1$, if and only if the largest proper value of $T'(\bar{z})T(z)$ is less than one. In particular, to show the existence of the inverse (37), one need only prove that the largest proper value on the unit circle of the following Hermitian matrix is below one:

$$(38) \qquad [A_d(z) + A_d'(\bar{z})]^{-1} C_v(\bar{z}) C_v(z) [A_d'(\bar{z}) + A_d(z)]^{-1}$$

This is like the situation where we have an arbitrary matrix, M, with complex elements for which we seek the largest proper value. If M^* denotes the conjugate transpose of M, then, as is well known, $M = \sqrt{[\rho(M^*M)]}$, where $\rho(\cdot)$ denotes the spectral radius of the matrix, that is, the proper value of the largest modulus (Varga 1962, sec. 1.3).

With these results we can investigate the positive definiteness of certain $n \times n$ Hermitian matrices and obtain inferences about the linear operator in the pertinent infinite dimensional space. In particular the conclusion about the existence of a Cournot-Nash equilibrium is a consequence of the behavior of certain kinds of positive definite matrices described in the next two lemmas.

LEMMA 6. *Let $K = [k_{ij}]$ denote a symmetric $n \times n$ matrix of real scalars with positive diagonal entries and nonpositive off-diagonal entries. If K is positive definite, then also*

$$\begin{bmatrix} k_{11} & -\beta_{12}k_{12} & \cdots & -\beta_{1n}k_{1n} \\ -\beta_{12}k_{12} & k_{22} & \cdots & -\beta_{2n}k_{2n} \\ & \cdots & & \\ -\beta_{1n}k_{1n} & -\beta_{2n}k_{2n} & \cdots & k_{nn} \end{bmatrix}$$

is positive definite for all β_{ij} with $0 \leq \beta_{ij} \leq 1$, where $k_{ij} \geq 0$ for all i, j, $i \neq j$ and $k_{ii} > 0$ for all i.

Further, using the convenient notation $K(\beta)$ to denote the dependence of the matrix on the scalars β_{ij}, it is also true under these hypotheses that

$$\rho(K_d(\beta)^{-1} K_v(\beta)) < 1$$

for all β_{ij} such that $-1 \leqslant \beta_{ij} \leqslant 1$, where
$$K(\beta) = K_d(\beta) - K_v(\beta),$$

and $K_d(\beta)$ denotes the diagonal matrix with the same diagonal elements as $K(\beta)$.

Proof. First, observe that the diagonal matrix K_d is positive definite and independent of β. We are first to show that $K(\beta)$ is positive definite for all β_{ij} with $0 \leqslant \beta_{ij} \leqslant 1$. Without loss of generality we may assume that $K(1)$ is irreducible. Because of the symmetry of K, this means that the rows and columns of $K(1)$ cannot be permuted so that $K(1)$ can be represented as follows:

$$\begin{bmatrix} K_{11} & 0 \\ 0 & K_{22} \end{bmatrix},$$

where the submatrices K_{11} and K_{22} are irreducible. In particular we may not set $\beta_{ij} = 0$ in the following argument though this constitutes no loss of generality since, if the theorem is true for positive β, then it is also true for zero β's, by application of these results to the appropriate submatrices. In our application, irreducibility means that none of the cross slopes in the demand equations, a_{ij} and a_{ji}, are *both* zero.

Proceeding now with the proof, by corollary 2 of theorem 3.11 (Varga 1962) $K(1)$ is positive definite and irreducible if and only if $K(1)^{-1} > 0$. Hence, $K(1)^{-1} > 0$. Therefore, from Varga's theorem 3.11, $\rho(K_d^{-1} K_v(1)) < 1$.

$$(39) \qquad\qquad |K_d^{-1} K_v(\beta)| \leqslant |K_d^{-1} K_v(1)|,$$

which means that the absolute value of each element of the matrix on the left hand side of (39) is not greater than the corresponding element on the right hand side. Inequality (39) implies that

$$(40) \qquad\qquad \rho(K_d^{-1} K_v(\beta)) \leqslant \rho(K_d^{-1} K_v(1)) < 1,$$

appealing to Varga's lemma 2.3. Hence Varga's corollary 2, theorem 3.11 implies $K(\beta)^{-1} > 0$ so that $K(\beta)$ is positive definite for all β_{ij} with $0 < \beta_{ij} \leqslant 1$.

To complete the proof we observe that (39) is valid for all β_{ij} such that $-1 \leqslant \beta_{ij} \leqslant 1$. Hence, (40) holds for the same range of β's. QED.

We now wish to extend these results to certain matrices whose elements are power series. Consider the matrix power series

$$\begin{aligned} A(z) &= A_0 + A_1 z + A_2 z^2 + \cdots \\ &= [a_{ij}(z)], \; i, j = 1, \ldots, n, \end{aligned}$$

which is the $n \times n$ matrix giving the generating function for $A(E)$. We may also write $A(z)$ as follows:

$$A(z) = A_d(z) - A_v(z) \qquad \text{[cf. 2(12)],}$$
$$A_d(z) = A_{0d} - A_{1d}z - A_{2d}z^2 - \ldots,$$

where the diagonal matrices A_{jd} satisfy

(41) $$A_{jd} \geqslant 0, \qquad \text{if } j > 0 \text{ and } A_{0d} > 0.$$

Likewise,

$$A_v(z) = A_{0v} + A_{1v}z + A_{2v}z^2 + \ldots$$
(42) $$A_{jv} \geqslant 0, \qquad \text{for all } j.$$

These properties express the assumption that the n goods are all gross substitutes for each other and that a given good at one point in time competes with itself at other points in time. The $n \times n$ Hermitian matrix $B = A(z) + A'(\bar{z})$ has a similar representation. We shall now prove the following:

LEMMA 7. *Let $B = A(z) + A'(\bar{z})$ be positive definite for all $|z| = 1$, the A_{jd} satisfy (41), and A_{jv} satisfy (42). Then*

(43) $$\rho(B_d^{-1}B_v) < 1, \qquad \text{for all } |z| = 1.$$

Let the generating functions of the forecasting operators satisfy the conditions as follows:

   i. $|w_{ij}(z)| \leqslant 1$ for all $|z| = 1$.
   ii. $w_{ij}(z)$ has nonnegative scalar coefficients,
   iii. $w_{ij}(z) \geqslant 0$ if $0 \leqslant z \leqslant 1$ and $w_{ij}(1) = 1$.[15]

Then

(44) $$\rho(B_d^{-1}(C_v + C_v')) < 1,$$
where

$$\begin{bmatrix} 0 & a_{12}(w_{12})w_{12} & \cdots & a_{1n}(w_{1n})w_{1n} \\ & \cdots & & \\ a_{n1}(w_{n1})w_{n1} & & \cdots & 0 \end{bmatrix}$$

---

15. Functions with these properties are known as characteristic functions in the theory of probability. If X is a random variable and $dF(x)$ its cumulative density function then the characteristic function is defined as follows:

$$\phi(t) = \int e^{-ixt} dF(x).$$

For the necessary and sufficient conditions that a given function be a characteristic function see Feller (1966, chap. 19, sec. 2). Actually, this result is known as Bochner's theorem. See lemma 5. For a good discussion of the properties of characteristic functions see Gnedenko and Kolmogorov (1954, secs. 11 and 14). It is easy to see that max $|w_{ij}(z)|$ occurs where $z = 1$.

The generating functions of these forecasting operators have another interesting property. Recall that the generating function of the lead operator E is given by the complex scalar z. Hence for $|z| \leqslant 1$, $w_{ij}(z)$ behaves just like the generating function of E. Thus it preserves the same analytic properties as the lead operator.

Proof. We begin by showing that the Hermitian matrix B_d^{-1} can be expanded in a power series in z and $\bar{z}$, which has nonnegative matrix coefficients. Next we show that the formal power series expansion of $(I - B_d^{-1} B_v)^{-1}$ also has positive matrix.coefficients. Appealing to lemma 6 we can prove (43) and, finally, with a similar argument, prove (44).

Since B is positive definite so is B_d so that B_d^{-1} exists. On the unit circle, $z + \bar{z} = 2\cos\theta$, $z^2 + \bar{z}^2 = 2\cos 2\theta$, ..., so that

$$(45) \qquad B_d = 2(A_{0d} - A_{1d}\cos\theta - A_{2d}\cos 2\theta - ...).$$

$A_{0d} > 0$ by hypothesis so that A_{0d}^{-1}, a diagonal matrix, satisfies the same condition. A typical element of B_d in (45) is a function $f(\theta)$ of the following form:

$$f(\theta) = a_0 - a_1\cos\theta - a_2\cos 2\theta - ....$$

We are given that the a_j's are all nonnegative and that

$$\min_\theta f(\theta) > 0, \qquad \text{for all } \theta \text{ with } -\pi \leqslant \theta \leqslant \pi.$$

The power series expansion

$$1/f(\theta) = a_0[1 - a_0^{-1}(a_1\cos\theta + a_2\cos 2\theta + ...)]^{-1}$$

can be justified by appealing to the Neumann series theorem if

$$\max_\theta a_0^{-1}[a_1\cos\theta + a_2\cos 2\theta + ...] < 1, \qquad \text{for } -\pi \leqslant \theta \leqslant \pi.$$

But this is clearly true because all of the a's are nonnegative and $\min f(\theta)$ is at $\theta = 0$ so that

$$a_0 - a_1 - a_2 - ... > 0,$$

giving the desired conclusion. Hence B_d^{-1} can be expanded in a power series with positive matrix coefficients.

Consider the formal power series expansion as follows:

$$(I - B_d^{-1} B_v)^{-1} = I + P_1 z + P_2 z^2 + ... + R_1 \bar{z} + R_2 \bar{z}^2 + ...$$

By carrying out the multiplications involved in computing the powers of $B_d^{-1} B_v$ and collecting the matrix coefficients of like powers of z and $\bar{z}$, we can verify that

$$(46) \qquad R_j \quad \text{and} \quad P_j \geqslant 0.$$

More simply, observe that the matrix coefficients of B_d^{-1} are all positive and that the same is true of the matrix coefficients of B_v. Since products of positive matrices can only be positive, this gives (46). However, this does not yet justify the power series expansion which depends on (43) being true. Let us now prove (43).

Consider B for $z = 1$. B is then a real, symmetric, and positive definite matrix with positive diagonal terms and nonpositive off-diagonals. Hence, assuming it is irreducible, which entails no loss of generality, Lemma 6 implies

$$\rho(B_d(1)^{-1}B_v(1)) < 1.$$

[As in lemma 6 let $B(z)$ denote the dependence of the matrix on the complex scalar z.] But also

(47) $$\left|B_d(z)^{-1}B_v(z)\right| \leqslant \left|B_d(1)^{-1}B_v(1)\right|,$$

where the right-hand side is a nonnegative matrix so that it has nonnegative real elements. Therefore,

$$\rho(B_d(z)^{-1}B_v(z)) \leqslant \rho(B_d(1)^{-1}B_v(1)) < 1$$

applying Varga's lemma 2.3. This proves (43).

To prove (44) observe that

$$\begin{aligned}\left|B_d(z)^{-1}[C_v(z)+C_v'(z)]\right| &\leqslant \left|B_d(z)^{-1}C_v(z)\right| + \left|B_d(z)^{-1}C_v'(z)\right| \\ &\leqslant B_d(1)^{-1}C_v(1) + B_d(1)^{-1}C_v'(1) \\ &\leqslant B_d(1)^{-1}B_v(1),\end{aligned}$$

where we appeal to the properties of $w(z)$, (41) and (42). Hence

$$\rho(B_d(z)^{-1}[C_v(z)+C_v'(z)]) \leqslant \rho(B_d(1)^{-1}B_v(1)) < 1,$$

which gives (44). QED.

We are now ready for the main result.

THEOREM 14. *Under the hypotheses of lemma 7 there is a Cournot-Nash equilibrium for the perceived demand equations as follows:* $q = b - (A_d - C_v)p$.

Proof. It is sufficient to show that for all $|z| = 1$

$$\rho(B_d(z)^{-1}C_v(z)) < 1 \qquad\qquad \text{[see (37)],}$$

since the Cournot-Nash equilibrium is the solution of

$$(A_d + A_d' - C_v)Y = G,$$

according to (25). Since

$$\left|B_d(z)^{-1}C_v(z)\right| \leqslant B_d(1)^{-1}[C_v(1)+C_v'(1)]$$

and by lemma 7

$$\rho(B_d(1)^{-1}[C_v(1)+C_v'(1)]) < 1,$$

we obtain the desired conclusion. QED.

This means that if the n goods are gross substitutes and, given forecasting equations with the properties as described in lemma 7, there is a Cournot-Nash equilibrium, although every firm's information is restricted to knowing only its own cost and demand parameters together with past prices of its competitors. The properties of the forecasting equations allow more general expectations than the ones described above in chapter IV. Thus the forecasting equations as follows:

$$\Pi_1^J(1-\delta_j)/\Pi_1^J(1-\delta_j L) \qquad (j=1,\ldots,J \text{ and } 0 \leqslant \delta_j < 1)$$

would also satisfy the hypotheses of lemma 7. Of course, one should not count on the existence of an equilibrium for arbitrary sets of forecasting equations nor should one expect there to be an equilibrium even for well-behaved expectations if the goods are not gross substitutes. Though the conditions of theorem 14 are merely sufficient, it is hard to see how they can be weakened in economically meaningful ways and still imply the existence of a Cournot-Nash equilibrium. Hence, they may well be close to being necessary as well as sufficient. It should also be noted that collusion would require the exchange of a considerable amount of information among the colluding firms in addition to prices. In fact, it is hard to see how a full monopoly equilibrium among the n colluding firms could be implemented without centralized management. Although, in contrast to the Cournot-Nash equilibrium, there would be no need for conjecture about the future values of prices but only about the future values of the sequence $\{g_t\}$ [cf. (36)], the optimal monopoly price policy also requires information about all of the demand parameters. This is also true under static conditions, but there is need for more information under dynamic conditions to secure the maximum present value of the joint net return. Since the cost of information rises with the amount, the larger is the amount of information needed under dynamic conditions, the larger the cost of collusion. It would appear that collusion is less likely to dominate competition (the Cournot-Nash equilibrium) under changing market conditions.

Before studying the situation for durable goods, let us derive some implications of a symmetric demand function that has the current rate of demand depending on both past and future prices as follows:

$$q_t = b_t - [A(E) + A'(L)]p_t.$$

A symmetric demand arises if it is assumed that the customers choose their goods to maximize a utility functional subject to a budget constraint [see Telser and Graves (1971, chap. 2)]. To derive the necessary conditions for the monopoly solution in this case, we require additional identities like those given in 4(17) to 4(19). The new identities are as follows:

$$\Sigma \alpha^t p_t' A'(L)p_t \equiv \Sigma \alpha^t p_t' A(\alpha E)p_t,$$
$$\Sigma \alpha^t p_t' A'(L)p_t \equiv \Sigma \alpha^{t/2} p_t' A'(\alpha^{1/2}L)\alpha^{t/2}p_t,$$
$$\Sigma \alpha^t p_t' A(\alpha E)p_t \equiv \Sigma \alpha^{t/2} p_t' A(\alpha^{1/2}E)\alpha^{t/2}p_t.$$

The monopoly equilibrium given by the necessary condition for maximum PV corresponding to (10) is

(48) $\quad g_t - [A(\alpha^{-1/2}E) + A'(\alpha^{-1/2}L) + A(\alpha^{1/2}E) + A'(\alpha^{1/2}L)]x_t = 0.$

A sufficient condition for the solvability of this equation and the existence of a maximum PV corresponding to (11) is that for some $m > 0$ and for all square summable $\{x_t\}$,

(49) $\Sigma\, x_t'[A(\alpha^{-1/2}E) + A'(\alpha^{-1/2}L) + A(\alpha^{1/2}E) + A'(\alpha^{1/2}L)]x_t \geqslant m\,\Sigma\, x_t'x_t.$

It is interesting to observe how the discount factor enters this version of the problem. First, as shown in section 4,

(50) $$|A_{-j}| < M\alpha^{|j|/2},$$

or the derivative of PV with respect to $\{x_t\}$ will not exist since (50) is necessary for a bounded derivative. Moreover, (50) expresses the nature of the customers' effective horizon, that is, the time span over which there is a perceptible effect on the current decision. The higher is the discount rate, so that the smaller is α, the shorter is the effective horizon because A_{-j} approaches zero more rapidly as j increases the smaller is α. However, with a symmetric demand, since positive powers of $\alpha^{1/2}$ appear as shown in (48) and given $0 < \alpha \leqslant 1$, there is no additional restriction imposed by symmetrical demands.

Second, if the sufficient condition (49) is satisfied by some positive discount rate, σ^*, it is also satisfied for all smaller discount rates, $\sigma \leqslant \sigma^*$, but it will not be satisfied in general by higher discount rates $\sigma > \sigma^*$ unless the matrix of price coefficients satisfies another condition, namely,

(51) $A_{-j} = \alpha^{j/2}R_{-j}, \quad A_j = A'_{-j} = \alpha^{j/2}R_j.$[16]

Define the linear operator $R(E)$ as follows:

$$R(E) = R_0 + R_{-1}E + R_{-2}E^2 + \ldots.$$

---

16. By lemma 5, a sufficient condition for the existence of a monopoly equilibrium if the demand operator is $A(E)$ is that the Hermitian matrix

$$A(\alpha^{-1/2}z) + A'(\alpha^{-1/2}\bar{z})$$

be positive definite for all z on the unit circle, that is, $|z| = 1$. It can be shown that if this condition is satisfied for all z on the unit circle, then it also holds for all z inside the unit disk $|z| < 1$. Thus, if it holds for a given $\alpha = \alpha_0$, then it also holds for all α such that

$$|\alpha^{-1/2}z| \leqslant |\alpha_0^{-1/2}z|, \ |z| = 1.$$

Hence the condition is satisfied for all $\alpha \geqslant \alpha_0$, which is the assertion in the text.

The relation between the positive definiteness of the quadratic form on the unit circle to its behavior within the unit disk relies on the following facts. First,

(i) $$\eta^*\,[A(\alpha^{-1/2}z) + A'(\alpha^{-1/2}z]\eta$$

is a harmonic function of z. Hence it attains its extrema on the boundary of the region where it is defined. Since it is positive definite for $|z| = 1$ it must remain so within the unit disk. The property of harmonic functions is described in Ahlfors (1953, chap. 5, thm. 3). The vector η is a complex $n \times 1$ column vector. That the function given in (i) is harmonic follows from the fact that it is the real part of the scalar valued complex function $\eta^*\,A(\alpha^{-1/2}z)\eta$ (η^* denotes the conjugate transpose of the complex vector η.)

If $R+R'$ is positive definite and if the demand coefficients satisfy (51), then there is a monopoly equilibrium for *any* discount factor. The conditions expressed by (51) are quite reasonable. They mean that both future and past prices are made equivalent to present prices with a suitable discount factor. If (51) is satisfied, then the higher the discount rate, the less do events remote from the present affect the current rate of demand. Moreover, if $R+R'$ is positive definite, the discount factor in (51) can even be set equal to 1, or, equivalently, the discount rate set equal to zero, and there will still be a monopoly equilibrium. Therefore, the necessary and sufficient conditions for the existence of a maximum PV can be put in terms of $\alpha = 1$. Hence, if a maximum exists for $\alpha = 1$ and if (51) holds, then a maximum exists for any positive discount rate. Since the existence of a Cournot-Nash equilibrium is a consequence of the existence of a monopoly equilibrium, this proves theorem 15.

THEOREM 15. *If the price coefficients of the demand equation satisfy (51) then a necessary and sufficient condition for the existence of a monopoly equilibrium for any positive discount rate is that $R+R'$ be positive definite. Hence a Cournot-Nash equilibrium also exists under the same hypotheses.*

All the preceding analysis assumes $G = 0$, which can be interpreted as the choice of very long time periods such that all goods may be regarded as perishable. We now consider the problem for time periods of arbitrary length so that some goods are durables. If so, the demand relations become

$$(52) \qquad s_t = b_t - A(E)r_t,$$

$$(53) \qquad r_t = (I - GE)p_t,$$

$$(54) \qquad q_t + d_t = (I - GL)s_t.$$

The first relation gives the desired stock as a function of current and expected rents, r_t; the second defines the rents as a function of the depreciation rates, the current price and the expected price; and the third relation is the inventory balance equation relating stocks to purchases and consumption.

Instead of (52) suppose that desired stocks were a direct function of expected prices as follows:

$$(55) \qquad s_t = b_t - A(E)p_t.$$

The corresponding demand equation relating purchase rates to prices would be

$$(55') \qquad q_t + d_t = (I - GL)b_t - (I - GL)A(E)p_t.$$

However, even if $A+A'$ were positive definite and $0 \leqslant \gamma_i < 1$, one could not guarantee a finite maximum PV.[17] Put differently, if the desired stocks satisfy (55), so that customers ignore the implicit rent in the price sequence, then a monopolist could exploit their ignorance and become indefinitely rich, which is impossible. To prevent this, we are led to postulate a demand relation representing rational consumer behavior. Assuming constant depreciation rates, the desired demand relation is given by the combination of (52) and (53).

LEMMA 8. *If $0 \leqslant \gamma_i < 1$ and if $A + A'$ is positive definite, then*

$$(I-G)'A(I-G)+(I-G)A'(I-G)'$$

is positive definite.

Proof. We must show that there is a positive constant $m > 0$ such that for all square summable X,

(56) $(X, [(I-G)'A(I-G)+(I-G)A'(I-G)']\cdot X) \geqslant m(X, X).$

The left-hand side of (56) equals

$$((I-G)X, A(I-G)X)+((I-G)'X, A'(I-G)'X) \geqslant \delta \|(I-G)X\|^2,$$

where $\delta > 0$ since $A + A'$ is positive definite. The hypothesis about γ_i ensures the invertibility of $I - G$. Hence, there is a $\sigma > 0$ such that for all X,

$$\|(I-G)\cdot X\|^2 \geqslant \sigma \|X\|^2.$$

Choose $m = \delta\sigma$. QED.

---

17. The following example does not quite fulfill the claim in the text though it does show that the Hermitian matrix for (55′) need not be positive definite. Consider a situation with two products such that $A(E) = I + A_1 E$, where

$$A_1 = \begin{bmatrix} 0 & \alpha_{12} \\ \alpha_{21} & 0 \end{bmatrix},$$

and the α's are real scalars. Assume

$$G = \begin{bmatrix} 0 & 0 \\ 0 & \gamma \end{bmatrix},$$

giving a very simple situation. The Hermitian matrix for the quadratic form is given as follows:

$$(I - G\bar{z})(I + A_1 z)+(I + A_1'\bar{z})(I - Gz)\dot{} =$$
$$\begin{bmatrix} 2 & \alpha_{12}z + \alpha_{21}\bar{z}(1 - \gamma z) \\ \alpha_{12}\bar{z} + \alpha_{21}z(1 - \bar{z}) & 2 - \gamma(z + \bar{z}) \end{bmatrix}$$

By hypothesis the matrix is positive definite if $\gamma = 0$. Also, the diagonal terms are positive for all γ with $0 \leqslant \gamma < 1$, and $|z| = 1$ as is necessary for positive definiteness, but even for $z = 1$, there are values of $\alpha_{12} = \alpha_{21} = \alpha$ such that the determinant of the matrix would be negative. Thus with a common value for the α's, the determinant is given by the following expression:

$$\delta(\alpha,\gamma) = 4(1-\gamma) - \alpha^2(2-\gamma)^2$$

Clearly, there are admissible values of α and γ such that $\delta(\alpha,\gamma)$ would be negative so that the Hermitian matrix of the form could not be positive definite. An admissible value of α in this case is simply $|\alpha| < 1$.

We can now obtain necessary and sufficient conditions for a monopoly equilibrium in the case of durable goods.

THEOREM 16. *Let the demand relations satisfy (52) to (54). For a maximum monopoly PV it is necessary that the price sequence $\{\alpha^{t/2}p_t\} = \{x_t\}$ satisfy*

(57)
$$\{[I-\alpha^{1/2}L]A(\alpha^{-1/2}E)[I-\alpha^{-1/2}GE]$$
$$+[I-\alpha^{-1/2}GE]A'(\alpha^{-1/2}L)[I-\alpha^{1/2}GL]\}x_t = f_t,$$

where

$$f_t = [I-\alpha^{-1/2}GL]A'(\alpha^{-1/2}L)[I-\alpha^{1/2}GE]\alpha^{t/2}c_t-\alpha^{t/2}d_t+[I-\alpha^{-1/2}GL]b_t.$$

The necessary condition (57) is solvable and a maximum exists if

$$A(\alpha^{-1/2}E)+A'(\alpha^{-1/2}L)$$

is positive definite for some $\alpha < 1$ and if for this α

$$\alpha^{1/2}\gamma_i < 1, \qquad\qquad \text{for all } i.$$

Proof. First, observe that the necessary condition for a maximum, though complicated in appearance, is merely a straightforward application of the general condition that the derivative of the functional *PV* with respect to the control variable, $\{\alpha^{t/2}p_t\}$ must equal zero. Thus the argument yielding (10) applies equally well here, given that the demand equation relating purchase rates to prices satisfies

(58) $$q_t = (I-GL)b_t+(I-GL)A(E)(I-GE)p_t-d_t.^{18}$$

The hypothesis about positive definiteness and lemma 8 gives the desired conclusion about the solvability of the necessary condition and the existence of a maximum. QED.

Finally, by means of theorem 13 we may set

$$B = (I-G)'\cdot A\cdot(I-G)+(I-G)\cdot A'\cdot(I-G)'$$

and conclude that:

COROLLARY. *If a monopoly equilibrium exists under the hypotheses of theorem 16 then a Cournot-Nash equilibrium also exists.*

With these necessary and sufficient conditions for the existence of a monopoly equilibrium we can see if it makes a difference to the monopoly return whether the durable is rented instead of being sold outright.

---

18. $PV = \Sigma \alpha^t(p_t-c_t)'q_t$, where, as above, c_t denotes the t-period per unit costs. To derive the expression for f_t we may proceed as follows:
$$\Sigma \alpha^t c_t'q_t = \Sigma \alpha^t c_t'[(I-GL)b_t-d_t-(I-GL)A(E)(I-GE)p_t].$$

$$\Sigma \alpha^t c_t'(I-GL)A(E)(I-GE)p_t = \Sigma \alpha^t p_t'[I-\alpha^{-1}GL]A'(\alpha^{-1}L)[I-\alpha GE]c_t$$
$$= \Sigma x_t'[I-\alpha^{-1/2}GL]A'(\alpha^{-1/2}L)[I-\alpha^{1/2}GE]\alpha^{t/2}c_t.$$
Taking the derivative with respect to the control variable $x_t = \alpha^{t/2}p_t$ gives us the major portion of the expression for f_t in (54). The terms in f_t involving b_t and d_t are derived similarly.

In accordance with (53), let r denote the rental so that

$$(59) \qquad\qquad r = (I-G)p.$$

Without loss of generality and for the sake of simplifying the notation, we may assume the constant per unit costs, c_t, equal zero.[19] Hence the monopolist chooses r to maximize the present value of the stream of rents where

$$PV = \Sigma \, \alpha^t r_t' s_t$$

and

$$s_t = b_t - A(E)r_t.$$

Therefore, the rental version of the monopoly problem is equivalent to selling perishables and all of the results pertaining to perishables apply herein. In particular the necessary condition for a maximum PV is given by

$$(60) \qquad\qquad b - (A+A')r = 0,$$

and the sufficient condition is that $(A+A')$ be positive definite.

With outright sale of the durable, the PV of the monopoly return satisfies

$$PV = \Sigma \, \alpha^t p_t' q_t = \Sigma \, \alpha^t p_t' \{(I-GL)b_t - (I-GL)A(E)(I-GE)p_t - d_t\}.$$

since the applicable demand equation refers to purchase rates (58). The necessary condition for maximum PV for the outright sale policy is

$$(61) \qquad (I-G)'b - [(I-G)' \cdot A \cdot (I-G) + (I-G) \cdot A' \cdot (I-G)']p = 0.$$

It is a routine calculation to show that the maximum PV is the same for rental as for outright sale. This can be done by calculating the maximum PV for outright sale, using the necessary condition shown in (61), and verifying that it gives precisely the same maximum PV as for rental, using (60) and the relation between p and r given by (59). Consequently, it would appear that a monopolist would be indifferent between rental and outright sale, but there is a further consideration, the nonnegativity of the equilibrium sequences $\{s_t\}$ and $\{q_t\}$. We need not be concerned about the nonnegativity of p_t and r_t since it may well be true that profit maximization dictates temporary sales below cost. Thus to induce customers to try new goods, these are sometimes initially sold below cost. But it would not make sense to allow negative values of s_t and q_t. Of course, the solutions of the necessary maximum conditions can give nonnegative values of these variables. If so, then the monopolist would be indifferent between the rental and outright sales

---

19. The PV for the net return is given by
$$PV = (s,r-c) = (b-Ar,r-c) = (b,r) - (r,Ar) - (c,b-Ar).$$
Thus, by setting the per unit costs equal to zero, we ignore the term in the present value given by $(c,b-Ar)$. This has no material effect on the subsequent analysis.

policies. Since (61) might not give nonnegative s_t and q_t, it is advisable to study nonnegativity more systematically. To this end we prove lemma 9.

LEMMA 9. *Under the hypotheses of lemma 8, $A_d^{-1} > 0$, and*

$$(62) \qquad \rho(A_d^{-1}A_v) = \rho(A_d^{-1}A_v') \leqslant \rho(B_d^{-1}B_v) < 1.$$

Proof. The hypothesis of lemma 8 asserts that $B = A(z) + A'(\bar{z})$ is a positive definite Hermitian matrix on $|z| = 1$. Since

$$A_d(z) = A_{0d} - A_{1d}z - A_{2d}z^2 - \ldots,$$

where the A_{jd} are nonnegative diagonal matrices, $A_{0d} > 0$, the same argument that implies $B_d^{-1} > 0$ for $|z| = 1$ now yields $A_d(z)^{-1} > 0$. Thus the positive definiteness of $A(z) + A'(\bar{z})$ implies B_d is positive definite so that

$$A_{0d} > \sum_{j=1}^{\infty} A_{jd},$$

which implies in turn that

$$\rho(A_{0d}^{-1} \sum_{j=1}^{\infty} A_{jd}) < 1.$$

Therefore,

$$[1 - A_{0d}^{-1}(A_{1d}z + A_{2d}z^2 + \ldots)]^{-1}$$

is a power series in z with positive matrix coefficients.

Because of the hypotheses about A_{jd} and A_{jv}, the largest proper values of $B_d^{-1} B_v$ and $A_d^{-1} A_v$ are attained on the unit circle, where $z = 1$. Recall that $B = B_d - B_v$ and $A = A_d - A_v$. At the point $z = 1$, $A_d(1) > 0$, and $B_d(1) > 0$. Since B has negative off-diagonal elements and positive diagonal elements and since it is a real positive definite matrix if $z = 1$, $[A(1) + A'(1)]^{-1} > 0$ (Varga 1962, cor. 2, thm. 3.11). Hence

$$\rho((1/2)A_d(1)^{-1}[A_v(1) + A_v'(1)]) < 1$$

(Varga 1962, thm. 3.11). Therefore, there is a strictly positive proper vector x such that

$$A_d^{-1}(A_v + A_v')x = 2\rho \cdot x.$$

(For convenience we may now drop the notation to indicate $z = 1$ since it will be understood for the remainder of the proof.) Let σ denote the maximal proper value of $A_d^{-1}A_v$ and let τ denote the maximal proper value of $A_d^{-1}A_v'$. Since both $A_d^{-1}A_v > 0$ and $A_d^{-1}A_v' > 0$, σ and $\tau > 0$. Also, there are strictly positive proper vectors, $y > 0$ and $u > 0$ such that $A_d^{-1}A_v y = \sigma y$ and $A_d^{-1}A_v' u = \tau u$. We now show that $\sigma = \tau$. $u'A_v y = \sigma u'A_d y$ and $y'A_v'u = \tau y'A_d u$ imply

$$y'A_v'u = \sigma y'A_d u = \tau \cdot y'A_d u.$$

Since $y'A_d u > 0$, $\sigma = \tau$, proving

$$\rho(A_d^{-1}A_v) = \rho(A_d^{-1}A_v').$$

Also,

$$y'(A_v+A_v')x = 2\rho \cdot y'A_d x = (2\rho/\sigma)y'A_v'x.$$

Therefore,

$$y'A_v x = [(2\rho/\sigma)-1]y'A_v'x \geqslant 0.$$

Hence,

$$(2\rho/\sigma)-1 \geqslant 0 \quad \text{or} \quad 2\rho \geqslant \sigma.$$

Finally, we must show that $\sigma \leqslant \rho$. By the extremal property of the proper values associated with the spectral radius (Varga 1962, pp. 27–8) $A_v x \geqq \sigma A_d x$ and $A_v'x \geqq \sigma A_d x$. Adding, this gives

$$(A_v+A_v')x \geqq 2\sigma A_d x.$$

Hence by Debreu and Herstein (1953, lemma*, sec. 4) $2\sigma \leqslant 2\rho$, which means that $\rho(A_d^{-1}A_v) \leqslant \rho(B_d^{-1}B_v)$. QED.

This result provides important information about the positivity of A^{-1}.

THEOREM 17. *Under the hypotheses of lemma 8, $A^{-1} > 0$.*

Proof. Consider $A_d(I-\lambda^{-1}A_d^{-1}A_v)$, which equals A if $\lambda = 1$. By lemma 9, $A_d^{-1} > 0$ and by Debreu and Herstein (1953, thm. 3*), $(I-\lambda^{-1}A_d^{-1}A_v)^{-1} > 0$ if and only if $\lambda < \rho(A_d^{-1}A_v)$. Hence, $A^{-1} > 0$ if $\rho(A_d^{-1}A_v) < 1$, which is true by lemma 9. QED.

$A^{-1} > 0$ makes good sense since it implies that a nonnegativity constraint on stocks imposes an upper bound on the admissible rentals. Thus, from (52) $s = b-Ar \geqslant 0$ implies $b \geqslant Ar$. Therefore, $A^{-1} > 0$ means that $s \geqslant 0$ implies $r \leqslant A^{-1}b$. Moreover, if $A^{-1} > 0$, then

$$[(I-G)' \cdot A \cdot (I-G)]^{-1} > 0,$$

provided $0 \leqslant \gamma_i < 1$ because with these restrictions on γ_i it readily follows that both $(I-G)^{-1} > 0$ and $(I-G)'^{-1} > 0$. Therefore, $A^{-1} > 0$ also implies there is an upper bound on prices given as follows:

$$p \leqslant [(I-G)' \cdot A \cdot (I-G)]^{-1}(I-G)'b.$$

It is easier to study the maximum PV for the rental policy with nonnegativity constraints on s and to choose s as the control variable instead of r. The demand equation becomes

$$r = A^{-1}b-A^{-1}s,$$

and for the PV we have

$$PV = (r, s) = (s, A^{-1}b)-(s, A^{-1}s).$$

It is easily shown that if $A+A'$ is positive definite, then also $A^{-1}+A'^{-1}$ is positive definite (and conversely.)[20] Hence the sufficient conditions for the existence of a maximum *PV* are satisfied. The necessary conditions are as follows:

(63) $$w + A^{-1}b - (A^{-1} + A'^{-1})s = 0,$$

where $w \geqq 0$ is a square summable sequence of Lagrangian multipliers such that $(s, w) = 0$. We can readily find the value of the maximum *PV* from (63).

$$(s, A^{-1}b) - (s, [A^{-1} + A'^{-1}]s) = 0$$

implies that

$$\max PV = (1/2)\,(s, [A^{-1} + A'^{-1}]s) > 0.$$

There is a similar development for the outright sales policy subject to a nonnegativity constraint on q.

With nonnegativity constraints on s and q, it is no longer true that the maximum *PV* is necessarily the same for the outright sales policy as for the rental policy. To see why, suppose that one policy has a natural nonnegative solution of its necessary condition while the other does not. A natural nonnegative solution is one for which the corresponding Lagrangian multipliers are zero. Thus, the rental policy has a natural nonnegative solution if $w = 0$ implies that $s \geqslant 0$ satisfies (63). If one policy has a natural nonnegative solution while the other does not, then the one with the natural nonnegative solution must yield the larger maximum *PV* (in degenerate cases there can be equality). That is, if one must *impose* binding constraints in order to achieve nonnegativity, then this must reduce profit. In fact, the outright sales policy can never be more profitable than the rental policy.

To prove this, consider the inventory balance equation (54) rewritten herein:

(64) $$s = (I - G)'^{-1}(q + d).$$

By hypothesis, $d \geqq 0$. As shown above, $0 \leqslant \gamma_i < 1$ implies that $(I - G')^{-1} > 0$. Hence, if the outright sales policy has a natural nonnegative solution, then the rental policy must also have one. Consequently, the rental policy must be at least as profitable as the outright sales policy. Moreover, $(I - G')$ itself is not a positive operator. That is,

(65) $$q + d = (I - G')s$$

does not have a nonnegative solution $q \geqslant 0$ whenever the rental policy has a natural nonnegative solution $s \geqslant 0$. Put differently, $s \geqslant 0$ does not imply

---

20. Here is a brief proof. First, observe that $A^{-1} + A'^{-1} = A^{-1}(A' + A)A'^{-1}$. Assume that $A + A'$ is positive definite implying that A and A' are both invertible. Hence there is a positive m such that

$$(X, A^{-1}(A' + A)A'^{-1}X) = (A'^{-1}X, (A + A')A'^{-1}X)$$
$$\geqslant m \|A'^{-1}X\|^2$$
$$\geqslant mn \|X\|^2$$

for some positive n by the invertibility of A'^{-1}. The converse, namely, that the positive definiteness of $A^{-1} + A'^{-1}$ implies the positive definiteness of $A + A'$, is similarly proved.

$q \geqslant 0$ in (65) given $d \geqq 0$. Therefore, the maximum PV for the rental policy can exceed the maximum PV for the outright sales policy. Perhaps this argument explains why certain companies such as IBM, United Shoe Machinery, and Xerox preferred rental to outright sales.

There is an intuitively appealing interpretation of these mathematical results. In order that the outright sales policy yield the monopolist the same net revenue as the rental policy, it would be necessary to drop the non-negativity constraint on q. This would mean that the monopolist would not only have to sell but also that he would have to be able to buy from his customers at the prices that are implicit in the optimal rental program where the price is determined from the relation $p = (I-G)'^{-1}r$. Because the rental policy is more flexible than the outright sales policy, it is more lucrative.

6. Conclusions

Before summarizing the results of this chapter, it is well to observe how its subject matter relates to the analysis in the preceding chapter. Chapter V studies competition among goods of different quality and uses an example from spatial economics as a concrete instance. There the problem is to predict the location of n firms, prices, and outputs under various competitive conditions. Since space is a continuum, this example frees one of the arbitrary restriction to n goods although, as it turns out, the presence of fixed costs leads to the prediction of a provision of a finite number of goods. Goods available at different points in time provide another instance in the general rubric of quality variation. Like the spatial problem, there is in principle a continuum of products, one at each point of time. Although one could analyze this explicitly, it is easier to divide time into an infinite number of discrete periods of arbitrary calendar length. In this way one can still have the essential features of the dynamic problem and can use simpler mathematics. Even so, one must acquire some familiarity with functional analysis to provide an adequate understanding of competitive and monopoly equilibria under dynamic conditions. A more general analysis with more powerful techniques of functional analysis would permit a fairly complete treatment of quality variation but would not differ in many essential ways from our study of the dynamic equilibrium.

The device that serves us well is common in mathematical analysis. We convert an economic problem into a maximum problem and then derive the resulting properties of the solution. In this case, starting from a set of arbitrary demand relations for n related goods subject to mild restrictions, we seek necessary and sufficient conditions for the existence of a maximum present value of the stream of net returns assuming there is monopoly control over the n goods. This imposes a number of conditions on the demand relations if the goods are durables. Indeed the assumption of durability is virtually one of necessity since the calendar length of a time period is arbitrary in economic analysis and all goods are durable for short enough periods. In particular we observe that, to ensure the existence of a monopoly equilibrium, one may postulate a relation between desired stocks and prices such that the latter become a function of the present value of the rent of the durable. Hence the price of the good is viewed as the present value of the services which it yields. Otherwise, one could construct cases in which a monopoly could obtain an infinite present value. Moreover, the postulate that a monopoly equilibrium exists implies that a Cournot-Nash competitive equilibrium exists. Finally, this approach illuminates the question of whether a monopolist prefers rental to outright sales of a durable good. Taking into account nonnegativity constraints on rates of sales and on stocks, it is shown that a rental policy is never dominated by an outright sales policy and there are circumstances such that rental is definitely more lucrative. This approach does not require consideration of other factors sometimes used to explain the use of rental instead of outright sale such as the possibility of extracting a higher return by price discrimination if the good is rented instead of sold.

A very interesting class of problems arises from an examination of the consequences of price expectations. Assuming that the forecasts depend on past observations both because of the inherent properties of real economic systems and because this is justified by experience, we can see whether a given class of forecasting equations combined with an economic structure has an equilibrium. It is hardly surprising to learn that only some kinds of expectations imply an equilibrium under dynamic conditions. In other words, given an economic structure that would have an equilibrium with perfect foresight, there may not be an equilibrium with arbitrary expectations, because some expectations are destabilizing. A class of forecasting equations that has an equilibrium for n gross substitutes under dynamic demand conditions, is a generalization of first order adaptive expectations which includes the Cournot expectations model as well as arithmetic mean expectations as a limiting case. The proof of the existence of an equilibrium for this class of expectations makes one suspect that if all goods were not gross substitutes or if some past prices receive negative weights in forming expectations, then there would not be an equilibrium. Hence although gross substitutes and the generalized price expectations given above are only sufficient but not necessary conditions for an equilibrium, it may well be that they are close to providing necessary and sufficient conditions for the existence of an equilibrium.

The following two chapters are empirical. Chapter VII gives some estimates of demand elasticities for certain branded and advertised goods. These elasticities permit a judgment about the excess of price over the marginal cost of these goods. Chapter VIII studies the relation between rates of return in four-digit manufacturing industries and the relative size of firms in these industries.

VII

Estimates of Demand, Price Policy, and the Ratio of Price to Marginal Cost by Brand for Selected Consumer Goods

1. Introduction

Using the preceding analysis as a guide, we now examine the pertinent evidence for three established products in moderately concentrated markets (frozen orange juice concentrate, instant coffee, and regular coffee) and one new product (instant mashed potatoes) in order to obtain a detailed picture of demand conditions and price policies by brand. In these markets a few firms account for more than one-third of total sales, and it is believed that consumers are somewhat brand loyal. Hence one may regard each firm as in possession of some monopoly power.

Therefore, a firm's ability to obtain more than the competitive return is limited by the availability of competing brands that are more or less close substitutes for the given brand as measured by the size of the own and cross price elasticities. Lower limits for the ratio of price to marginal cost are simple functions of these price elasticities.

Assuming the firms choose prices to maximize their net return and depending on whether or not there is collusion, there follow certain relations between price and marginal cost. Thus, assume the demand for n competing goods is given by the linear relation as follows:

$$(1) \qquad\qquad q = b - Ap,$$

where q, p, and b are $n \times 1$ vectors, and A is an $n \times n$ matrix of real scalars such that $a_{ii} > 0$ and $a_{ij} \leqslant 0$, for $i \neq j$. Assume there are constant per unit costs denoted by the $n \times 1$ vector c. The joint net return, denoted by β, is

$$(2) \qquad\qquad \beta = (p - c, q) = (p, b - Ap) - (c, b) + (p, A'c).$$

If the firms collude to maximize the joint net return so that prices are chosen to maximize β, then the price policy equations are given by

$$(3) \qquad\qquad 0 = b - (A + A')p + A'c,$$

274

which can also be written as follows:

(4)
$$q - A'(p - c) = 0.$$

By definition the price elasticity for product i with respect to the price of product j is

(5)
$$\eta_{ij} = \frac{\Delta q_i}{\Delta p_j} \frac{p_j}{q_i} = a_{ij} \frac{p_j}{q_i}.$$

Substituting these elasticities into the price policy equations in place of a_{ij}, the elements of the matrix A', we obtain the equations as follows:

(6)
$$q_1 = \eta_{11}(p_1 - c_1)q_1/p_1 + \ldots + \eta_{n1}(p_n - c_n)q_n/p_1,$$
$$\ldots$$
$$q_n = \eta_{1n}(p_1 - c_1)q_1/p_n + \ldots + \eta_{nn}(p_n - c_n)q_n/p_n.$$

However, the ith component of β, denoted by β_i, is the net return of firm i and $\beta_i = (p_i - c_i)q_i$ so that (6) implies

(7)
$$p_1 q_1 = \eta_{11}\beta_1 + \eta_{21}\beta_2 + \ldots + \eta_{n1}\beta_n,$$
$$\ldots$$
$$p_n q_n = \eta_{1n}\beta_1 + \eta_{2n}\beta_2 + \ldots + \eta_{nn}\beta_n.$$

The Cournot-Nash equilibrium is the solution of the set of equations as follows:

(8)
$$b - Ap - A_d(p - c) = 0,$$

where A_d is a diagonal matrix formed from the diagonal elements of A, and (8) may also be expressed in the form

(9)
$$q = A_d(p - c),$$

which is a special case of (4), obtained by setting the off-diagonal elements of A' in (4) equal to zero. Thus in the Cournot-Nash equilibrium for firm i

(10)
$$q_i^C = a_{ii}(p_i^C - c_i),$$

where the superscript C reminds us that price and quantity differ under the Cournot-Nash equilibrium from those prevailing under collusion. Let γ_i^C denote the gross revenue and β_i^C the net revenue of firm i in the Cournot-Nash equilibrium. Therefore, using (10),

(11)
$$\gamma_i^C = \eta_{ii}\beta_i^C$$

relates gross and net receipts via the own price elasticity. Equivalently,

(12)
$$\beta_i^C/\gamma_i^{\,C} = 1/\eta_{ii} > 0,$$

recalling our sign convention that makes $a_{ii} > 0$ (and $a_{ij} \leqslant 0$, if $i \neq j$). In words, the reciprocal of the own price elasticity gives the percentage excess of the price over marginal cost in the Cournot-Nash equilibrium. Hence equation (12) gives good reason for wanting estimates of the price elasticity. However, the estimate shown in (12) is too low if there is collusion as we shall now see. The system (7) is pertinent for collusion:

$$(13) \qquad \beta_i/\gamma_i - 1/\eta_{ii} = -\sum_j (\eta_{ji}/\eta_{ii})(\beta_j/\gamma_i) \geq 0.$$

It follows that with collusion the ratio of the net to the gross revenue exceeds the reciprocal of the price elasticity, which gives a lower limit to the sought-for ratio.

Unfortunately, one cannot determine from a study of the demand relations alone whether or not the firms are in collusion. Hence without knowledge of cost conditions as well, which would in principle give direct estimates of the ratios β_i/γ_i, one cannot tell whether (12) or (13) is the correct formula for describing the situation. That is, on the basis of demand estimates one cannot deduce the exact relation between net and gross revenue even if one is willing to assume that firms seek a maximum net return, although (12) gives a plausible lower limit to the ratio of net to gross revenue. Before presenting some simple numerical estimates of the elasticities, it is necessary to describe the data and engage in some other preliminaries.

The basic data used for estimating demand are monthly series of prices and market shares by brand covering somewhat more than six years from the beginning of 1954 to about the middle of 1960. The data are from the Consumer Panel of the Market Research Corporation of America (MRCA), a sample of about 10,000 families designed to represent the population of customers for the products sold by MRCA clients. A household member of the panel records transaction data in a diary which is sent weekly to MRCA. These data describe the transactions fairly completely and include the quantity purchased, the amount paid, the brand name, the place of purchase, and so on. Panel members receive some remuneration for their participation.

A panel, since it contains the same families for some time, permits the study of changes in consumer behavior over time. There is the risk that households willing to participate in a panel behave differently than households who are unwilling to join. For example, if panel members are more habit prone than nonpanel members, then this might bias estimates of the price elasticities toward zero. Possibly, panel members become more self-conscious and expert in their buying habits which may bias the price estimates in the opposite direction. Unfortunately, these questions cannot be answered, though one can take some simple precautions. Thus one can check the characteristics of panel members with census figures for all households. In most respects there is tolerably good agreement. However, in 1957 certain kinds of households were under-represented. Thus, households with housewives under 30, working wives, and renters were among these (Telser 1962a). In his doctoral dissertation Sudman (1962) studied these and related questions in detail for the MRCA panel. His findings for the revised panel are reassuring.

Fortunately, for our purposes we impose no heavy demands since we require only market share and price figures. There is in addition an important redeeming feature.

The operator of the panel is a profitable enterprise that has survived for many years despite competition from others. Its clients purchase the data for their own purposes and can compare the figures with their own records of shipments and prices. Since we only require price and market share data and since many of these clients are of long standing, it is safe to conclude that the panel data are of sufficient accuracy for our purposes.

The market share and price data do depart from the model's prescriptions in two respects. First, the model speaks of quantity and not market share. Second, the producers' and not retail prices are referred to in the model. These departures have their advantages as well as some disadvantages.

The use of market share instead of absolute quantities imposes a lighter burden on the information needed for reasonably good estimates of the important demand parameters. For example, those factors which affect the sales of all brands to the same extent would have to be included in a regression where absolute sales are the dependent variable, but they may be safely ignored if market shares are dependent. Thus, market shares are free of seasonal change although absolute sales figures are not.

The use of retail prices avoids difficult problems that would arise in attempting to measure the relevant price at the level of manufacture. Obviously, the empirical work would require transaction and not list prices, but, unfortunately, the latter are more often readily available at the manufacture level. Since the price data come from the buyers who record the actual amount they pay and who specify in detail the nature of the good they buy, it is even true that these price data are more accurate than, say, the prices collected for the Consumer Price Index (see, for instance, Kruskal and Telser 1960). Moreover, if there is competition among retailers, then the prices we observe reflect the complex terms of sale involved in the dealings between retailers and manufacturers or their distributors.

There are, nevertheless, undeniable disadvantages in the use of these prices. The retail price of an item depends on distribution costs as well as on the price from the manufacturer. Therefore, some of the changes in the retail price are explained by changes in distribution costs. It is as if the retail price were a biased estimator of the manufacturer's price that is also subject to measurement error. However, the measurement error and bias may well be larger for estimates of transactions prices at the manufacture level and certainly the cost of collecting such data would be larger.

Another aspect of the price figures is worth noting. Families record prices only for the brand they actually buy and do not record prices of the available competing brands. Hence we are in ignorance of the prices of all of the alternative brands simultaneously available to a given household. Therefore, with these data it would be impossible in principle though hardly desirable in practice to estimate individual household demand. Fortunately, however, families do buy different brands, and the panel includes clusters of families who are exposed to many competing brands. As a result we have readings on the prices of all of the brands simultaneously available to a group of households.

Certain other aspects of the data will be considered below. For now, in order to relate the ratio of net to gross revenue to the price elasticity with the aid of market share demand functions, we must show how the price elasticities derived from the latter relate to the price elasticities for ordinary demand functions in terms of absolute quantities. Let

q_i = absolute quantity of brand i,
$Q = \Sigma q_i$,
m_i = market share of brand i,
p_i = price of brand i,
$\bar{p}$ = average price of all brands = $\Sigma m_i p_i$,
p_i^* = average price of all brands excluding brand i.

Equation

$$p_i^* = (\bar{p} - m_i p_i)/(1 - m_i)$$

implies

(14) $$\bar{p} = (1 - m_i)p_i^* + m_i p_i.$$

By definition $q_i = m_i Q$ so that

(15) $$\Delta q_i = Q\Delta m_i + m_i \Delta Q.$$

Denote the quantity price elasticity of brand i by η_q, where

$$\eta_q = (\Delta q_i/\Delta p_i)(p_i/q_i)$$

$$= \frac{Q}{q_i} p_i \frac{\Delta m_i}{\Delta p_i} + \frac{p_i m_i}{q_i} \frac{\Delta Q}{\Delta p_i}$$

(16) $$= \frac{\Delta m_i}{\Delta p_i} \frac{p_i}{m_i} + \frac{\Delta Q}{\Delta p_i} \frac{p_i}{Q}.$$

On the right-hand side of (16), the first term is the market share elasticity, and the second is the aggregate quantity elasticity with respect to the price of brand i. Hence (16) implies that the market share elasticity, denoted by η_m, is smaller in magnitude than the absolute quantity elasticity since

$$\eta_q - \eta_m = \frac{p_i}{Q} \frac{\Delta Q}{\Delta p_i} < 0.$$

To approximate $\Delta Q/\Delta p_i$, one may assume that $\Delta \bar{p}/\Delta p_i = m_i$ in accordance with (14). Therefore,

$$\frac{\Delta Q}{\Delta p_i} = \frac{\Delta Q}{\Delta \bar{p}} \frac{\Delta \bar{p}}{\Delta p_i} = m_i \frac{\Delta Q}{\Delta \bar{p}}$$

$$(17) \qquad = m_i \frac{Q}{\bar{p}} \eta_Q,$$

where η_Q denotes the elasticity of the total quantity Q with respect to the average price $\bar{p}$. It follows that

$$(18) \qquad \eta_q \approx \eta_m + (m_i\, p_i/\bar{p})\eta_Q.$$

Let w_i denote brand i's proportion of total revenue so that $w_i = (p_i q_i)/(\bar{p}Q)$. This allows us to express (18) as follows:

$$(19) \qquad \eta_q \approx \eta_m + w\eta_Q.$$

With (19) one can form a reasonable guess at the size of the bias due to the use of the market share elasticity instead of the absolute sales elasticity. The bias is smaller, the smaller is the brand's proportion of total receipts and the smaller in magnitude the aggregate quantity elasticity. The latter is probably

TABLE 7.1
Products, Brands, and Sample Periods Analyzed in the Following Tables

PRODUCT AND BRAND	IDENTIFICATION OF SAMPLE PERIODS
Frozen orange juice concentrate (*OJ*): Birds Eye, Libby's, Minute Maid, Snow Crop	1 = April 1954 to December 1959 2 = April 1954 to March 1957 3 = April 1957 to December 1959
Instant coffee (*IC*): Chase and Sanborn, Maxwell House	1 = April 1954 to June 1960 2 = April 1954 to March 1957 3 = April 1957 to June 1960
Regular coffee (*RC*): Chase and Sanborn, Folger's, Hills Brothers, Maxwell House	1 = April 1954 to June 1960 2 = April 1954 to March 1957 3 = April 1957 to June 1960

NOTE: Brands (alphabetically arranged here) are given the following symbols in arbitrary order in the tables: OJ—A, B, C, E; IC—A, B; RC—A, B, C, D.

less than one, and w is slightly larger than m because the brands under study command above average prices. Therefore, the bias approximately equals the market share, about 10 percent for these brands.

Let us now consider some simple estimates of the market share price elasticities and compare them with accounting measures of the ratio of profits to sales. The basic data are monthly time series of prices and market shares by brands for frozen orange juice concentrate, *OJ*, instant coffee, *IC*, and regular coffee, *RC*. Prices are deflated by the monthly Consumer Price Index (*faute de mieux*, see Kruskal and Telser 1960). A condition of using these data is that individual brands may not be identified. Table 7.1 lists the brands, which are subsequently identified by letter though not in the same order as in table 7.1. Table 7.1 also shows the dates of the sample periods. For

TABLE 7.2
Summary Statistics for Frozen Orange Juice Concentrate (OJ),
Regular Coffee (RC), and Instant Coffee (IC)

PRODUCT AND BRAND	MARKET SHARE (MS)		DEFLATED OWN PRICE (PD)		DEFLATED OTHER PRICE (POD)	
	Mean	S.E. × 10	Mean	S.E. × 10	Mean	S.E. × 10
OJ A1	0.1279	0.2239	17.51	2.175	14.72	2.252
2	0.1432	0.1932	16.32	0.7306	13.69	0.5931
3	0.1158	0.1654	18.46	2.465	15.44	2.710
B1	0.05639	0.2546	17.06	2.283	15.07	2.233
2	0.07976	0.1992	15.92	0.6909	14.05	0.5938
3	0.03791	0.0831	17.96	2.681	15.88	2.689
C1	0.05627	0.1349	16.91	2.432	15.07	2.233
2	0.06523	0.1121	15.33	0.557	14.05	0.5938
3	0.05133	0.1318	18.16	2.615	15.88	2.689
E1	0.05247	0.2155	15.18	2.338	15.07	2.233
2	0.07194	0.1393	13.72	0.5154	14.05	0.5938
3	0.03707	0.1177	16.32	2.574	15.88	2.689
IC A1	0.3494	0.2893	39.35	9.029	37.49	8.884
2	0.3672	0.2872	47.52	4.865	42.25	6.096
3	0.3338	0.1834	32.22	4.686	30.72	4.045
B1	0.08975	0.2291	35.29	8.243	38.12	8.909
2	0.09650	0.2836	42.55	5.608	46.05	5.602
3	0.08387	0.1482	28.97	3.628	31.21	4.231
RC A1	0.1261	0.1637	77.45	13.76	71.46	13.00
2	0.1150	0.0685	88.93	7.149	82.17	7.471
3	0.1358	0.1614	67.44	9.674	62.12	8.929
B1	0.04067	0.07335	74.80	13.41	72.20	13.02
2	0.03644	0.06832	86.05	7.208	82.94	7.426
3	0.04436	0.05608	64.98	9.149	62.83	8.952
C1	0.1214	0.1441	75.52	13.92	72.20	13.02
2	0.1111	0.1059	87.11	7.294	82.94	7.426
3	0.1303	0.1090	65.42	9.792	62.83	8.952
D1	0.06259	0.07853	75.19	13.81	72.20	13.02
2	0.06418	0.08667	86.53	6.8303	82.94	7.426
3	0.06120	0.06883	65.30	10.28	62.83	8.952

NOTE: In those cases where a given brand has a small market share, *POD* is the average price of all the brands including the given brand.

some brands the time series are long enough to yield meaningful estimates of demand for two subsamples. This is true, for example, for orange juice, brand *A*, *OJA*.

The simplest estimates of market share responsiveness to price derive from the summary figures of table 7.2. An estimate of the slope of market share with respect to price is given by the ratio of the change in market share to the change of $(PD - POD)$, the difference between the deflated price of the given brand and the deflated price of all other brands, calculated between the two subsample averages. To illustrate consider OJA. The average deflated price of $OJA2$ was 16.32 cents per unit while the deflated average price of all other brands was 13.69 cents per unit giving a difference of 2.63 cents. From the first to the second subsample the average market share of OJA fell from 0.1432 to 0.1158 while the average price difference rose by 0.39 cents per unit. Hence the estimated slope of market share with respect to the own price is -0.0703. At the sample mean price and market share the implied price elasticity is -9.62. Table 7.3 presents the results for all those brands such that the available data allow this simple procedure to be used.

TABLE 7.3
Simple Estimates of the Slope of Market Share with respect to Price, and Own-Price Elasticities

PRODUCT AND BRAND		Δm	$\Delta(PD\text{-}POD)$	SLOPE	η
OJ	A	-0.0274	$+0.39$	-0.0703	-9.62
	B	-0.04185	$+0.21$	-0.199	-6.17
	C	-0.0112	$+1.00$	-0.0112	-3.38
	E	-0.0349	$+0.77$	-0.0453	-13.11
IC	A	-0.0334	-0.77	$+0.0413$	$+$
	B	-0.0126	$+1.26$	-0.0100	-3.93
RC	A	0.0208	-1.44	-0.0144	-8.85
	B	0.00792	-0.96	-0.00823	-15.13
	C	0.0192	-1.58	-0.01215	-7.56
	D	-0.00298	-1.12	$+0.00267$	$+$

Table 7.3 shows positive price slopes in two cases and negative slopes in the eight remaining cases. The latter estimates imply own price elasticities in the range -3.4 to -15.1. Hence the estimated lower limits for the ratios of profits to gross receipts range from 29 to 6.6 percent. One must remember that these results refer to individual brands of narrowly specified products. Also, these are successful brands, and their implied profit to sales ratios can be expected to exceed the figures for a random sample of brands.

Let us compare these results with estimates of the ratio of accounting profits to sales. Some pertinent data are given in table 7.4, the source being the *Fortune* list of the 500 largest U.S. corporations. The median ratios over all industries vary between 4.2 and 5.6 percent for the period 1961–67. There is, of course, more variation among industries. Thus, the ratio is about 10 percent for pharmaceuticals; 8–9 percent for tobacco; 4–7 percent for soaps and cosmetics; 3.4–4.8 percent for motor vehicles; and 2.4–3.2 percent for food and beverages. The latter figure is most pertinent for our purposes. In addition let us examine the figures for some of the individual corporations who make the brands in the sample. These figures, shown in table 7.4, show as much variation among the companies as there is among the industries with about the same order of magnitude for both. Hardly surprising, the upper range of the profit to sales ratio is considerably below the upper limit implied by the reciprocal of the price elasticities. There are good reasons for

TABLE 7.4

Ratio of Profits to Gross Receipts for Selected Industries and Corporations in Selected Years

A. INDUSTRY MEDIANS

	1961	1962	1963	1964	1965	1966	1967
Pharmaceuticals	10.5	10.5	10.6	10.8	10.3	10.2	9.6
Tobacco	8.9	8.9	8.9	9.1	8.6	8.1	8.2
Soaps, cosmetics	4.3	6.1	7.0	5.6	6.5	6.7	7.2
Chemicals	5.3	6.1	6.0	6.6	7.5	6.8	5.9
Textiles	3.0	3.6	3.3	3.9	4.4	4.4	3.1
Motor vehicles, parts	3.4	3.4	4.2	4.5	4.4	4.8	4.4
Food, beverages	2.8	2.7	2.4	2.7	2.8	3.2	2.4
Apparel	3.0	2.5	2.8	3.7	4.1	4.1	3.3
All industry	4.2	4.2	4.4	5.0	5.5	5.6	5.0

B. SELECTED CORPORATIONS

	1962	1963	1964	1966	1967
American Home Products	10.6	10.7	10.8	10.3	10.5
Beatrice Foods	2.2	2.3	2.4	3.1	3.3
Borden	3.1	3.1	3.5	3.7	3.5
Coca Cola (Minute Maid)	8.2	8.2	7.8	9.1	9.6
General Foods	6.1	6.5	6.2	6.0	6.0
Proctor and Gamble	6.8	7.0	6.8	6.7	7.1
Standard Brands	3.9	4.0	3.8	4.1	4.1

SOURCE: *The Fortune Directory: The 500 Largest U.S. Industrial Corporations.*

believing that this must be because the brands in the sample are the successful ones whose profits offset the losses of the unsuccessful ones.

In one important respect accounting measures of profit do not correspond to the economic concept of profit. The latter views profits as a random variable equal to the difference between the actual and the expected return to the firm owners (Knight 1921). The accountant includes the owners' return in his profit measure so that it would be positive on average. In contrast, the measure afforded by the reciprocal of the price elasticity is closer to the economic concept. Moreover, it measures the return at the margin while the accounting measure is the ratio of the total net receipts to the total gross receipts. If total costs are a rising function of output, then this gives an additional reason for the accounting measure to exceed the economic measure given by the ratio of the price elasticity. Consequently, the finding of a higher return at the margin than the accounting measure would only seem consistent with the view that the brands in the sample are the more successful ones. In this chapter we shall attempt to refine the price elasticity estimates. In the next chapter we shall examine in detail the results of an empirical study of the returns to manufacturing industries.

Given the more than six years of monthly data available and our curiosity, we shall subject the figures to a more searching scrutiny which can confirm the results of the simple exercise described above and thereby lend confidence to similar approaches for other products where more detailed figures are unavailable. The results of our investigation may also increase our confidence in the simple procedure by showing that more complicated methods give either the same or unreasonable results. These are good reasons for a more detailed examination of the time series evidence, the task of the next section.

Section 3 examines the relations among brand prices. It attempts to throw light on the nature of firm price policy and to see how well a firm can predict competing prices using past prices. Section 4 studies a special situation that arose during the introduction of a new product, instant mashed potatoes. During the early stage the market was dominated by a single brand and the sample period covers some of this period as well as the subsequent time when the entry of new brands led to a loss of the first brand's leading position giving us the opportunity of estimating the demand during a time of temporary monopoly.

2. Estimates of the Demand Relation between Market Share and Prices

The main reason for estimating the relation between market share and prices instead of between quantity and prices is simplicity, because the forces affecting absolute quantity fall into one of two mutually exclusive classes, those which affect all competitors to the same extent and those which do not. Thus, in terms of absolute quantity the demand relation is

$$(1) \qquad q = f(p, p^*)g(z),$$

where p is the price of the given brand, p^* the average price of all other brands, and z represents all those variables which affect all brand sales to the same extent. The demand for all the competing brands can be represented by the same kind of demand relation as (1) thereby giving

$$(2) \qquad q^* = f^*(p, p^*)g(z).$$

Therefore, market share, m, is independent of z since by definition $m = q/(q + q^*)$.

This formulation has another advantage. It can justify the treatment of price as exogenous in the market share demand equation. Let us see why. Assume that unit costs, denoted by c, are independent of output rates. Let the firm choose its price to maximize its net return. Hence its price policy is the function defined by the necessary condition for a maximum net return, which is as follows:

$$[(p-c)/q]\frac{\partial q}{\partial p}+1 = 0.$$

The function $g(z)$ divides out, and we obtain

(3) $$[(p-c)/f]\frac{\partial f}{\partial p}+1 = 0.$$

Therefore, p is independent of the determinants of $g(z)$. Moreover, even those z's which are policy variables and, consequently, optimally chosen by

TABLE 7.5
Autoregressions of Market Share for OJ, IC, and RC

PRODUCT AND BRAND		CONSTANT	COEFFICIENT OF		R	S.E.
			M-1*	M-2*		
OJ	A1	0.03493	0.6807	0.04378	0.7061	0.01607
			5.86	0.37		
	B1	0.003359	0.8376	0.09250	0.9284	0.009586
			7.23	0.80		
	C1	0.01508	0.8328	−0.1009	0.7594	0.008894
			7.16	0.87		
	E1	0.005822	0.7791	0.1009	0.8714	0.01072
			6.73	0.87		
IC	A1	0.08845	0.5783	0.1702	0.7312	0.0201
			4.91	1.51		
	B1	0.01320	0.8356	0.01591	0.8485	0.01229
			7.19	0.14		
RC	A1	0.01147	0.7403	0.1706	0.8779	0.007950
			6.42	1.46		
	B1	0.008347	0.8780	−0.07991	0.8069	0.00434
			7.42	0.67		
	C1	0.01650	0.6089	0.2626	0.8708	0.007185
			5.49	2.37		
	D1	0.02888	0.5071	0.02913	0.8223	0.006791
			4.09	0.23		

*The number below the coefficient is the t-ratio.

the firm do not affect p. Only those policy variables entering f in addition to price can lead to biased estimates of the price coefficients if they are omitted from the market share demand relation, which can occur if the outside investigator excludes any market share demand variable used by the firm in forming its price policy. For example, the regressions below exclude two such variables, the number of retail outlets carrying the given brand and advertising outlays. Both of these variables are probably positively correlated with market share and with the firm's own price. As a result the least squares estimates of the own price coefficients are biased upward, that is, the true price slope is probably algebraically smaller than the estimate. It was argued in the preceding section that the hypothesis of competition implies that the reciprocal of the price elasticity gives a lower limit to the estimated ratio of net to gross return. In addition we saw that the market share price elasticity is too low since it excludes the effect of price on aggregate sales. We now have an additional reason for believing that the market share price elasticity is too low because the regressions exclude some pertinent variables that are probably positively correlated with both price and market share. It is perhaps superfluous to add that if data were available for the measurement of advertising outlays and the extent of retail distribution then these variables would certainly be included in the regressions.[1]

A salient fact about these data is the autocorrelation of market share. Table 7.5 shows this clearly; current market share varies directly with market

---

1. It is of some interest to derive in detail the bias in the simple case for which the firm's demand is linear as follows:

(i) $$q_t = b - ap_t + kx_t + u_t.$$

The term $\{x_t\}$ is a sequence of exogenous variables that does not affect all firms to the same proportionate extent. The term $\{u_t\}$ is a sequence of identically distributed random variables that are mutually independent and are independent of the policy variable p_t and are independent of the exogenous variable x_t. There are constant per unit costs. The firm's t-period net revenue is $\beta_t = (p_t - c_t)q_t$.

Assume that the firm knows the demand parameters, a, b, and k, the values of x_t and c_t, and that it chooses p_t to maximize the expected value of its net return. Hence it sets the unknown value of the random variable u_t equal to its mean value, that we may assume is zero. The necessary condition for the maximum is given by

(ii) $$p_t = (1/2a)(b + kx_t + ac_t).$$

Obviously, p and x are linearly related so that if the outside investigator estimates the equation as follows:

(iii) $$q_t = b - ap_t + v_t,$$

instead of (i), then the least squares estimate of a will be biased where

(iv) $$v_t = kx_t + u_t.$$

Denote the least squares estimate of a by a'. Hence

$$a' = \operatorname{cov}(q_t, p_t)/\operatorname{var} p_t$$

(v) $$= -a + \operatorname{cov}(v_t, p_t)/\operatorname{var} p_t$$

If $\operatorname{cov}(x_t, c_t) = 0$, then

(vi) $$\operatorname{cov}(v_t, p_t) = (k^2/2a)\operatorname{var} x_t.$$

Therefore, (v) reduces to

(vii) $$a' = (-2a^2 \operatorname{var} p_t + k^2 \operatorname{var} x_t)/2a \operatorname{var} p_t$$

But

$$\operatorname{var} p_t = (1/4a^2)(k^2 \operatorname{var} x_t + a^2 \operatorname{var} c_t).$$

Therefore, (vii) simplifies to

(viii) $$a' = a(k^2 \operatorname{var} x_t - a^2 \operatorname{var} c_t)/(k^2 \operatorname{var} x_t + a^2 \operatorname{var} c_t).$$

We can draw several conclusions from (viii). First, $a' \leqslant a$. Second, the larger is $k^2 \operatorname{var} x_t$ relative to $a^2 \operatorname{var} c_t$, the smaller is the bias. Third, the bias does not depend on the sign of either k or a.

share lagged one period and in some cases with market share lagged two periods. The problem is to see whether this is due to autocorrelation of the factors explaining market share or whether it is due to an intrinsic feedback relation in the demand. The latter can result from price expectations or the durability of the goods as is shown in the preceding chapter. Although the goods in the sample are not commonly regarded as durables, for time periods as short as one month they may exhibit some characteristics of a durable good because consumers may hold stocks of these goods. Moreover, there can be a delayed response to price change for many reasons so that one should not rule out a priori the possibility of an intrinsic dynamic demand relation. Therefore, there is the empirical problem of measuring the extent to which the autocorrelation of market share is due to the intrinsic dynamics and to what extent it is due to the autocorrelation of the factors determining market share.

We shall consider the class of demand relations such that the current market share is a linear function of the deflated price of the given brand, the deflated average price of all other brands, lagged prices and lagged market shares. There are complications if the residuals are serially correlated, which may lead us to believe that there is an intrinsic dynamic demand relation when this is false. Serial correlation of residuals can impart serial correlation of market share not easily disentangled from that serial correlation of market share due to the intrinsic dynamic structure of the system. We now examine alternative forms of the demand relation and methods of using the evidence to discriminate among the alternative models.

Let

$$p_{t-j} = PD-J = \text{the deflated price of the given brand at time } t-j,$$

$$p^*_{t-j} = POD-J = \text{the deflated average price of all other brands at time } t-j,$$

$$m_{t-j} = MS-J = \text{market share at time } t-j,$$

where $j = 0, 1, 2$, and the deflator is the monthly Consumer Price Index.

The simplest regression equation is given by

(4) $$m_t = b_0 + b_1 p_t + b_2 p^*_t + u_t,$$

where u_t denotes the random disturbance.[2] We expect a negative b_1 and a positive b_2. (It is convenient to drop the sign conventions of the preceding

---

2. Varients of least squares estimates are used throughout. However, none of these take into account the restriction imposed by the fact that market shares must sum to 1. Observe, however, that we do not estimate the market share regressions for *all* of the brands in a given commodity class so that the shares of the brands actually estimated do not sum to 1. In effect we treat each brand one at a time and view it against all of its competitors. It is an easy exercise to derive the restrictions on the demand parameters that are imposed by use of market share as the dependent variable. Let

(i) $$m_{it} = b_i + \sum_j a_{ij} p_{jt} + k_i m_{i, t-1} + u_{it}$$

material and not display an explicit minus sign in front of the own price slope.) Autocorrelation of the residuals can produce two complications. First, it makes simple least squares estimates of the coefficients of (4) inefficient. There is also a more serious possibility. The residual represents the effects of omitted variables as well as the effects of random shocks. If the omitted variables are serially correlated, then they may also be correlated with price, making least squares estimates of b_1 inconsistent. This point deserves elaboration.

It follows from (1) and (2) that the market share depends on p and p^*. Suppose that the demand equation (1) is actually of the form

$$(5) \qquad q = f(p, p^*, x)g(z),$$

where x is a variable that has a different effect on each brand's sales. It follows that the market share function would be

$$(6) \qquad m = \phi(p, p^*, x).$$

Regardless of whether x itself is a policy variable, that is, regardless of whether the firm chooses the level of x to maximize its net return, it follows from (5) that p would depend on x because p is a policy variable. Now if x is not one of the variables explicitly included in the regression equation (4), then least squares estimates of b_1 would be inconsistent. Moreover, if the residual, u_t, is autocorrelated then this may result from autocorrelation of x_t, a component of u_t. Specifically, assume that the u's satisfy a first-order Markov scheme so that

$$(7) \qquad u_t = \alpha u_{t-1} + \varepsilon_t,$$

---

denote the market share relation for firm i, where $i, j = 1, \ldots, n$.

$$(ii) \qquad \sum_i m_{i,t} \equiv 1$$

is an identity. Summing (i) over i we must have the restrictions as follows:

$$(iii) \qquad \sum_i u_{it} \equiv 0,$$

$$(iv) \qquad \sum_{i,j} a_{ij} \equiv 0,$$

$$(v) \qquad \sum_i k_i \equiv n(1 - \sum_i b_i).$$

It is plausible to rewrite (iv) as follows:

$$(iv') \qquad \sum_j \sum_i a_{ij} \equiv 0$$

and to require that

$$(vi) \qquad \sum_i a_{ij} = 0, \qquad\qquad \text{for all } j.$$

If the market is dichotomized into the given brand and all others, as is shown by (4) in the text, then the complement of (4) is given by

$$1 - m_t = \beta_0 + \beta_1 p_t + \beta_2 p_t^* + v_t.$$

Hence $u_t + v_t = 0$, for all t. Restriction (vi) above implies that $b_1 + \beta_1 = 0$ and $b_2 + \beta_2 = 0$. By using these restrictions one could also devise more efficient estimation methods than simple least squares, but these would be more complicated, and it is doubtful whether the improvement in the estimates would be worth the effort.

where ε_t is independent of u_{t-1}, and is independently distributed over time with mean zero and constant variance. Then (4) becomes

(8) $$m_t = b_0 + b_1 p_t + b_2 p_t^* + \alpha u_{t-1} + \varepsilon_t.$$

But then we can replace u_{t-1} with its value implied by (4) to obtain

(9) $$m_t = b_0(1-\alpha) + b_1 p_t + b_2 p_t^* - \alpha b_1 p_{t-1} - \alpha b_2 p_{t-1}^* + \alpha m_{t-1} + \varepsilon_t.$$

Although the new residual in (9) may still include the effect of the omitted variable, x_t, it is plausible to believe that least squares estimates of (9) will suffer less from inconsistency than least squares estimates of (4).

An important rival to (4) and (9) is a distributed lag relation as follows:

(10) $$m_t = k_0 + k_1 p_t + k_2 p_t^* + a m_{t-1} + v_t,$$

where v_t is a serially uncorrelated random disturbance. Equations (9) and (10) are similar. They differ in that lagged prices do not appear in (10) and in that the coefficients of (9) obey certain restrictions. One can hardly avoid error if one tries to discriminate between (9) and (10) solely by reliance on the statistical significance of the lagged price coefficients in (9). A better way to choose is to see how close the unconstrained least squares estimates of the coefficients of (9) are to the theoretical values as predicted by a first-order Markov process of the residuals.

Equations (9) and (10) have different economic implications. The former implies an equal short- and long-term response to price while the latter implies a greater long- than short-term price response. Thus, in (10), the long-term price response equals the short-term response divided by $1-a$, which is a positive number larger than one, if $0 < a < 1$. However, if the good has durable features, then a can be negative and the short-run price response might exceed the long-term response. For example, a price decrease might induce customers to stock up so that in subsequent periods they lower purchases and run down inventories.[3]

---

3. To see why a negative lagged quantity coefficient is compatible with a durable good, suppose that the current quantity bought varies inversely with the stock on hand so that the demand equation is given by

(i) $$q_t = f_t - b p_t - a s_{t-1},$$

where s_{t-1} = stocks at the end of period $t-1$ = stocks at the beginning of period t. The inventory balance equation is

(ii) $$s_t = \gamma s_{t-1} + q_t \qquad\qquad (0 \leqslant \gamma < 1).$$

Hence

(iii) $$s_{t-1} = (1 - \gamma L)^{-1} q_{t-1}.$$

Substituting (iii) into (i), we obtain

$$q_t = f_t - b p_t - a(1 - \gamma L)^{-1} q_{t-1}.$$

We may multiply through by $(1 - \gamma L)$ to obtain

(iv) $$(1 - \gamma L) q_t = (1 - \gamma L) f_t - b(1 - \gamma L) p_t - a q_{t-1},$$

which reduces to

(v) $$q_t = (1 - \gamma L) f_t - b(1 - \gamma L) p_t + (\gamma - a) q_{t-1}.$$

Hence the coefficient of lagged q is negative if and only if $\gamma < a$. The form of (v) closely resembles (9) in the text (with certain obvious modifications). It differs from (9) with respect to the restric-

Even if the market share completes its response to price within one period so that (4) is the appropriate regression, the residuals may satisfy a second order Markov process as follows:

(11) $$u_t = \alpha_1 u_{t-1} + \alpha_2 u_{t-2} + \varepsilon_t',$$

where $\{\varepsilon_t'\}$ is a sequence of mutually independent and identically distributed random variables with zero means such that every ε_t' is independent of the lagged u's. Equations (4) and (11) imply the market share regression as follows:

(12) $$m_t = b_0(1-\alpha_1-\alpha_2)+b_1 p_t+b_2 p_t^* -\alpha_1 b_1 p_{t-1}-\alpha_1 b_2 p_{t-1}^* \\ -b_1\alpha_2 p_{t-2}-b_2\alpha_2 p_{t-2}^* +\alpha_1 m_{t-1}+\alpha_2 m_{t-2}+\varepsilon_t'.$$

The difference between (9) and (12) hinges on the presence of prices lagged two periods in (12) and their absence from (9).

Thus one can continue postulating higher and higher order autoregressive schemes for either or both the residuals and the dependent variables. Regardless of the assumption concerning the generation of the residuals, one may interpret unconstrained least squares estimates of regressions such as (9) or (12) as dynamic representations of demand which allow a more complicated path of market share in response to price. These equations are special cases of

(13) $$A(L)m_t = b_0+B_1(L)p_t+B_2(L)p_t^* +\text{random shock},$$

where $A(L)$, $B_1(L)$, and $B_2(L)$ are polynomials in the lag operator L and are not necessarily related by autoregressions that are assumed to generate the residuals like the ones given in (7) or (11). One should not be sanguine about the chances of estimating general schemes even with the availability of moderately long time series. We shall not attempt estimation of autoregressions involving more than two lags.[4]

Turning to the evidence, we consider only those brands for which the longest time series are available, thereby amplifying the power of the tests to discriminate among the rival hypotheses. For these brands, regressions are

---

tions imposed on the coefficients. If q_{t-1} has a negative coefficient, then the short-run price response will exceed the long-run price response. Thus, assume there is a once and for all price change in period 0. Then

$$\Delta q_0 = -b\Delta p_0,$$
$$\Delta q_1 = b\gamma\Delta p_0+(\gamma-a)\Delta q_0$$
$$\cdots$$
$$\Delta q_t = (\gamma-a)^{t-1}\Delta q_1, \qquad \text{for } t > 1.$$

For stability it is necessary that $|\gamma-a| < 1$. There is an oscillatory approach to equilibrium if the lagged quantity coefficient is negative. Observe that a price decrease initially raises purchases and subsequently lowers purchases by an amount equal to $-a\Delta q_0$. Thus the initial price response exceeds the subsequent response which is then followed by a return to equilibrium.

4. For a thorough discussion of estimation problems for this class of time series models, see Griliches (1967). The bias due to the presence of serially correlated errors in a distributed lag model is discussed also in Griliches (1961).

VII. Demand and Price Policy by Brand

also estimated for two subperiods in order to see whether the parameters are
stable. It is reassuring to discover that this is indeed the case.

To save space these results are not shown in table 7.16 which includes only
the 5 basic regressions for each of the 10 brands for which the longest time
series are available. The 5 regressions correspond to the 5 alternative hypotheses. Thus, I assumes a simple market share regression (4); II assumes a
simple first-order distributed lag; III allows a test of the hypothesis that the
residuals of I are autocorrelated; IV does the same thing assuming a second
order autocorrelation of the residuals; and the rationale for V is given below
in (14) and (15).[5]

The results are as follows. The current price coefficients have the expected
signs and respectable t-ratios. The multiple R's are about the same and range

TABLE 7.6

Estimates of α_1 and α_2 Assuming a Second-Order Markov Process Generates the Residuals
of Equation (4) Using Coefficients from IV in Table 7.16

PRODUCT AND BRAND		ESTIMATES OF α_1			ESTIMATES OF α_2		
		PD	*POD*	*MS-1*	*PD*	*POD*	*MS-2*
OJ	A1	0.556	0.327	0.732	0.091	0.258	0.024
	B1	0.696	0.823	0.826	0.541	0.160	0.115
	C1	0.284	0.246	0.673	0.110	0.145	−0.068
	E1	0.520	0.430	0.630	0.139	0.601	0.134
IC	A1	0.674	0.701	0.552	0.180	0.095	0.243
	B1	0.522	0.683	0.821	0.161	0.651	−0.025
RC	A1	0.585	0.469	0.626	0.087	0.168	0.037
	B1	0.243	0.230	0.461	0.043	0.058	0.018
	C1	0.875	0.977	0.506	0.226	0.260	−0.053
	D1	0.437	0.215	0.368	0.183	0.339	−0.018

upward from 0.8. A comparison of I and III indicates the presence of autocorrelated residuals in I. The simple first order distributed lag equation II
does not give as good predictions as III–V judged by the criterion of the standard error of prediction. Moreover, the direct estimates of II do not adequately guard against the possibility that the residuals in the distributed lag
equation are serially correlated. For these reasons we focus our attention on
regressions III–V.

The pertinent statistics are shown in tables 7.6–7.9. Assuming the true

5. Observe that the average price of all other brands is derived by dividing total outlays on
these brands by the total number of units bought. Hence each brand price is weighted by its
current market share. As is well known, this price average is lower than one which would give
each brand a constant weight during the sample period because there is an inverse relation
between price and market share.

equation is (4) and there is a second order Markov process generating the residuals, regression IV implies three estimates of α_1 and α_2 as follows:

For α_1 the 3 estimates are
    coefficient of PD-1/coefficient of PD-0,
    coefficient of POD-1/coefficient of POD-0,
    coefficient of MS-1.
For α_2 the 3 estimates are
    coefficient of PD-2/coefficient of PD-0,
    coefficient of POD-2/coefficient of POD-0,
    coefficient of MS-2.

The numerical results are shown in table 7.6. There is no agreement at all among the estimates of α_2 and considerable disagreement among the estimates of α_1. Hence one can safely reject the hypothesis that a second order Markov process generates the residuals.

This narrows the contest to two candidates based on III and V. Regression III assumes a first order Markov process generates the residuals for the basic equation (4). Hence, III corresponds to (9); V assumes the basic equation is

TABLE 7.7

First-Order Distributed Lag and First-Order Markov Process of Residuals Selected Estimates from Regression V in Table 7.16

PRODUCT AND BRAND		ESTIMATES OF δ FROM			ESTIMATE OF a	COEFFICIENT OF MS-2	
		PD	POD	Mean		Predicted	Actual
OJ	A1	0.580	0.532	0.556	0.207	−0.115	−0.046
	B1	0.695	0.765	0.730	0.091	−0.066	0.113
	C1	0.318	0.338	0.328	0.357	−0.117	−0.106
	E1	0.353	0.333	0.343	0.314	−0.108	0.100
IC	A1	0.849	0.791	0.820	−0.202	0.166	0.184
	B1	0.501	0.352	0.426	0.358	−0.153	0.031
RC	A1	0.643	0.666	0.655	0	0	0.003
	B1	0.260	0.260	0.260	0.207	−0.054	−0.001
	C1	0.695	0.751	0.723	−0.282	0.020	0.033
	D1	0.538	0.495	0.517	−0.121	0.062	−0.110

(10), first order distributed lag, and that in addition the residual, v_t, obeys a first order Markov process given by

(14) $$v_t = \delta v_{t-1} + \xi_t.$$

Combining (14) with (10), we obtain the following regression:

(15) $$m_t = k_0(1-\delta) + k_1 p_t + k_2 p_t^* - \delta k_1 p_{t-1} - \delta k_2 p_{t-1}^* + (a+\delta)m_{t-1} - \delta a m_{t-2} + \xi_t.$$

Observe that the coefficient of m_{t-1} is the sum of the structural parameter, a, and the autocorrelation coefficient of the residual, δ. We can try to distinguish between (10) and (15) by seeing whether the estimates of (15) obey the restrictions implied by (14). Table 7.7 gives the estimates of δ. Since there are two explanatory variables, p_t and p_t^*, there are two estimates of δ, one derived from the p coefficients and the other from the p^* coefficients. Table 7.7 shows close agreement between these two estimates in all but one case. The exception is *IC B*1 for which the difference between the two estimates is 0.15. Since the two sets of estimates, while close together, are not the same, it is reasonable to derive a single estimate by averaging the two. The coefficient of m_{t-1} minus this estimate of δ gives an estimate of a. We can test the hypothesis of a first-order distributed lag combined with a first order Markov process on the residuals by comparing the actual coefficient of m_{t-2} with its hypothetical value. The last two columns of table 7.7 give the two sets of figures. There is tolerable agreement especially in view of the rather high standard errors attached to the estimates of the coefficients of m_{t-2}.

There is a second test based on regression III [equation (9)]. This regression explains the presence of lagged m and lagged p's as the result of serially correlated residuals. Table 7.8 compares the actual price coefficients of III with

TABLE 7.8

A Comparison of Actual and Predicted Coefficients of PD-1 and POD-1 Derived from Regression III in Table 7.16

PRODUCT AND BRAND		COEFFICIENT OF $(PD\text{-}1) \times (-10)$			COEFFICIENT OF $(POD\text{-}1) \times 10$		
		Actual	Predicted	Difference	Actual	Predicted	Difference
OJ	*A*1	0.159	0.204	0.045	0.121	0.172	0.051
	*B*1	0.0614	0.0755	0.0141	0.0883	0.0998	0.0115
	*C*1	0.0312	0.0594	0.0284	0.0347	0.0624	0.0277
	*E*1	0.0418	0.0864	0.0446	0.0393	0.0812	0.0419
IC	*A*1	0.138	0.120	−0.018	0.107	0.100	−0.007
	*B*1	0.0325	0.0512	0.0187	0.0180	0.0382	0.0202
RC	*A*1	0.393	0.400	−0.007	0.378	0.372	−0.006
	*B*1	0.058	0.102	0.044	0.053	0.096	0.043
	*C*1	0.351	0.227	−0.124	0.355	0.209	−0.146
	*D*1	0.237	0.153	−0.084	0.234	0.160	−0.074

NOTE: See equation (9) for the formulas relating the coefficients. For example, coefficient of $PD\text{-}1 = -$coefficient $MS\text{-}1 \times$ coefficient $PD\text{-}0$.

those predicted on the hypothesis of a first order Markov process generating the residuals. That is, (9) predicts that the coefficient of a lagged price is minus the product of the current price coefficient and the lagged market share coefficient. The agreement between the actual and predicted coefficients is very good in most cases and in only one case, *RC C*1, is there a difference of any magnitude. Since these results follow from the hypothesis of no distributed lag mechanism so that they assume $a = 0$, we see that the time series

evidence supports this belief. This is also confirmed by the direct estimates of
a shown in table 7.7, which assumes a combination of first order distributed
lag and a first order serially correlated residual [equation (15)]. The estimates
of the *a*'s are tolerably close to zero. Finally, there is the direct evidence
obtained from a comparison of the price coefficients of III and V. In all 10
cases the coefficients of *PD*-0 are virtually the same in III and V, and, likewise,
the *POD*-0 coefficients are nearly equal in the two regressions. Therefore, it

TABLE 7.9

Estimates of Market Share Price Elasticities by Brand Evaluated at Sample Means from Regression V in Table 7.16

PRODUCT AND BRAND		PD ELASTICITY		POD ELASTICITY		NAÏVE ESTIMATES
		Short Run	Long Run	Short Run	Long Run	
OJ	A1	−3.89	−4.90	2.78	3.51	− 9.62
	B1	−2.55	−2.80	2.90	3.19	− 6.17
	C1	−2.96	−4.60	2.77	4.31	− 3.38
	E1	−3.38	−4.93	3.10	4.52	−13.11
	Mean	−3.26	−4.31	2.89	3.88	
IC	A1	−1.72	−1.43	1.44	1.20	+
	B1	−2.49	−3.88	1.97	3.07	− 3.93
	Mean	−2.10	−2.65	1.70	2.14	
RC	A1	−3.74	−3.74	3.21	3.21	− 8.85
	B1	−4.09	−5.16	3.63	4.58	−15.13
	C1	−3.16	−2.46	2.80	2.18	− 7.56
	D1	−4.97	−4.43	4.98	4.44	+
	Mean	−3.99	−3.95	3.85	3.85	
Instant mashed		−1.04	−10.4	−	−	
potatoes*		−2.24	−11.2	0.56	2.80	

* The first set of estimates correspond to II in table 7.14 and the second to III in table 7.14.
The naïve estimates are from table 7.3.

makes little difference for estimates of the short-run price elasticities derived
from time series regressions whether we use III or V. Nor is this all. If there is
a distributed lag mechanism, the estimated coefficients of lagged market
share, the *a*'s, are fairly small, indicating that the difference between the short-
and long-run price elasticity is also fairly small. Finally, there is strong evidence that the residuals satisfy a first order Markov process.

In summarizing this work it is best to give the estimates of the price elasticities for V. The short-run price elasticities are virtually the same as they would
be for III, and we gain some notion of the estimated long-run price elasticities.
These results are given in table 7.9. The own and the cross price elasticities
are quite close together. Thus, with constant relative prices among the brands,
there would also be constant market shares. Let us also compare these results
with the naïve estimates of price elasticities. Plainly, there is one striking fact.
The naïve estimates imply much larger price elasticities than the more

sophisticated ones. It may well be that the naïve estimates, since they derive from two subsamples of about $3\frac{1}{2}$ years each, come closer to the long run elasticities. It is also true that in the two cases where the naïve estimates fail by giving positive price slopes, the regression estimates yield slopes with the "right" signs. This experience nevertheless suggests that even the naïve estimates, which use no regressions and can be done on the back of an envelope, may give reasonable estimates of the price elasticities.

3. Estimates of the Relations among Competing Prices

The results of the preceding section give lower limits for the own price elasticities. Therefore, the reciprocals of these elasticities estimate the upper limits for the ratio of net to gross receipts, assuming Cournot-Nash competition. Unfortunately, we cannot tell whether this assumption is justified from the demand parameters. Nor can we discover the state of competition from the relations among prices since common cost and demand factors enter all prices in any case. Whether these common forces explain positive relations among competing prices or whether the firms consciously follow a collusive price policy is not easily answered. It is plausible, however, to expect a looser relation among prices under competition than under collusion because of the very nature of collusion. This section gives empirical estimates of price relations among competing goods. The brand prices are divided by the Consumer Price Index to eliminate common forces affecting all prices.

Before examining the relations among prices, it is advisable to see how well prices can be predicted from their own past behavior. Table 7.10 has the results. There are two autoregressions for every brand, a first order autoregression and the best fitting autoregression from among those of degrees 2 to 4 as judged by the smallness of the standard error of estimate. The evidence is plain. Inclusion of prices lagged more than one period reduces the standard error of estimate by very little. Moreover, the sum of the coefficients of lagged prices in the higher order autoregression is almost the same as the coefficient of p_{t-1}. Finally, in the first order autoregression the coefficient of p_{t-1} is always well above 0.9. Hence prices closely conform to a random walk and the best predictor of next month's price is this month's price. These findings support the Cournot expectations model. Recall from chapter VI that for this model if the goods are gross but not perfect substitutes then an equilibrium exists.

A good way to examine the price relations is by means of autoregressions of the price differentials, $p_t - p_t^*$. In this way we partially control for common cost and demand forces affecting price and can see how well competing prices are related while using the minimum number of degrees of freedom. Denote the deflated price differential lagged J periods as follows:

$$(1) \qquad\qquad DPD\text{-}J = p_{t-j} - p_{t-j}^* \qquad\qquad (j, J = 0, 1, 2, ...).$$

TABLE 7.10
PD Autoregressions for OJ, IC, and RC

Product and Brand	Constant	PD-1	Coefficients of PD-2	PD-3	PD-4	S.E.	R	Σ Coefficients
OJ A1 I	0.7784	0.9562	...	...	...	0.6384	0.9575	0.9562
	1.308	28.378						
II	1.1505	1.2381	−0.1695	−0.1338	...	0.6029	0.9633	0.9348
	1.994	10.683	0.948	1.201				
B1 I	1.4650	0.9141	...	...	...	0.9452	0.9139	0.9141
	1.789	19.231						
II	1.7564	0.9864	−0.1942	0.4236	−0.3190	0.9064	0.9245	0.8968
	2.110	8.720	1.245	2.736	2.872			
C1 I	0.6077	0.9647	...	...	...	0.6500	0.9650	0.9647
	1.1591	31.451						
II	1.0627	1.1415	−0.0936	0.1277	−0.2385	0.6133	0.9702	0.9371
	2.0696	9.682	0.509	0.696	2.049			
E1 I	0.7392	0.9520	...	...	...	0.7203	0.9529	0.9520
	1.355	26.849						
II	1.0161	1.0941	−0.1444	0.1676	−0.1840	0.7120	0.9560	0.9333
	1.819	9.336	0.827	0.964	1.590			
IC A1 I	0.3468	0.9799	...	...	...	0.8374	0.9953	0.9799
	0.754	85.758						
II	0.3093	1.0243	−0.1974	0.1528	...	0.8392	0.9955	0.9797
	0.670	8.547	1.151	1.303				
B1 I	0.7627	0.9673	...	...	...	1.2893	0.9866	0.9673
	1.098	50.279						
II	0.5613	0.8700	−0.2189	0.0680	0.2468	1.2058	0.9888	0.9659
	0.860	7.468	1.393	0.448	2.276			
RC A1 I	1.6451	0.9695	...	...	...	1.6341	0.9922	0.9695
	1.433	66.268						
II	1.7181	1.1555	−0.1854	...	...	1.6137	0.9925	0.9701
	1.514	10.232	1.661					
B1 I	1.8446	0.9660	...	...	...	2.1486	0.9859	0.9660
	1.231	48.875						
II	1.5889	1.1405	−0.1475	−0.2224	0.1989	2.1148	0.9869	0.9695
	1.0564	9.1589	0.809	1.220	1.709			
C1 I	1.0618	0.9764	...	...	...	1.7039	0.9918	0.9764
	0.917	64.625						
II	1.1734	1.1608	−0.1843	...	...	1.6850	0.9921	0.9765
	1.023	9.977	1.599					
D1 I	1.1338	0.9757	...	...	...	1.8748	0.9901	0.9757
	0.892	58.609						
II	1.2260	1.2976	−0.5337	0.2116	...	1.784	0.9913	0.9755
	0.997	11.268	2.922	1.836				

Table 7.11 gives two autoregressions of DPD; the first order autoregression and the best fitting autoregression chosen from among four with lags up to 4 periods. There are four major conclusions. First, the price differentials are well described by a first order process since the introduction of lagged differentials before last month's contributes little toward reducing the standard error of prediction. Second, there is much less collinearity among the lagged price differentials than among the lagged prices themselves, as is shown by comparing the sum of the lagged price differential coefficients with the coefficient of $p_{t-1} - p_{t-1}^*$ in the first order autoregression. Third, the lagged price differential coefficient tends to be lower than the lagged price coefficient. Fourth, the constant terms of the first order autoregressions in table 7.11 have a larger t-ratio than the constant terms of the first order autoregressions in table 7.10. This means that there is a stable set of price differences among the brands describing their long run relations.

There are some notable differences in the results by commodity. For regular coffee the lagged price differential coefficients range from 0.68 to 0.80. For three orange juice brands, the coefficients are 0.65, 0.66, and 0.73; but one, brand B, has a markedly lower coefficient, 0.40. This is hard to explain because this brand does not seem to differ from the others. For example, like brand E it experienced a 50 percent drop in its market share from the time of the first to the second subsample. The $p_t - p_{t-1}^*$ coefficients also differ for the instant coffee brands. For the leading brand this coefficient is 0.84 and for the second brand it is 0.52. One is tempted to conclude that the price of the leading brand approaches its normal level more slowly than its closest competitor in this case. In general, one would like to say that the coefficient of $p_{t-1} - p_{t-1}^*$ measures the rapidity of the convergence of a brand's price to its long run level.

To see whether this interpretation is defensible, consider the evidence of a closely related set of regressions as shown in table 7.12. These regressions are estimates of the following equation:

(2) $$p_t = a_0 + a_1 p_t^* + b_1 p_{t-1} + b_2 p_{t-1}^* + \text{residual.}$$

The autoregressions of the price differentials in table 7.11 give estimates of

(3) $$p_t - p_t^* = c_0 + c_1(p_{t-1} - p_{t-1}^*) + \text{residual,}$$

which is the special case of (2) derived by setting $a_1 = 1$ and $b_2 = -b_1$. In fact, by comparing the standard error of prediction for corresponding regressions in table 7.11 and 7.12, we see that (3) gives almost as good a fit as (2). This is equivalent to performing an F-test to see whether (3) can be accepted as a statistical explanation of these data.[6] Suppose on the basis of these

---

6. Although the dependent variable in (2) and (3) is different, it is nevertheless valid to compare the standard error of the predictions in the two equations to test the hypothesis that the restrictions imposed on the parameters in (3) are valid. This is because (2) is equivalent to the equation as follows:

TABLE 7.11
Autoregressions of DPD by Brand for OJ, IC, and RC

PRODUCT AND BRAND			CON-STANT	DPD-1	COEFFICIENTS OF DPD-2	DPD-3	DPD-4	S.E.	R	Σ COEFFICIENTS
OJ	A1	I	0.9564	0.6598	...	...	...	0.3928	0.6659	0.6598
			3.901	7.627						
		II	1.5059	0.6311	0.1440	−0.0944	−0.2186	0.3783	0.7107	0.4621
			4.804	5.598	1.102	0.729	1.964			
	B1	I	1.1906	0.3989	...	...	...	0.5351	0.3981	0.3989
			5.351	3.707						
		II	1.0736	0.4489	−0.1347	0.3072	−0.1629	0.5231	0.4784	0.4585
			3.336	3.805	1.087	2.485	1.332			
	C1	I	0.5005	0.7306	...	...	...	0.4842	0.7289	0.7306
			3.188	9.098						
		II	0.3644	0.6338	−0.0328	0.3409	−0.1380	0.4688	0.7607	0.8039
			2.086	5.422	0.241	2.504	1.135			
	E1	I	0.0472	0.6499	...	...	...	0.4769	0.6518	0.6499
			0.844	7.344						
		II	0.0504	0.5151	0.2788	0.0506	−0.2068	0.4658	0.6883	0.6377
			0.920	4.385	2.104	0.383	1.779			
IC	A1	I	0.3029	0.8367	...	...	...	0.9409	0.8346	0.8367
			1.803	12.584						
		II	0.4031	0.9823	−0.2879	0.3602	−0.2670	0.9140	0.8521	0.7876
			2.556	8.306	1.764	2.212	2.253			
	B1	I	−1.3173	0.5165	...	...	...	1.0192	0.5226	0.5165
			4.235	5.092						
		II	−0.6835	0.4486	−0.0017	−0.0454	0.3297	0.9752	0.6028	0.7312
			1.708	3.900	0.0133	0.358	2.872			
RC	A1	I*	1.5323	0.7419	...	...	...	0.8748	0.7415	0.7419
			3.086	9.179						
	B1	I	0.8463	0.6778	...	...	...	1.2637	0.6781	0.6778
			3.063	7.663						
		II	0.7278	0.6828	0.1138	−0.2748	0.1990	1.2512	0.7024	0.7208
			2.232	5.658	0.795	1.925	1.661			
	C1	I	0.6939	0.7864	...	...	...	0.8973	0.7665	0.7864
			2.409	9.915						
		II	5.584	0.6702	0.1556	...	...	0.8934	0.7727	0.8258
			1.824	5.534	1.266					
	D1	I	0.6023	0.8034	...	...	...	1.1094	0.8027	0.8034
			2.344	11.181						
		II	0.2762	0.6266	0.0834	−0.1279	0.3246	1.0524	0.8330	0.9067
			1.0215	5.359	0.610	0.951	2.811			

* In this case the first-order autoregression has the lowest standard error.

(2′) $p_t - p_t^* = a_0 + (a_1 - 1)p_t^* + b_1 p_{t-1} + b_2 p_{t-1}^* + \text{residual}$,

and the residual in (2) and (2′) is precisely the same. Hence one may compare the standard errors of residuals in (2) and (3) to test whether (3) is an adequate representation of the data.

TABLE 7.12

Estimates of Price Policy Regressions for OJ, IC, and RC by Brand, PD-0 Dependent

PRODUCT AND BRAND		CONSTANT	POD-0	COEFFICIENTS OF PD-1	POD-1	S.E.
OJ	A1	1.1454	0.7754	0.6891	−0.4828	0.3669
		2.761	12.153	8.159	5.083	
	B1	1.3626	1.0928	0.3944	−0.4980	0.5379
		2.813	11.505	3.643	3.514	
	C1	0.2561	0.7219	0.7659	−0.4759	0.4513
		0.699	8.910	9.950	4.586	
	E1	−0.0058	0.8220	0.6715	−0.4901	0.4680
		0.016	9.876	7.671	4.300	
IC	A1	0.2158	0.3586	0.8884	−0.2548	0.7885
		0.493	3.081	15.728	2.153	
	B1	0.1511	0.8920	0.2978	−0.2460	0.9554
		0.230	5.452	2.549	1.321	
RC	A1	0.5621	1.0414	0.5862	−0.6005	0.8473
		0.918	13.673	5.910	5.068	
	B1	−0.1227	1.3709	0.5907	−0.9414	1.1804
		0.146	12.671	6.777	7.899	
	C1	−1.5156	1.1211	0.5252	−0.6023	0.8152
		2.497	15.223	5.431	5.308	
	D1	−1.1627	1.1775	0.7004	−0.8473	1.0686
		1.501	12.049	8.835	7.693	

TABLE 7.13

Tests for the Presence of First Order Serial Correlation of Residuals of Regressions in Table 7.12

PRODUCT AND BRAND		COEFFICIENT OF PD-1 FROM TABLE 7.12	−[COEFFICIENT OF POD-1]/ [COEFFICIENT OF POD-0]
OJ	A1	0.689	0.623
	B1	0.394	0.456
	C1	0.766	0.659
	E1	0.671	0.596
IC	A1	0.888	0.710
	B1	0.298	0.276
RC	A1	0.586	0.577
	B1	0.591	0.687
	C1	0.525	0.537
	D1	0.700	0.720

results that we do accept (3). There remains the question of how to interpret c_1. Does c_1 measure the rapidity of convergence of a given brand's price to its long-term level? To answer this question, let us assume a still simpler model than (3), namely,

$$(4) \qquad\qquad p_t - p_t^* = K + u_t,$$

where

$$(5) \qquad\qquad u_t = \delta u_{t-1} + e_t,$$

so that u_t, the residual of (4), satisfies the familiar first order Markov process. Together equations (4) and (5) imply that $b_2 = -a_1\delta$. Hence,

$$(6) \qquad\qquad b_1 = -b_2/a_1 = \delta.$$

Table 7.13 gives the figures to show how well the estimates satisfy (6). We can see close agreement between the PD-1 coefficient and the predicted coefficient in the adjacent column. Therefore, the coefficient of $p_{t-1} - p_{t-1}^*$ should be interpreted as resulting from the serial correlation of the residuals assumed in (4) instead of as a measure of the rapidity of approach to the long-term price difference K. Nor is this all. As we shall see in the next section, equations such as (2) or (3) can be interpreted as estimates of the brand's price policy. Hence the coefficient of p_{t-1} is related to the lagged market share coefficient in the demand relation under the hypothesis that the firms choose prices in order to maximize the present value of their net return [see 4(3)]. If the lagged market share coefficient in the demand equation is small, then the lagged price coefficient in the price policy equation must be still smaller according to the hypothesis that prices are chosen optimally. Since we have found the coefficient of m_{t-1} in a distributed lag mechanism to be small, we should also find price policy equations of the form (4). In the next section which analyzes the experience for the leading brand of a new product, instant mashed potatoes, we shall encounter a case where the lagged price coefficient in (2) and the lagged market share coefficient in the demand equation are closely related by the equation describing the optimal policy.

4. Competition in a New Product

Instant mashed potatoes was a new product during the sample period from June 1958 to September 1962. During this time the leading brand adhered to a relatively stable price while it experienced a drop in its market share from 75 percent at the beginning to 33 percent at the end. This decline was not the result of a fall in its absolute sales. It resulted from growth in total sales as new brands entered. The entrants initially asked low prices to attract clientele, and those new brands which were successful subsequently raised their prices to a level not much below the leading brand.

The entrants' initial low price presumably induced customers to try the new product at a small cost, and those customers who liked it continued buying after prices rose to their normal level. These circumstances offer a good opportunity to study a temporary monopoly.

The first statistical sign of unusual market conditions is given by the simple regression of market share on prices in table 7.14. Regression I shows a positive p_t coefficient and a negative p_t^* coefficient, just the opposite of our

TABLE 7.14

Instant Mashed Potatoes, Estimates of the Demand for the Leading Brand with Market Share (MS-0) as the Dependent Variable

	CONSTANT	PD-0	POD-0	PD-1	POD-1	PD-2	POD-2
I	0.4735	0.02868	−0.03532				
	0.766	1.374	6.093				
II	0.6212	−0.01714	−0.005886				
	1.832	1.406	1.408				
III	0.5751	−0.03701	0.007593	0.02553	−0.01687		
	1.685	2.911	1.417	2.118	3.163		
IV	0.6939	−0.03714	0.01070	0.01826	−0.01793		
	2.030	2.980	1.923	1.453	3.405		
V	0.4186	−0.03146	0.01077	0.00095	−0.01741	0.02228	−0.0001
	1.064	2.468	1.838	0.059	2.279	1.769	0.016
VI	0.6736	−0.03157	0.009840	−0.00070	−0.01783	0.02955	0.003272
	1.508	2.457	1.647	0.043	2.272	1.974	0.397
VII	0.7075	−0.03026	0.008886	−0.001487	−0.01577	0.02512	0.002937
	1.562	2.310	1.435	0.091	1.852	1.417	0.353

	PD-3	POD-3	M-1	M-2	M-3	S.E.	R
I	...	...	...	...	...	0.08559	0.7579
II	...	...	0.8766	...	...	0.04691	0.9353
			10.838				
III	...	...	0.8139	...	...	0.04100	0.9530
			11.219				
IV	...	...	0.6367	0.2348	...	0.04020	0.9558
			5.041	1.698			
V	...	...	0.5485	0.2954	...	0.03971	0.9588
			3.714	2.070			
VI	−0.01374	−0.003866	0.5797	0.2620	...	0.03994	0.9603
	1.094	0.629	3.841	1.706			
VII	−0.01104	−0.004772	0.5489	0.1884	0.1025	0.04021	0.9607
	0.829	0.752	3.448	0.982	0.648		

expectations. Moreover, the negative p_t^* coefficient is more than 6 times its standard error. Finally, the Durbin-Watson statistic is 0.86, indicating the presence of a high positive serial correlation of the residuals.

The addition of the lagged market share in regression II, table 7.14 greatly improves the fit, changes the sign of the p_t coefficient from positive to negative, and converts the serial correlation from a large positive to a small negative number (the new Durbin-Watson statistic is 2.3), but the p_t^* coefficient remains negative though it moves closer to zero.

The addition of lagged prices in III improves the fit, but, beyond regression III, from IV to VII the standard error of estimate stays about the same. The p_t coefficient stays in the range -0.037 to -0.030 and the p_t^* coefficient hovers around 0.01. The corresponding market share price elasticities are between -1.9 and -2.3 for the own price and only 0.56 for the cross price.

The difference between III and IV in table 7.14 is slight. Moreover, it does not seem possible to interpret the latter in the same way as in section 2. Thus the brands in section 2 have market share equations that are first order distributed lag with first order autocorrelated residuals, but for instant mashed potatoes the residuals are serially independent since the estimated autocorrelation of residuals derived from the ratio of the coefficients of p_t and p_{t-1} is about 0.5 but that derived from the ratio of p_t^* and p_{t-1}^* is about

TABLE 7.15

Estimates of Price Policy Regressions for the Leading Brand of Instant Mashed Potatoes, PD-0 Dependent Variable

	CON-STANT	PD-1	PD-2	PD-3	COEFFICIENTS OF PD-4	POD-0	POD-1	S.E.	R
I	8.08	0.6870	...	...	...	...	...	0.490	0.6944
	3.133	6.892							
II	7.218	0.7545	-0.1918	0.1089	0.0486	...	...	0.497	0.7064
	2.288	5.231	1.071	0.608	0.345				
III	12.88	0.5618	...	...	...	-0.07014	...	0.472	0.7280
	3.938	5.068				2.253			
IV	13.52	0.5419	...	...	...	-0.03829	-0.03786	0.474	0.7305
	3.925	4.675				0.644	0.630		

1.7, a sizeable difference. Moreover, for the three established goods after correcting for serially correlated residuals there is a much smaller lagged market share coefficient than for the leading brand in the new product, instant mashed potatoes.

Consider now the evidence from the price regressions shown in table 7.15. The autoregression of p_t seems to be first order but with a smaller p_{t-1} coefficient, 0.69, than for the established goods' brands in the preceding section. Moreover last month's price is not the best forecast of this month's price for the new product. The autoregression of p_t^*, not shown in the table, is also of first order but has a much higher lagged price coefficient, 0.86, more nearly like the values of the comparable coefficients for the established commodity brands. Regression IV, corresponding to 3(2) is unlike the corresponding regressions of section 3 since it gives no evidence of serially correlated residuals. Indeed, IV has a most interesting interpretation, as we shall see.

To develop this interpretation we must briefly review some analysis from chapter VI in order to derive the necessary condition for the maximum present value of the monopoly net return. Assume that the leading brand regards the prices of competing brands as exogenous and that its demand equation is given by

(1) $$A(L)q_t = b_t - p_t.$$

Thus, for the subsequent development it is convenient to set the price coefficient equal to -1 resulting in no loss of generality. $A(L)$ is a scalar polynomial in the lag operator L. Assume there is a constant per unit cost c_t that may change over time. The present value of the firm's net revenue is given by

$$PV = \Sigma \, \alpha^t q_t (p_t - c_t) = \Sigma \, \alpha^t q_t [b_t - c_t - A(L)q_t].$$

The necessary condition for maximum PV with quantity as the policy variable is as follows:

(2) $$b_t - c_t - [A(L) + A(\alpha E)]q_t = 0$$

(E is the lead operator, see sec. VI.2). To convert this into the necessary condition in terms of price, observe that (2) implies

$$-c_t + p_t - A(\alpha E)q_t = 0.$$

Since

$$q_t = A(L)^{-1}(b_t - p_t),$$

and recalling that the operators $A(L)$ and $A(\alpha E)$ are scalar polynomials so that products commute, $A(L)A(\alpha E) = A(\alpha E)A(L)$, we obtain

$$-c_t + p_t - A(\alpha E)A(L)^{-1}(b_t - p_t) = 0,$$

which implies the price policy equation as follows:

(3) $$[A(L) + A(\alpha E)]p_t = A(L)c_t + A(\alpha E)b_t.$$

The characteristic equation for (3), z a complex scalar, is given by

$$A(z^{-1}) + A(\alpha z) = 0.$$

We may set $\alpha = 1$ because with monthly data, even if the annual discount rate is as high as 18 percent, the monthly discount rate would only be 1.5 percent and the discount factor α would be $\alpha = 1/(1+0.015)$, which is close to one. For annual discount rates below 18 percent, the monthly discount factor α is even closer to 1. Hence the characteristic equation for the price policy equation becomes

(4) $$A(z^{-1}) + A(z) = 0.$$

For the original demand equation (1), the characteristic equation is $A(z^{-1}) = 0$, which is obviously related to (4). Our goal is to relate the coefficient of p_{t-1} in the price regression III, table 7.15 to the coefficient of m_{t-1} of the market share demand equation in regression II, table 7.14 on the grounds that the former estimates the price policy equation.

Assume that the current and lagged p_t^* represent the right-hand side of (3) because the price of all other brands depends on roughly the same cost and demand factors as does the given brand. Moreover, with $\alpha = 1$, the operator $A(L) + A(E)$ can be factored so that it can be represented as follows:

(5) $$A(L) + A(E) = M(E)M(L)$$

[see lemma 4, sec. VI.5 and eq. VI.5(35)]. Hence we may represent (3) empirically by the following expression:

(6) $$M(L)p_t = M(E)^{-1}P(L, E)p_t^*,$$

where

$$A(L)c_t + A(E)b_t = P(L, E)p_t^*.$$

In the simple cases we shall consider, we can determine the factors $M(E)$ and $M(L)$ from the stable roots of the characteristic equation (4). $M(L)$ is composed of those roots of (4) that are inside the unit circle (see the examples in sec. VI.4).

Assume that the market share regressions in table 7.14 provide estimates of $A(L)$. For regression II, $A(L) = 1 - 0.88L$. Therefore, equation (4) becomes

$$(1 - 0.88\,z^{-1}) + (1 - 0.88\,z) = 0,$$

for which the roots are 0.60 and 1.67. Hence the coefficient of p_{t-1} in the price regression III, table 7.15 should be 0.60. It is actually 0.56 in III and 0.54 in IV. This is very good agreement.

We may also try more complicated models. Thus suppose the demand equation is given by

(7) $$A(L)q_t = b_t - C(L)p_t,$$

instead of (1), where $C(L)$ is a scalar polynomial in L. This converts (7) into a form like (1) using the relation

$$C(L)^{-1}A(L)q_t = C(L)^{-1}b_t - p_t.$$

The necessary condition for maximum PV requires the quantity sequence to satisfy

(8) $$[A(L)C(L)^{-1} + A(\alpha E)C(\alpha E)^{-1}]q_t = C(L)^{-1}b_t - c_t.$$

The corresponding price policy equation is

(9) $$[A(L)C(\alpha E) + A(\alpha E)C(L)]p_t = A(L)C(\alpha E)c_t + A(\alpha E)b_t.$$

The characteristic equation for (9) is more complicated and is given by

$$(10) \qquad\qquad A(z^{-1})C(\alpha z) + A(\alpha z)C(z^{-1}) = 0.$$

This is the appropriate price policy equation for a market share demand relation such as III, table 7.14. For III, table 7.14,

$$A(L) = 1 - 0.81\, L, \quad C(L) = 1 - 0.7\, L.$$

The latter operator is derived from the coefficients of p_t and p_{t-1} as follows:

$$-0.0371[1 - (0.0255/0.037)L] = -0.037(1 - 0.7\, L).$$

The equation corresponding to (10) is

$$(11) \qquad (1 - 0.8z^{-1})(1 - 0.7z) + (1 - 0.8z)(1 - 0.7z^{-1}) = 0$$

for which the roots are 0.73 and 1.37. Hence the coefficient of p_{t-1} in III, Table 7.15 is predicted to be 0.73. It is actually 0.55. Hence the more complicated model gives poorer results than the simpler one.

Still more complicated market share demand relations imply higher order autoregressive price policy equations. The estimates of these equations shown in table 7.15 conform less well to the predictions than the two simpler ones above.

A similar interpretation explains why the price relations for the three established products, frozen orange juice concentrate, instant coffee, and regular coffee are so simple. In the preceding section we accept the hypothesis that

$$(12) \qquad\qquad p_t - p_t^* = K + u_t,$$

where

$$u_t = \delta u_{t-1} + e_t,$$

a first order Markov process. Equation (12) is implied by a market share demand relation in which the coefficient of lagged market share is zero so that $A(L) = 1$. But this is in fact nearly true of the estimated market share demand relations as shown in table 7.6. Therefore, a single theory can explain the price relations for all four products in terms of the size of the m_{t-1} coefficient in the market share demand relations. The closer this coefficient is to zero the less related are the current and the lagged price differentials.

Demsetz (1962) reports on frozen orange juice concentrate in a period of its history like the one for instant mashed potatoes. Using monthly data for 1950 to 1957 from the *Chicago Tribune* Consumer Panel, he estimates demand parameters for frozen orange juice concentrate. Demsetz finds that the group of leading and more heavily promoted brands declined in market share from 70 to 30 percent and only slightly reduced their price. His estimated own price elasticity is -1.6, and his cross price elasticity is 0.6. Using regression II, table 7.14 the own and cross price elasticities are -1 and -0.3, respectively.

For III the estimated elasticities are −2.2 and 0.4. Apparently, similar forces were at work in the early stages of both products.

5. Conclusions

A simple procedure gives reasonable estimates of price elasticities for 8 out of 10 brands in 3 established consumer products. These elasticities are derived from the means of two subsamples, which can be used to measure the slope of market share with respect to the deflated price. The results given in table 7.3 exceed the price elasticities derived from monthly time series regressions, table 7.9. It is, therefore, of interest to mention two other cases where equally simple methods give estimates of own price elasticities. Both are derived from episodes described in Telser (1962a). In the first, a certain brand of margarine experienced a sharp price change during the sample period and comes as close to offering a controlled experiment in price effects as one could wish for. The implied price elasticity is −8.23 (see Telser 1962a, p. 323, and chart 5, p. 316). The second episode occurs in beer. Professor Aaron Director brought to my attention the pertinent data. Anheuser-Busch reduced their price of beer per case in St. Louis in two steps while their competitors did not change their prices. The Anheuser-Busch price was $2.93 in October 1953, $2.68 in January 1954, and $2.35 in June 1954. Their market share rose from 12.5 percent at the end of December 1954 to 16.55 percent at the end of June 1954, say in response to the first price reduction. For this six month period the implied price elasticity is −3.8. From the end of June 1954 to the beginning of March 1955 their market share rose from 16.55 percent to 39.3 percent. During this eight-month period the price elasticity is −11.2. For the entire fourteen-month period the elasticity is −10.8. The data can be found in FTC v. Anheuser-Busch, Inc., 80 S. Ct. 1267 (1960) (Telser 1962a, n. 23, p. 324).

Using the data for a new product we are able to deduce a relation between the market share price regression and a certain autoregression of prices for the leading brand. This evidence supports the view that the price policy of the leading brand is as if it were a temporary monopoly. Thus, assuming a first order distributed lag market share demand relation, the implied coefficient of lagged price in the price policy equation on the hypothesis of a profit maximizing monopoly is very close to the actual estimate.

For the three established products the data are consistent with the hypothesis that the market share demand relation is a first order distributed lag with residuals generated by a first order Markov process. The lagged market share coefficient is quite small and, therefore, implies an even smaller coefficient of lagged price in the price policy equation. In fact the lagged price coefficient is virtually zero, according to the evidence presented in section 3. This evidence indicates that there are constant price differentials among the brands subject to an additive random disturbance obeying a first order Markov process. One may interpret these results in terms of brand loyalty. Thus the leading brand of the new product, instant mashed potatoes, commanded more brand loyalty than any of the brands in the established products where we take the lagged market share coefficient in the market share demand relation as a measure of brand loyalty.

TABLE 7.16
Estimates of Market Share Demand Regressions by Brand of Frozen Orange Juice Concentrate, Instant Coffee, and Regular Coffee

	CONSTANT	PD-0	POD-0	PD-1	POD-1	PD-2	POD-2	M-1	M-2	S.E.	R
					OJ A1 (n = 77)						
I	.2438	-.02721	.02449	...	...	...	...	...	...	.01777	.6220
		6.730	6.274								
II	.1273	-.01948	.01821	...	...	...	...	.5746	...	.01309	.8197
		6.222	6.109					7.963			
III	.07852	-.02814	.02377	.01588	-.01215	...	...	.7247	...	.01209	.8503
		7.457	6.473	3.556	2.771			8.773			
IV	.06834	-.02814	.02226	.01566	-.007285	.002575	-.005735	.7320	.02439	.01221	.8566
		7.317	5.637	2.615	1.190	.527	1.202	5.907	.198		
V	.07877	-.02844	.02415	.01649	-.01284	...	...	.7631	-.04576	.01226	.8509
		7.398	6.399	3.536	2.770			6.643	.484		
					OJ B1						
I	.1611	-.01698	.01228	...	...	...	...	...	...	.02152	.5516
		3.940	2.785								
II	.02157	-.005909	.005673	...	...	...	...	.8799	...	.009121	.9363
		3.073	2.983					18.41			
III	.008392	-.008122	.01073	.006145	-.008830	...	...	.9297	...	.008517	.9463
		4.272	4.240	3.099	3.539			19.53			
IV	.006105	-.008465	.01118	.005895	-.009197	.0004582	.0001788	.8258	.1152	.008623	.9473
		4.341	4.043	2.560	2.122	.210	.059	6.316	.897		
V	.008405	-.008438	.01085	.005869	-.008303	...	...	.8214	.1133	.008512	.9471
		4.386	4.285	2.936	3.264			7.204	1.045		

Table 7.16 continued

	CONSTANT	PD-0	POD-D	PD-1	POD-1	PD-2	POD-2	M-1	M-2	S.E.	R
					OJ C1						
I	.07776	-.01389 8.685	.01416 8.129	...	...	...	...	...	...	.009603	.7116
II	.03848	-.008443 5.775	.008686 5.572	...	...	...	...	.5273 7.054	...	.007456	.8405
III	.03249	-.009839 5.065	.01034 5.360	.003116 1.337	-.003471 1.566	...	...	.6039 6.671	...	.007432	.8464
IV	.03238	-.01015 5.005	.01022 5.010	.002887 1.072	-.002517 .870	.001118 .459	-.001481 .588	.6732 5.613	-.0685 5.577	.007511	.8500
V	.03404	-.009850 5.079	.01035 5.375	.003128 1.344	-.003497 1.580	...	...	.6850 5.867	-.1061 1.097	.007422	.8492
					OJ E1						
I	.1008	-.02272 8.363	.01967 6.915	...	...	...	...	...	...	.01474	.7379
II	.02662	-.009987 4.633	.009365 4.461	...	...	...	...	.6850 10.19	...	.009534	.9011
III	.02059	-.01161 4.711	.01091 4.268	.004180 1.548	-.003928 1.346	...	...	.7444 9.580	...	.009507	.9046
IV	.02240	-.01053 4.367	.007053 2.558	.005479 1.934	-.003036 .822	-.001465 .544	-.004238 1.490	.6308 5.372	.1340 1.188	.009056	.9176
V	.01860	-.01167 4.731	.01079 4.213	.004115 1.522	-.003597 1.223	...	...	.6571 5.456	.1003 .949	.009513	.9059

Table 7.16 continued

	CONSTANT	PD-0	POD-0	PD-1	POD-1	PD-2	POD-2	M-1	M-2	S.E.	R
					IC A1 (n = 73)						
I	.3081	−.005722	.007107	...	...	...	...	...	...	.02467	.5413
		3.302	4.035								
II	.1219	−.004307	.004987	...	...	...	...	.6023	...	.01834	.7840
		3.308	3.724					7.588			
III	.06494	−.01563	.01301	.01376	−.01066	...	...	.7696	...	.01512	.8635
		6.528	5.518	5.544	4.327			10.74			
IV	.05407	−.01482	.01330	.009996	−.009329	.002667	−.001268	.5525	.2431	.01485	.8750
		6.159	5.583	2.467	2.413	.911	.454	4.396	2.106		
V	.05346	−.01527	.01344	.01297	−.01063	...	...	.6181	.1843	.01473	.8732
		6.533	5.832	5.301	4.432			6.248	2.159		
					IC B1						
I	.06842	−.01224	.01189	...	...	...	...	...	...	.01902	.5745
		5.282	5.547								
II	.01779	−.005926	.005624	...	...	...	...	.7420	...	.01108	.8809
		4.077	4.140					11.72			
III	.01296	−.006298	.004702	.003253	−.001799	...	...	.8127	...	.01089	.8887
		4.376	2.135	2.060	.836			11.43			
IV	.01570	−.006743	.003937	.003517	.002689	−.001083	−.002562	.8212	−.02529	.01075	.8969
		4.577	1.782	2.143	.900	.647	1.277	6.752	.213		
V	.01288	−.006340	.004647	.003174	−.001637	...	...	.7847	.0307	.01096	.8889
		4.352	2.087	1.967	.731			6.454	.285		

Table 7.16 continued

RC A1 (n = 73)

	CONSTANT	PD-0	POD-0	PD-1	POD-1	PD-2	POD-2	M-1	M-2	S.E.	R
I	.2127	-.005968 7.903	.005257 6.578	...	...	...	...	...	...	.006802	.9123
II	.1282	-.004405 6.662	.004007 5.915	...	...	...	...	.4187 6.166	...	.005501	.9444
III	.07642	-.006100 8.569	.005677 6.823	.003929 4.317	-.003785 3.907	...	...	.6556 7.938	...	.004938	.9567
IV	.07476	-.006123 8.434	.005504 6.298	.003580 3.215	-.002581 2.088	$.5358 \times 10^{-3}$ .562	$-.9268 \times 10^{-3}$ .867	.6263 5.631	.03736 .348	.005013	.9574
V	.07632	-.006097 8.470	.005674 6.749	.003923 4.226	-.003778 3.820	...	...	.6528 6.284	.003354 .044	.004975	.9567

RC B1

	CONSTANT	PD-0	POD-0	PD-1	POD-1	PD-2	POD-2	M-1	M-2	S.E.	R
I	.06672	-.002866 10.86	.002608 9.591	...	...	...	...	...	...	.003729	.8653
II	.04260	-.002068 6.949	.001911 6.586	...	...	...	...	.3662 4.373	...	.003324	.8962
III	.03570	-.002227 6.547	.002046 3.981	$.5798 \times 10^{-3}$ 1.386	$-.5322 \times 10^{-3}$ 1.000	...	...	.4666 4.231	...	.003324	.8994
IV	.03484	-.002249 6.547	.002070 3.627	$.5476 \times 10^{-3}$ 1.192	$-.4767 \times 10^{-3}$ .628	$.9613 \times 10^{-4}$ .211	$-.1193 \times 10^{-3}$ .210	.4607 3.472	.01920 .140	.003399	.8895
V	.03573	-.002226 6.311	.002045 3.881	$.5789 \times 10^{-3}$ 1.348	$-.5313 \times 10^{-3}$ .979	...	...	.4673 3.670	-.001083 .011	.003349	.8994

Table 7.16 continued

RC C1

	CONSTANT	PD-0	POD-0	PD-1	POD-1	PD-2	POD-2	M-1	M-2	S.E.	R
I	.1899	−.002573	.001741	...	...	...	...	...	...	.005577	.9243
		4.215	2.670								
II	.1455	−.002593	.001998	...	...	...	...	.2278	...	.005423	.9296
		4.369	3.100					2.246			
III	.1019	−.005030	.004664	.003513	−.003555	...	...	.4470	...	.004749	.9481
		6.917	5.314	4.791	4.222			4.455			
IV	.1010	−.004941	.004749	.004324	−.004641	−.001117	.001233	.5059	−.05294	.004793	.9495
		6.552	5.236	4.416	3.603	1.280	1.210	4.395	.455		
V	.09662	−.005077	.004705	.003529	−.003536	...	...	.4409	.03324	.004781	.9482
		6.827	5.281	4.772	4.164			4.304	.356		

RC D1

	CONSTANT	PD-0	POD-0	PD-1	POD-1	PD-2	POD-2	M-1	M-2	S.E.	R
I	.04467	−.003108	.003485	...	...	...	...	...	...	.004946	.7837
		9.389	9.928								
II	.04098	−.002931	.003281	...	...	...	...	.08121	...	.004955	.7864
		7.526	7.755					.863			
III	.02737	−.004138	.004343	.002371	−.002338	...	...	.3700	...	.004575	.8271
		8.535	6.831	3.733	3.216			3.176			
IV	.02840	−.004466	.004350	.001951	$-.9341 \times 10^{-3}$	$.8170 \times 10^{-3}$	$-.1477 \times 10^{-3}$	.3678	−.01758	.004499	.8416
		8.817	6.669	2.539	.973	1.162	1.782	2.892	.140		
V	.03107	−.004141	.004321	.002227	−.002141	...	...	.3956	−.1097	.004559	.8312
		8.572	6.818	3.457	2.883			3.352	1.213		

VIII

Some Determinants of the Returns to Manufacturing Industries

1. Introduction

A basic premise of this empirical research is that the returns to an industry depend on the size of its capital stock. It follows that some of the differences in the measured rates of return among industries are explained by the omission of certain components of their "true" capital. An important empirical task is to determine and, if possible, to measure all of the components of an industry's capital stock. The data used in this study are from two sources; the 1963 Census of Manufactures and the employment turnover statistics of the Bureau of Labor Statistics. The reasons for using the former in a study of this nature are obvious, and we shall soon see the relevance of the labor force turnover figures.

Since the data from the Census of Manufactures refer to establishments instead of companies, several important capital stock components only available at the company level cannot be measured. Nevertheless, the establishment data have certain advantages. First one can use these data to infer the existence of one capital component capable of explaining some of the differences among industry rates of return. This component is the firm's *specific human capital*. The adjective "specific" refers to the hypothesis that this kind of human capital results from the firm's outlays on its work force to impart skills and knowledge specific to the firm. Second, analysis of the data at the establishment level facilitates the study of whether the higher the concentration ratio the higher is the rate of return. Finally, these data permit some inferences about the relation between company size and the industry rate of return.

The data on work force turnover in manufacturing industries enable the testing of certain implications of the hypothesis that specific human capital is a component of the total firm capital. For example, the higher the quit rate the higher is the depreciation rate of specific human capital. There are also

312

some algebraic relations deduced from the basic regressions explaining the returns among industries from the Manufacturing Census data that can be investigated with the BLS turnover data.

A basic problem at the outset is the difficulty of measuring the relevant capital. The figures entered as capital on the books of a firm show the outlays on only those items that are conventionally regarded as capital by accepted accounting practices. Although they include expenditures on such tangibles as plant, equipment, and inventories, they exclude such intangible capital items as the cost of the good will created by advertising and other marketing activities as well as the cost of the technical knowledge resulting from research and development outlays. There are also difficulties from the lack of comparable data at the establishment and the company levels. Some of these problems are insurmountable even in principle because certain firm activities are carried on at its establishments while others occur at its central offices for the benefit of the whole company. Examples of the latter include research and development outlays and part of the salaries paid to the highest levels of management. However, some central office activities can be allocated to establishments using data available from *Enterprise Statistics*. This is done in the subsequent analysis and the details are given in appendix 1. Moreover, some central office activities in connection with marketing and finance could be allocated to the appropriate establishments were the pertinent data available. This deficiency could be remedied in future censuses. Because of the decision to use data at the establishment level, I must perforce exclude that portion of the capital stock that is measurable only at the company level. This imparts an upward bias to the estimates of the rates of return for larger companies relative to smaller ones unless some of the return from their capital is also lodged at the company level.

In view of these difficulties with the measurement of capital at the establishment level, the decision to study the returns at the four-digit level of industrial classification needs some defense. The four-digit level is the most aggregative one at which the concentration ratios are available. Since I wish to estimate the relation between concentration ratios and rates of return, I am a prisoner of the figures available. The concentration ratio represents one aspect of the structure of an industry while the rate of return depends on its performance. Hence to estimate the relation between the concentration ratio and the rate of return is to relate *structure* and *performance*. It it be true that a high concentration ratio is associated with a high rate of return, *ceteris paribus*, then this would suggest that the concentration ratio is a proxy for the degree of competition in the industry, or, at least, it is consistent with the view that competition varies inversely with the concentration ratio.

In estimating the relation between the concentration ratio and the rate of return, the measured stock of capital should exclude the present value of the monopoly return, if any. This is for two reasons. First, economists do not regard the present value of the monopoly return as a legitimate cost to society although it is a cost to any person who wishes to acquire the monopoly from its owner. According to this view, the capital stock we seek to measure is a component of the social capital. In particular we wish to measure the replacement cost of the firm's capital stock. The second reason for excluding the present value of the monopoly return is, paradoxically, to measure its effect. To understand this reason, consider an alternative approach to measuring the capital of the firm. Thus, suppose instead of measuring the capital of the firm from its book value, which gives the original cost of the capital items, that we would measure the capital of the firm from the current market value of its debt and equity, assuming, say, that equity shares are traded on a stock exchange. This procedure would include the present value of the monopoly return and, therefore, should not be expected to show any relation between the implied rate of return and the concentration ratio. Thus the difference between an ideal measure of the replacement cost of the firm and the market value of its debt and equity is the present value of the monopoly return. In principle this is the goal to which we aspire if we wish to measure the prevalence of monopoly in the economy. The present study is a step toward this goal.

If in fact at a given point of time a cross-section of industries shows that high concentration ratios are associated with high rates of return, then we should expect to see high entry rates in the highly concentrated industries. Eventually this ought to lower both the concentration ratio and the rate of return. In other words, if a cross-section of industries shows a positive association between these two variables, then time series for a given industry should also show a positive relation between the two, assuming entry takes time and that the maximum present value of the return by the firms existing in the industry dictates a strategy leading to a lower return as entry occurs. One such strategy implies that it is less profitable to lower prices at the outset in an effort to discourage entry than to do so after competitors are actually present. At least one must admit that a positive association between rates of return and the concentration ratios in the cross-section implies that the most lucrative policy over time is the one that does not constantly forestall entry. To test this hypothesis would require data for a sample of comparable industries over time giving the capital stock, rates of return, concentration ratios, and other relevant variables. Unfortunately, such a study is beyond our means with the available data. First, the earliest comprehensive collection of statistics giving the concentration ratio by industry is for the year 1935. Second, figures on the stock of depreciable assets at the four-digit level of classification were first made available for a subset of industries in the 1958 Census of Manufactures. These data include about half of the four-digit industries based on the 1945 system of classifications. Although similar data are available for the 1963 Census for all four-digit industries, these are not comparable to the 1958 figures because of extensive changes in the composition of the industries resulting from the new classification system adopted in

1957, which was applied to the regular census data and not to the special sample. One cannot construct comparable samples for the two years 1958 and 1963 with the published data. Nor is this all. It may well be that a study of the relation between rates of return and concentration ratios over time needs data spanning several decades because concentration ratios change slowly. The autoregression coefficients for the concentration ratios between census years is not far below one. A study of the relation between concentration ratios and rates of return over time would be most welcome though we are not in complete ignorance on this subject. There are some findings for a few industries, which do not constitute a random sample and, therefore, do not afford a solid basis for generalization.[1]

These remarks reflect the belief that the relation between the concentration ratio and the rate of return represents a more or less long-lived state of disequilibrium. Without special legal protection of some sort, how can a high concentration ratio yielding a high return persist? Moreover, a positive association between the two variables may result from unconscious forces and does not imply a deliberate attempt by the firms in an industry to secure a noncompetitive rate of return. There are many ways that a firm can secure an advantage over its rivals enabling it to grow relatively and to obtain a higher rate of return. For instance, a firm may find how to reduce costs, consequently, raising its sales and its profits. Product changes and the introduction of new products can have similar effects on profits and sales. These considerations can explain the observed positive association between concentration ratios and rates of return.

The concept of specific human capital also deserves some discussion. This concept, though far from novel, is still not so familiar as its importance deserves. Specific human capital is to be distinguished from nonspecific human capital. The latter includes those skills of value to a variety of employers. For example, a woman trained as a secretary has general skills making her useful to many kinds of employers. In addition, she has acquired certain skills and knowledge of benefit only to her current employer. If he bears the expense of her acquisition of the specific skills, then this expense is part of the firm's specific human capital. The readiness of an employer to invest in specific human capital depends on several factors, including the candidate's expected tenure with the company. Clearly, the employer risks the loss of his portion of the specific human capital should the employee quit. *Ceteris paribus* we expect a larger specific human capital per worker the lower is the work-force turnover. Moreover, it is in the employer's interest to have the employee share the cost of and the returns from the specific human capital. The larger is the share of the cost of the specific human capital that is borne

---

1. For evidence on the stability of concentration ratios over time, see Gort (1963). For an example of a study relating the rate of return to the market share in the steel industry see Stigler, "The Dominant Firm and the Inverted Umbrella" (1968, chap. 9).

by the employee, the less is his incentive to quit and the safer is the employer's share of the cost. However, the employee is unwilling to bear the whole cost without guarantees on the terms of employment ensuring the safety and yield of his investment.[2]

The presence of specific human capital has several implications. Over the business cycle it implies more variation in hours worked per employee than in the number of employed workers because it is cheaper for an employer to adapt to temporary variations in sales rates by changing hours worked than by changing the size of the labor force. Similarly, a firm expecting higher sales would rather increase overtime hours than hire new people unless the anticipated rise is expected to be too large and to last too long to be accommodated in this way.

There are also some interesting implications for output per man hour during various phases of business activity. Suppose the economy starts from a period of full employment and experiences a pervasive drop in demand. Initially, firms respond by reducing hours worked without reducing the number of their employees so that we may expect to observe a drop in output per man hour. With a large enough decline in demand expected to last for a while, the firms lay off some employees. The longer the duration of these layoffs, the stronger is the incentive for the workers to seek employment elsewhere. The firm's laidoff workers gradually disperse, and their firm specific skills decay through lack of use and obsolescence, resulting in a reduction in firm specific human capital. The longer the depression the more severe are these tendencies. Consequently, even after full recovery, output per worker will be lower than it was at a period of sustained full employment.

In the following material we shall study certain aspects of specific human capital primarily to see how specific human capital affects the firm's return. In addition we shall see whether specific human capital is the same for production as for nonproduction workers, the relation between specific human capital and labor earnings, the relation between concentration ratios and specific human capital, and the complementarity between physical and specific human capital. The fourth section contains a detailed analysis of the components of labor force turnover rates using the theory of specific human capital.

2. Description of the Data and Some Simple Summary Statistics

The sample consists of two sets of four-digit manufacturing industries for the years 1958 and 1963 for which the data are published by the Bureau of the

---

2. Alfred Marshall has many illuminating remarks on specific human capital (1920, bk. 6, chaps. 4 and 5). The topic was long neglected by economists. Perhaps the first recent study is Oi's dissertation (1961) and subsequent article (1962). Becker (1962, pp. 10–24) has a theoretical analysis. For a critical discussion of some of Oi's work see Mincer (1962, pp. 69–72).

Census. The first set is derived from a special census tabulation of 277 four-digit manufacturing industries based on the 1945 definitions of the industries. This set includes about half of all four-digit industries. My sample includes all but 12 of the industries in the special tabulation. The excluded industries and the reasons for their exclusion are given in appendix 2. The second set of data is derived from several sources. The basic source is the 417 four-digit manufacturing industries in the 1963 Census of Manufactures. Data for inventories and gross book value are from the *Annual Survey of Manufactures for 1963*. Data for employment and payrolls at the central offices and auxiliaries are from *Enterprise Statistics of 1963*. The *Enterprise Statistics* organize the primary census data by classifying companies into 112 manufacturing enterprise industries while the four-digit data group the establishments of the companies into more detailed categories. Appendix 1 gives a more detailed discussion of the relation between these two bodies of data. Essentially, I adapt some of the payroll data for the enterprise industries to a four-digit establishment level by imputing the central office and auxiliary payroll to the four-digit establishments comprising the company. This gives a more comprehensive measure of nonproduction worker payrolls. It was possible to use all but 18 of the total number of four-digit industries in the 1963 census. Industries in this sample are defined according to the 1957 system. Hence the 1963 and the 1958 data are not comparable. For more information about the data and the methods of collecting it, the reader should consult the original sources.[3]

Certain characteristics of these data deserve our attention here. The figures for gross book value refer only to depreciable assets at the establishment. Since land is not considered a depreciable asset, its value is excluded from these figures. As a result depreciable assets consist mostly of plant and equipment. Rented capital is excluded. The capital items included are valued at their original cost and not at their replacement cost. No information is available at the establishment level about financial assets such as cash balances, accounts receivable, holdings of government bonds, and so on. Such data are available only at the company level and cannot be imputed to establishments. We do have establishment data for one short-term asset, inventories. It is important to observe that these inventories include the value of goods in process and raw materials as well as the stocks of finished products. The inventory figures are from the *1963 Annual Survey of Manufactures*.

The figures for depreciation pose some special problems. We would like to have the replacement cost of the capital stock which is not necessarily the same as the net book value. The latter is derived from gross book value by means of an arbitrary formula for calculating the depreciation. One can argue that the contribution of an asset to overhead depends not on its net but on

---

3. The footnotes to table 8.1 and appendix 1 give the sources.

its gross book value because a machine's contribution to output does not decline according to the depreciation formula. For example, the marginal product of the famous one-horse shay remained constant until the very end, when it suddenly became zero. There is no all-purpose measure of depreciation. Sometimes the best measure is the outlay necessary to maintain a constant capital stock. However, a net revenue maximizing firm does not necessarily want a constant capital stock. If capital stock is like the one-horse shay, then it is better to use gross book value to represent the effective amount of the depreciable asset instead of net book value. Actually, the 1958 data suggest that it makes little difference for the empirical results whether gross or net book value is used.

Estimates of the return by industry are built-up from the flow figures. For 1958 one may calculate the return either including or excluding the depreciation allowance for that year. The return that includes the depreciation allowance is called the contribution to overhead (OH), and the figure excluding the depreciation allowance is called the net return (NR). Let D denote the depreciation allowance. Then

$$(1) \qquad\qquad NR = OH - D.$$

The contribution to overhead equals value added less payrolls at the establishment level. Let VA denote the value added and let PA denote the establishment payrolls.

$$(2) \qquad\qquad OH = VA - PA.$$

One more adjustment is made. Let $CAO\ PA$ denote the payrolls at central offices and auxiliaries imputed to the four-digit establishment (for the details see appendix 1). These include salaries of some central office personnel such as managers, clerks, and some engaged in research and development. The adjusted contribution to overhead (OHA) is defined as follows:

$$(3) \qquad\qquad OHA = OH - CAOPA.$$

OHA approximates the gross return to capital before taxes. It includes interest payments, dividends, retained earnings, corporate taxes, and depreciation. Certain other outlays are included as well, such as advertising expenditures and some research and development expenses in the form of services purchased from outside firms. The main sources of the firm's income excluded from this estimate are its returns on investments in other companies and on bonds. It should be noted that the census method of estimating payrolls leads to the inclusion of part-time help which is adjusted to a full-time equivalent.

Value added is derived from the value of shipments by subtracting the cost of materials. Let VS denote the value of shipments and let CM denote the cost of materials. Hence

$$(4) \qquad\qquad VA = VS - CM.$$

Measurement error is a serious problem in those industries in which there are interplant transfers of goods within the same company. In these cases it is difficult to value the shipments because no market transactions automatically record prices. In some especially troublesome cases the census refuses to publish estimates of the value of shipments because of their unreliability. Such industries are excluded from the sample used in this study. The value of shipments exclude freight charges (they are FOB), and they exclude excise taxes.

Concentration ratios are one of the variables used in the subsequent analysis. We shall use two different ones. The 1958 data include the four-firm concentration ratio based on the value of shipments while the 1963 analysis uses the four-firm concentration ratio based on value added. The differences are small according to the 1963 census where both measures are given for all industries, and the two can be compared by industry. With either concentration ratio it is assumed that an establishment makes a single group of products all within the same four-digit product class. Thus the four-firm concentration ratio is, say, the ratio of the value added by all of the establishments belonging to the four largest companies to the total value added by industry. One can also calculate the concentration ratio on a product basis instead of on an industry basis. On a product basis the concentration ratio is the ratio of the value added of the products in the four-digit class made by the four leading companies to the total value added of that product. The two measures would be the same if it were true that every establishment would specialize in making only the products within a single four-digit industry classification. If this is not true so that some establishments make products in different four-digit classes, then the concentration ratios based on product shipments would differ from the one based on industry shipments. A concentration ratio based on product shipments would give a more accurate picture if demand cross-elasticities were larger than supply cross-elasticities. Since the 1963 census gives both kinds of concentration ratios, we can compare the figures for the two approaches of measuring concentration. Some experiments indicate that the differences would be small and would consequently have little effect on the regression analysis. Since it is slightly more convenient to use the concentration ratio on the industry instead of on the product basis, the former is the variable in the regressions.

Another variable in the regressions is the number of companies operating at least one establishment in the given four-digit industry. Although there is a fairly strong inverse correlation between the concentration ratio and the number of companies by industry, it is far from perfect. Hence the combination of both variables in a regression equation conveys more information about industry structure than would either of them used in isolation. For example, there are times in which two industries have nearly the same concentration ratio but different numbers of companies and times where the industries have the same number of companies but different concentration

TABLE 8.1

Means and Standard Deviations for Selected Variables
by Four-Digit Manufacturing Industries, 1958 and 1963

Variable	Year	
	1958	1963
Payrolls (PA)	218.1	213.9
	(295.0)	(385.5)
Gross book	292.7	376.4
value (GBV)	(559.9)	(1122.7)
Number of		
companies (NO)	837	682
	(1815)	(1444)
Value added (VA)	408.8	447.0
	(495.8)	(845.9)
PA/GBV	121.9	102.3
	(114.6)	(84.9)
Concentration		
ratio (C)	37.9	37.3
	(22.7)	(22.1)
Contribution		
to overhead (OH)	190.6	233.1
	(231.6)	(485.5)
Inventory (INV)	not available	142.9
		(279.1)

SOURCES: 1958—267 four-digit industries from U.S.
Bureau of the Census, *U.S. Census of Manufactures: 1958.
Supplementary Employee Costs, Costs of Maintenance and
Repair, Insurance, Rent, Taxes, and Depreciation and Book
Value of Depreciable Assets;* 1957, Subject Report
MC58(1)-9. (Washington, D.C.: U.S. Govt. Printing Office:
1961).

1963—402 four-digit industries from U.S. Senate, Committee on the Judiciary, Sub-Committee on Antitrust and
Monopoly, *Concentration Ratios in Manufacturing Industry
1963: Part I.* (Washington, D.C.: U.S. Govt. Printing
Office, 1966). The figures for gross book value are taken
from the U.S. Bureau of the Census, *Annual Survey of Manufactures: 1964, Book Value of Fixed Assets and Rental Payments for Buildings and Equipment,* M64(AS)-6. (Washington, D.C.: U.S. Govt. Printing Office, 1967). Inventory
figures come from U.S. Bureau of the Census, *Annual Survey
of Manufactures: 1963, Value of Manufacturers' Inventories,*
M63(AS)-3 (Washington, D.C.: U.S. Govt. Printing Office,
1965).

NOTES: The figures in parentheses are the standard deviations. All units are in millions of dollars except for ratios, *C*
and *PA/GBV*, which are in percent and *NO*, which gives the
average number of companies per industry.

The concentration ratios give the share of the *value of
shipments* by the four leading companies in the four-digit
industry in 1958. For 1963 the concentration ratio gives
share of the *value added* by the four leading companies.

ratios. The main drawback to the use of the number of companies as a variable in the analysis is its susceptibility to large measurement error. It is especially hard to count the number of companies in industries having small establishments.

We now consider some gross aspects of the data. Table 8.1 gives the means and standard deviations for certain key variables appearing in the subsequent regression analysis. Since the 1958 data refer to a subsample of all four-digit manufacturing industries, this is reflected in some of the differences between the variables in the two years. The 10 percent rise in value added and value of shipments partly results from the price rise from 1958 to 1963. In contrast, the unadjusted contribution to overhead by industry rose nearly 22 percent largely because 1958 was a recession year with recovery under way by summer, while 1963 was a prosperous year. Since the contribution to overhead is a residual including profits, it typically rises more than sales during such periods. Now observe the concentration ratios. Despite the fact that the 1958 measure refers to value of shipments while the 1963 measure refers to value added, the two are sufficiently close together to afford a meaningful comparison between the two census years. According to these data, the average concentration ratio hardly changed over this five-year period. The labor figures do change. Payrolls are a smaller fraction of value added in 1963, 48 percent, than in 1958, 53 percent, which may also reflect the recession in 1958 if there was more specific human capital per worker in the more pros-

TABLE 8.2
Simple Correlation Coefficients for Selected Variables for the 1958 and 1963 Samples of Four-Digit Manufacturing Industries

	log *OH*	*C*	log *GBV*	log *PA/GBV*	log *NO*
			1958		
log *OH*	1	−0.07654	0.8612	−0.1982	0.4046
C		1	0.03853	−0.4487	−0.8066
log *GBV*			1	−0.5524	0.2400
log *PA/GBV*				1	0.3649
log *NO*					1
			1963		
log *OH*	1	0.0274	0.8951	−0.2667	0.3451
C		1	0.1005	−0.4167	−0.7483
log *GBV*			1	−0.5621	0.1901
log *PA/GBV*				1	0.3865
log *NO*					1

NOTES: All logarithms are to the base *e*, that is, natural logarithms. This sample includes the same 402 industries as are in table 8.1.
The 0.001 level of significance is 0.3211 for 100 observations.

perous year. Most of the variables appear in logarithmic form in the regressions to follow. Table 8.1 shows that the mean industry size is of the same order of magnitude as the standard deviation, but the mean of the logarithm of industry size is considerably larger than the standard deviation of the logarithm. Hence the distribution of the logs of the variables is much closer to

TABLE 8.3

Summary Statistics for 398 Manufacturing Industries in 1963 by Concentration Ratio Strata

VARIABLE	CONCENTRATION RATIO STRATA							
	Total		[0–25)		[25–50)		[50–100]	
	Mean	S.E.	Mean	S.E.	Mean	S.E.	Mean	S.E.
NO, number of companies[a]	690	1453	1486	2232	378	437	136	205
VA, value added	431	826	455	475	339	420	532	1407
C, concentration ratio[b]	0.370	0.221	0.151	0.059	0.352	0.067	0.675	0.128
INV, inventories	132	242	116	110	120	164	167	404
GBV, gross book value	349	1019	273	332	286	529	536	1809
TPA, total payrolls	225	414	252	255	176	232	262	688
PPA, production worker payrolls	142	277	165	158	109	149	159	468
MPA, nonproduction worker payrolls	84	151	88	123	67	93	103	229
AS, annual earnings per employee, all employees[c]	5.535	1.251	4.985	1.245	5.518	1.186	6.255	0.965
PS, annual earnings per employee, production workers[c]	4.846	1.179	4.417	1.186	4.793	1.129	5.463	0.969
MS, annual earnings per employee, nonproduction workers[c]	7.606	0.866	7.271	0.935	7.613	0.773	8.025	0.706
OH, contribution to overhead	225	476	214	261	178	233	307	819
OHA, contribution to overhead, adjusted	205	430	203	240	162	211	270	737
SPEC, specialization ratio	1.460	2.916	1.235	1.698	1.234	0.421	2.248	5.524

NOTES: Unless otherwise indicated all variables are in hundreds of millions of dollars.

Four industries are excluded from the 402 industry sample for this and the following tables because it was not possible to calculate *OHA*. The excluded industries are as follows: 2911, 3411, 3721, and 3871. See appendix 2.

a. Actual number   b. Decimal fraction   c. Thousands of dollars

normality than is the distribution of the original variables.

The simple correlation coefficients in table 8.2 provide additional information about how the variables are related. These are about the same in both years. Notice the large inverse correlation between C and log NO. Also observe the inverse correlation between C and log PA/GBV showing that in highly concentrated industries there is a higher ratio of gross book value to total payrolls. Table 8.2 does not readily show the simple relations among the variables because of common scale effects. To bring these out, we must resort to multiple regressions which control for common scale factors. Nor is this all. Simple correlations can easily mislead. For instance, consider the simple correlation between C and OH/GBV. The latter crudely estimates the rate of return. The simple correlation between these two variables is -0.22 for 1958 and -0.16 for 1963. However, as we shall see in the next section, the partial

TABLE 8.4
Some Simple Correlations for the 398 Four-Digit Manufacturing Industries and by Concentration Ratio Strata

	OHA/GBV	C	AS
ALL INDUSTRIES			
OHA/GBV	1	-0.2058	-0.3368
C		1	0.3952
$C \geqslant 50$ (109 INDUSTRIES)			
OHA/GBV	1	0.0178	-0.1886
C		1	0.1314
$25 \leqslant C < 50$ (154 INDUSTRIES)			
OHA/GBV	1	0.0072	-0.1748
C		1	0.0454
$0 \leqslant C < 25$ (135 INDUSTRIES)			
OHA/GBV	1	-0.1801	-0.3630
C		1	0.1132

correlation between C and the rate of return holding constant a set of other relevant variables is significantly positive.

We get a clearer picture of how the concentration ratio affects the industry returns by splitting the entire 1963 sample into three strata by concentration ratio. The 1958 sample is too small to follow the same procedure.

The three strata are as follows:

$$C \geqslant 50 \text{ percent,} \quad 109 \text{ industries;}$$
$$50 > C \geqslant 25, \quad 154 \text{ industries;}$$
$$25 > C \geqslant 0, \quad 135 \text{ industries.}$$

Table 8.3 shows the means and standard deviations for these three strata. Two facts stand out. First, the ratio of payrolls to gross book value is almost twice as high in the least concentrated strata as in the most concentrated one. Second, OHA/GBV is nearly 40 percent higher in the least than in the most concentrated strata. The latter findings in terms of the mean values of OHA/GBV confirm the negative value of the simple correlation between C and OHA/GBV for the entire sample.

One of the main reasons for this stratification is to see whether or not there is a nonlinear relation between rates of return and concentration. There is some implicit evidence of nonlinearity even in the simple correlations between C and OHA/GBV because it is negative in the least concentrated stratum and positive in the other two strata (table 8.4).

3. Multiple Regression Analysis of the Census Data

The main purpose of this analysis is to "explain" the rates of return by industries as a function of variables representing industry structure and capital components. Concentration ratios and the number of companies in an industry represent its structure. Specific human capital is part of the industry capital. I begin with a presentation of the argument which led me to choose the particular form of the regression equation.

The most straightforward regression equation relating the rate of return to its determinants is given by

$$(1) \qquad OHA/GBV = a_0 + a_1 C + a_2 TPA/GBV.$$

This form is vulnerable to several damaging criticisms. First, there is a possibility of spurious correlation due to the presence of a common denominator GBV on both sides of the equation. Second, the regression excludes other components of tangible capital in addition to plant and equipment, namely, inventories, raising the question of how to remedy this omission. One possible way is by redefinition of the capital stock as follows:

$$(2) \qquad K = GBV + INV.$$

K would be the capital measure in place of GBV. This is also a poor procedure because it assumes implicitly that both forms of capital, GBV and INV, affect OH in the same way without allowing the data to reveal whether this is so. There is no theoretical reason for making this stipulation. The reader may object that the hypothesis of profit maximization does imply that at the margin one dollar of inventories should contribute as much to the return as one dollar of plant and equipment. Though this be granted, it does not follow that the two forms of capital are perfect substitutes as is implied by (2). Moreover, since the dependent variable is *OHA* and not the net return to capital, differences in the rate of depreciation can explain differences in the marginal contribution to overhead of different kinds of capital. Finally, even if one accepts the view that firms maximize profits and equate the net return at the margin of all kinds of capital, it does not follow that in any given year all industries are in long-run equilibrium. To believe that they are in long-run equilibrium implies that it costs nothing to change the stock of capital and/or that all firms have excellent foresight. These are not useful assumptions in empirical work. Nevertheless consideration of profit maximization does lead to some helpful insights as we shall see below. There is one final point about the form of the regression (1) or its variant with K. In this form one cannot ascertain how the returns vary with size since the regression makes the implicit assumption that returns are proportional to size. Not only is it better to allow the data to tell us how returns do vary with size but also it is better to allow the data to reveal how different forms of capital contribute to overhead.

A simple regression meeting these objections is as follows:

$$(3) \qquad \log OHA = \sum_i a_{0i}\delta_i + a_1 \log GBV + a_2 \log INV + a_3 \log TPA$$

$$+ a_4 \log NO + a_5 C + a_6 SPEC.$$

This regression introduces two new factors to be discussed presently, δ_i, a set of industry dummy variables, and *SPEC*, a measure of specialization. Equation (3) closely resembles the Cobb-Douglas production function and has a similar interpretation. Since *OHA* is not divided by total capital, we avoid the problem of spurious correlation due to the presence of a common variable on both sides of the equation. Second, every component of capital appears separately so that the impact of each on *OHA* can be measured separately. Third, the sum of the coefficients of the capital components shows how the returns vary with size. Fourth, the effect of the concentration ratio on the *rate* of return is given by its coefficient a_5, because total capital is held constant by the inclusion of the other variables, *GBV*, *TPA*, and *INV*. Hence a_5 actually measures the effect of C on the rate of return.

Now we come to the dummy variables. In principle one may argue that every industry has its own relation such as (3) and that one should not expect the coefficients of different kinds of capital to be the same in different industries. Fitting (3) to a set of industries assumes the mobility of capital among industries and ignores differences in technology and market conditions. It is partly to meet this kind of criticism that the regression equation includes the concentration ratio C and the number of companies by industry *NO*. In addition one can partially measure industry effects by using dummy variables. The basic data consist of four-digit industries which can be aggregated into broader two-digit classes represented by the dummy variable δ_i. More precisely, if a given four-digit observation is within two-digit industry class i then δ_i equals 1 and otherwise δ_i equals 0. Hence the inclusion of the dummy variables shifts the entire regression plane according to the two-digit industry class that contains a given four-digit observation.

The variable *SPEC* is designed to measure the degree to which a given industry includes companies that specialize primarily in the products made in that industry. To explain this, it is necessary to describe certain facts about industrial organization from the point of view of the census statistics. Beginning with 1963, the census has classified companies into a set of 179 "enterprise industry categories." Each of these enterprise industries is composed of one or more four-digit SIC industries. Hence the enterprise industries permit a correspondence between companies and their component four-digit establishments. A company consists of one or more establishments where production occurs, and central offices and auxiliaries (*CAO*'s). The latter includes the company headquarters, warehouses, some of its research and development activities that do not take place at establishments, and some of its marketing activities with a similar proviso. Establishments roughly correspond to the economist's concept of a plant although an establishment often makes several products. Some companies have only one establishment so that they have no separate *CAO*'s and have the simplest kind of organization. Other companies have several establishments possibly in different four-digit SIC industries and one or more *CAO*'s; these are the diversified companies. A company is placed into an enterprise industry according to the primary activities of its four-digit SIC components. Hence, a company appearing in a given enterprise industry might operate establishments within four-digit SIC classes that belong to another enterprise industry. The specialization ratio is designed to measure the extent to which the output of a given four-digit SIC industry comes from companies specializing in the same four-digit category. To show this more precisely, it is helpful to consider an example. The enterprise industry 20*B*, prepared meats and dressed poultry, is made up of two four-digit SIC industries—2013, meat processing plants; and 2015, poultry dressing plants. The employment figures are as follows:

	Total 1963 Employment, All Establishments	Employees of Companies Classified in the Census, SIC Industry by Establishment, 1963
20*B*, prepared meats and dressed poultry	119,321	88,491
2013, meat processing plants	48,619	37,922
2015, poultry dressing plants	70,702	50,569

These figures mean that meat processing and poultry dressing are not the primary activities of many of the companies making these products. The specialization ratio for meat processing is 48,619/37,922 = 1.28. The ratio is 1.39 for poultry dressing. In some industries the given activity is primary, for example, meat packing plants with a ratio of 0.83. In general if the ratio is below one, then the four-digit SIC industry is the company's primary activity; and if the ratio is above one, then the four-digit SIC industry is not the primary activity of the companies operating in it. The average value of the specialization ratio for the 398 four-digit SIC industries in the sample is 1.46. Hence, on average, the companies do not specialize in a given four-digit industry.

The specialization variable is included in the regressions as a proxy for missing capital. Since both *GBV* and *INV* refer only to capital at the establishment level and since capital at the company level is excluded, the less specialized the company at the four-digit SIC level, the more capital is missed by the observed figure, assuming that the missing capital is proportional to employment. Since the higher is *SPEC* the less specialized is the company and the larger is the amount of missing capital, the coefficient of *SPEC* should be negative. As we shall see, the coefficient does in fact tend to be negative.

The problem of missing capital is obviously important in interpreting the coefficient of the concentration ratio. One can argue that a positive coefficient of the concentration ratio does not in fact indicate that rates of return vary directly with the concentration ratio but rather that missing capital is positively correlated with the concentration ratio. Thus, perhaps the highly concentrated industries are relatively less specialized than the unconcentrated industries so that *SPEC* varies directly with *C*. This seems to be true as is shown by table 8.3. Hence we correct for the missing capital as best as we can by introducing a proxy measure, *SPEC*, as one of the variables in the regression. Omission of this variable would lead to overestimating the effect of *C* on the rate of return.

Why should *TPA* appear as an explanatory variable in the regression? Since the dependent variable *OHA* represents the return to capital while *TPA* is the total payment to labor, it would not seem that *OHA* should depend on *TPA*. One possible explanation is spurious correlation due to a common scale effect, but this is implausible for two reasons. First, one would think that the contribution to overhead would depend on the variables directly representing capital such as *GBV* and *INV* leaving no room for *TPA*. Moreover, $OHA = VA - TPA$, so that there is no common variable on both sides of the equation to explain spurious correlation.

A hypothesis explaining the dependence of *OHA* on *TPA* is that the latter depends on firm-specific human capital which contributes to overhead. The specific human capital results from outlays by the firm in training its labor force in skills and knowledge primarily benefiting the firm. This raises the employee's marginal product to a level above his real wage rate, and specific human capital manifests itself by a significant positive coefficient of log TPA in (3). The argument can be put algebraically. Let MP_1 denote the marginal product of specifically trained workers and let W_1 denote their wage rate. Let H denote the specific human capital. The net return to the firm per dollar of specific human capital is given by

$$(4) \qquad r = [(MP_1 - W_1)E_1]/H - d_H,$$

where E_1 is the number of trained employees and d_H is the rate of depreciation of the specific human capital. Let

$$(5) \qquad MP_1 = \alpha W_1, \quad \text{with} \quad \alpha \geqslant 1,$$

so that (4) becomes

$$r = (\alpha - 1)TPA_1/H - d_H,$$

where $TPA_1 = W_1 E_1$. Therefore,

$$(r + d_H)H = (\alpha - 1)TPA_1,$$

so that

$$(6) \qquad H = [(\alpha - 1)/(r + d_H)]TPA_1$$

or in log form

$$(7) \qquad \log TPA_1 = \log [H(r + d_H)/(\alpha - 1)].$$

However, the total payroll is the sum of the payments to the untrained as well as the trained workers. During their training period the wage rate is W_0 and the marginal product is MP_0 with W_0 above MP_0. $TPA = TPA_1 + TPA_0$, where $TPA_0 = W_0 E_0$. Let $\beta = TPA_0/TPA$ so that $0 \leqslant \beta \leqslant 1$. Then

$$(8) \qquad H = [(1 - \beta)(\alpha - 1)/(r + d_H)]TPA$$

applies instead of (6), and instead of (7) we have

$$(9) \qquad \log TPA = \log [H(r + d_H)/(1 - \beta)(\alpha - 1)].$$

In this development it should be noted that E_0 refers to the net increment of personnel in the process of being endowed with specific human capital so that it is the gross increment less the training of the workers in replacement of depreciation. If there were no net investment of specific human capital, then E_0 would be zero. Hence formula (7) would be correct, and $TPA = TPA_1$ whenever H remains constant. This is to say that

$$\frac{dH}{dt} = (W_0 - MP_0)E_0.$$

The specific human capital resulting from the firm's outlays on recruiting and training its workers would be proportional to the total payroll if the three parameters, r, the rate of return; d_H, the rate of depreciation; α, the factor of proportionality between W_1 and MP_1 given by (5); and β were the same for all industries. If so, TPA would indeed be an ideal proxy for H. However, as we shall see, there is good reason to reject the hypothesis of proportionality and to give a more subtle interpretation of the results.

In one sense the presence in (3) of all three variables, GBV, INV, and TPA has a common rationale. Thus, one would hardly assert that GBV and INV are exact measures of the true capital they purport to represent although these variables are related to the true values which determine OHA. Similarly, the preceding analysis shows that TPA is related to H which is also a determinant of OHA. One can interpret these remarks in a formal statistical sense by using the model of "errors in the variables," but this would serve little purpose because without knowledge of the magnitude and relations among the errors it is not possible to deduce any useful conclusions from the empirical results.[4]

We gain some limited insight into the effects of measurement error from the 1958 sample which gives figures for accumulated depreciation, net book value and the depreciation allowance for 1957. Hence one can experiment with different measures of the return and capital to see how much they affect the results. Let

ACD = accumulated depreciation through 1956,
NBV = net book value.

Thus,

(10) $$NBV = GBV - ACD.$$

Let D = depreciation allowance for 1957. The net return NR is defined as follows:

(11) $$NR = OH - D.$$

This variable is closer to the accounting definition of profits before taxes, while OH can be regarded as a measure of the gross profit before taxes.[5] The two variables are related to each other in the same way that net and gross investment are related. One can use NR as the dependent variable instead of OH and NBV as an explanatory variable instead of GBV. In addition both variables NBV and ACD can be used as separate explanatory variables of either NR or OH. The empirical results show that it makes little difference which of the two variables, NR or OH, is the dependent variable for the same set of explanatory variables. In either case the coefficients of the same explanatory variables are about equal. Nor does it seem to matter whether we use

---

4. See Theil (1961, pp. 215 and 326–29).

5. No data are available for 1958 to estimate payrolls at the central offices and auxiliaries corresponding to the four-digit SIC industry establishments. Hence in these calculations only OH and not OHA appear.

GBV as an explanatory variable or the two separate variables, *NBV* and *ACD*. These results are due to the high correlations between corresponding pairs of variables. For example, the correlation between log *NR* and log *OH* is nearly one, 0.998 to be exact; and, similarly, the correlations between log *GBV* and log *NBV* and between log *GBV* and log *ACD* are nearly one. These high correlations explain why we find the same results whether log *OH* or log *NR* is the dependent variable or log *GBV* or the pair of separate variables, log *NBV* and log *ACD* are the explanatory variables.[6]

We now turn to the first of the main body of results, the multiple regression estimates of (3) and its relatives given in tables 8.5–8.7. First, consider the estimates of (3) shown in table 8.5. A number of empirical regularities appear. Thus, the relations are actually nonlinear in the sense that the coefficients vary systematically with the concentration ratio strata. For example, the coefficient of log *TPA* is 0.48 for the whole sample, and it rises from 0.12 in the most concentrated stratum to 0.63 in the least concentrated one. The coefficient of log *GBV* also changes regularly with stratum but inversely with log *TPA*. The coefficient of log *GBV* is 0.25 in the whole sample while it is 3.5 times as high in the most as in the least concentrated stratum. The behavior of the coefficient of log *INV* is the same as the coefficient of log *GBV*. Thus both

---

6. The following regressions support the assertions in the text.

$$\log NR = -0.2304 + 0.0111\ C + 0.6275 \log (PA/GBV) + 0.6753 \log NBV$$
$$\ \ 1.04 \quad\ \ 4.94 \quad\quad\ \ 11.0 \quad\quad\quad\quad\quad\quad\ \ 9.35$$
$$\ -0.1369 \log (GBV/NO) + 0.3470 \log ACD, \qquad R^2 = 0.841;$$
$$\ \ \ 3.62 \quad\quad\quad\quad\quad\ \ 4.77$$

$$\log OH = -0.0831 + 0.0101\ C + 0.5815 \log (PA/GBV) + 0.6594 \log NBV$$
$$\ \ 0.41 \quad\ \ 4.96 \quad\quad\ \ 11.2 \quad\quad\quad\quad\quad\quad\ \ 10.0$$
$$\ -0.1281 \log (GBV/NO) + 0.3637 \log ACD, \qquad R^2 = 0.868.$$
$$\ \ \ 3.73 \quad\quad\quad\quad\quad\ \ 5.51$$

The partial correlation between the concentration ratio, *C*, and the dependent variable is 0.29 in both regressions. It is also worth noting that

$$\log OH = 1.4764 + 0.3579 \log NBV + 0.4047 \log ACD, \qquad R^2 = 0.7517.$$
$$\ \ 12.44 \quad\ \ 4.22 \quad\quad\quad\ 4.56$$

This last regression clearly shows how the addition of the other explanatory variables contributes to the explanation of how log *OH* varies by industry. Finally, it should be noted that the 1963 sample includes 402 four-digit SIC industries.

Let us also observe the similarity between the 1963 and 1958 results in a multiple regression with the same explanatory variables.

1958 $\quad \log (OH/GBV) = 0.1542 + 0.01035\ C - 0.3352 \log GBV + 0.2886 \log NO,$
$$\ \ 0.55 \quad\ \ 4.08 \quad\quad\ \ 12.15 \quad\quad\quad\quad 7.38$$
$$\ R^2 = 0.395 \quad\quad\quad\quad \text{sample size} = 265,$$

1963 $\quad\quad\quad \log OH = 1.1361 + 0.01077\ C + 0.7161 \log GBV + 0.2885 \log NO,$
$$\ \ 4.85 \quad\ \ 6.39 \quad\quad\ \ 37.78 \quad\quad\quad\quad 10.67$$
$$\ R^2 = 0.848 \quad\quad\quad\quad \text{sample size} = 402.$$

To compare the coefficient of log *GBV* for the two regressions one should either add one to the coefficient of log *GBV* in the 1958 regression or subtract one from the log *GBV* coefficient in the 1963 regression. In the latter case this gives -0.2839 for 1963 which is close to -0.3352 for the 1958 coefficient. Thus, as is indicated by the simple correlations in table 8.2, the relations among the variables are very nearly the same in the two census years.

coefficients of the components of tangible capital rise with the concentration ratio stratum. The elasticity of C behaves similarly, although the coefficient itself does not change monotonically with concentration stratum. For the whole sample the elasticity of OHA with respect to C, which is actually the elasticity of the rate of return with respect to C since the amount of capital is held constant in this exercise, is 0.16. For the two least concentrated strata this elasticity is 0.083 and 0.092; and it is 0.66 in the most concentrated stratum. The variable $SPEC$, introduced to measure the specialization of the

TABLE 8.5

Selected Statistics for Regressions with log OHA Dependent, 398 Four-Digit SIC Manufacturing Industries, 1963, Stratified by Concentration, log of Total Payrolls an Explanatory Variable

		Strata		
	Total	[0, 25)	[25, 50)	[50, 100]
Variables	Coefficients and *t*-ratios			
C	0.4413	0.5480	0.2613	0.9776
	3.31	1.18	0.64	2.56
log NO	0.1018	0.0828	0.0811	0.1607
	4.28	2.12	1.98	3.40
log INV	0.2051	0.2033	0.2159	0.3643
	6.69	5.24	4.68	4.04
log GBV	0.2543	0.1266	0.2359	0.4456
	6.12	2.45	3.37	4.37
log TPA	0.4801	0.6308	0.5025	0.1177
	9.37	10.00	5.45	0.91
$SPEC$	−0.0032	−0.0367	−0.0552	−0.0015
	0.55	2.98	0.91	0.18
R^2	0.9284	0.9699	0.9492	0.9121
S.E.	0.3193	0.2082	0.2898	0.3923
Elasticity of C	0.163	0.083	0.092	0.660
Sum of coeff. of log INV, log GBV, log TPA	0.9395	0.9607	0.9543	0.9276

parent companies in the given four-digit industry, is defined so that specialization decreases as $SPEC$ increases. Hence the higher is $SPEC$, the lower ought to be the true rate of return because relatively more true capital is omitted from the measured capital figure. In every case the $SPEC$ coefficient is negative, as expected. Although the *t*-ratio of this variable is low except in the least-concentrated stratum, this fails to give an adequate idea of its importance. The effect of specialization or, equivalently, diversification, is made manifest by what happens to the coefficient of *C* if *SPEC* is omitted. In fact the omission of this variable tends to increase the coefficient of *C*, which confirms the view that part of the effect of *C* results from its role as a proxy for missing capital. Thus, the higher the concentration ratio, the more diversified the companies in the industry, and the more the measured rate of return tends to overstate the true rate by its failure to include all of the relevant capital.

Finally, it is worth noting that the coefficients of the variables, log *INV*, log *GBV*, and log *TPA* tend to vary inversely with the ratios *OHA/INV*, *OHA/GBV*, and *OHA/TPA*, respectively, where the latter are the ratios of the sample means. We shall return to these findings below.

TABLE 8.6

Selected Statistics for Regressions with log OHA the Dependent Variable 398, Four-digit SIC Manufacturing Industries, 1963, Stratified by Concentration, Payrolls divided into Production and Nonproduction Workers

| | TOTAL | STRATA | | |
		[0, 25)	[25, 50)	[50, 100]
VARIABLES	COEFFICIENTS AND *t*-RATIOS			
C	0.3305	−0.0122	0.2274	1.0103
	2.39	0.03	0.57	2.57
log *NO*	0.0959	0.0870	0.0696	0.1664
	4.00	2.28	1.74	3.32
log *INV*	0.1920	0.1851	0.2012	0.3606
	6.18	4.81	4.46	3.93
log *GBV*	0.2707	0.1506	0.2391	0.4389
	6.45	2.99	3.45	4.18
log *PPA*	0.1467	0.1640	0.0402	0.1191
	2.88	2.78	0.51	0.83
log *MPA*	0.3224	0.4323	0.4647	0.0086
	6.97	8.11	5.82	0.08
SPEC	−0.0004	−0.0373	−0.0574	−0.0019
	0.06	3.09	0.97	0.23
R^2	0.9289	0.9714	0.9526	0.9122
S.E.	0.3186	0.2037	0.2812	0.3943

NOTE: The coefficients and *t*-ratios of the dummy variables are shown in table 8.14, appendix 3.

The behavior of the coefficient of log *NO*, the number of companies in the four-digit SIC census industry, is of considerable interest. A priori, this coefficient can be either positive or negative without upsetting any standard economic theorems. If negative, then this would mean that given the share of the four leading companies, the larger is the total number of companies in the industry, the lower is the rate of return. This could be explained by saying that, given the concentration ratio, the more companies there are in the industry, the more competition and the lower is the rate of return. On the other hand, there is an equally plausible argument for a positive coefficient. Thus the higher is the rate of return, the more companies enter the industry attracted by the prospect of obtaining high profits. In fact, all of the coefficients of log *NO* are positive. Nor is this all. The coefficients tend to rise with concentration. One can argue that this finding means that the higher is the concentration ratio the greater is the inducement for a company to enter the industry, which assumes that the inducement to enter is measured by the size of the coefficient of log *NO*. This finding I interpret as being consistent with the positive coefficient of *C* itself. Moreover, it provides a clue in favor of a positive association between concentration and rates of return over time. Above I argue that a positive association between *C* and the rate of return in the cross-section implies a similar relation for time series. In particular a fall in the concentration ratio should be accompanied by a fall in the rate of return as a result of the entry and success of companies in the more concentrated industries. Hence the finding of a positive association between rates of return and the number of companies in the cross-section supports my argument about what time series data will show.

The total payroll can be divided into two parts, the payroll of the production workers, *PPA*, and the payroll of the nonproduction workers, *MPA*. The latter includes an estimate of the payroll at the central office and auxiliaries appropriate to the four-digit SIC census industry. One may argue that the specific human capital per employee differs between these two groups. Since the nonproduction workers include managers, foremen, supervisors, sales personnel, clerks, secretaries, and so forth, such employees probably have more firm specific skills than do the production workers. If this is so, then the coefficient of log *MPA* should be larger than the coefficient of log *PPA*.

To test this, consider table 8.6. For the whole sample the coefficient of log *MPA* is more than twice as high as the coefficient of log *PPA*. Moreover, in the two lowest concentration strata the difference is even more striking. For example, in the middle stratum the coefficient of log *PPA* is virtually zero and is less than 10 percent of the coefficient of log *MPA*. However, the results in the most concentrated stratum are disappointing. There both coefficients are small; and the coefficient of log *MPA*, in particular, is nearly zero, which is inconsistent with the findings for the two other strata.

Table 8.6 shows the coefficients of the concentration ratio in a different light than the corresponding coefficients in table 8.5. In particular the coefficient of *C* for the whole sample falls from 0.44 in table 8.5 to 0.33 in table 8.6. Moreover, in the least concentrated stratum the *C* coefficient is actually negative albeit very close to zero. Finally, the gradient of the *C* coefficient is much steeper in these regressions than in their predecessors. The remaining coefficients in these regressions are close to the values of their counterparts in table 8.5.

There is another aspect of the hypotheses about specific human capital that can be explored with these data. This is to relate specific human capital to the annual earnings per employee. Is it true that specific human capital per employee rises with annual earnings? Although the theory does not dictate

TABLE 8.7

Selected Statistics for Regressions with log OHA Dependent, 398 Four-digit SIC Manufacturing Industries, 1963, Stratified by Concentration, log TE, and log AS among Explanatory Variables

			STRATA	
	TOTAL	[0, 25)	[25, 50)	[50, 100]
VARIABLES	COEFFICIENTS AND *t*-RATIOS			
C	0.3913	0.4470	0.2761	0.9464
	2.93	0.97	0.67	2.47
log NO	0.1085	0.0790	0.0860	0.1628
	4.57	2.04	2.02	3.44
log INV	0.2124	0.2154	0.2180	0.3561
	6.96	5.56	4.69	3.94
log GBV	0.2318	0.1174	0.2309	0.4110
	5.52	2.29	3.25	3.83
log TE	0.4817	0.6238	0.5004	0.1400
	9.48	10.01	5.41	1.06
log AS	0.7969	0.8767	0.5866	0.5244
	6.21	6.31	2.81	1.29
$SPEC$	−0.0024	−0.0362	−0.0586	−0.0007
	0.42	2.97	0.95	0.08
R^2	0.9298	0.9854	0.9493	0.9133
S.E.	0.3167	0.2055	0.2907	0.3920

such a relation, it is, nevertheless, worth investigating. Table 8.7 gives the results. Let

$$TE = \text{total number of employees,}$$
$$AS = \text{average annual earnings per employee.}$$

Hence $TPA = AS \cdot TE$. Thus the regressions in table 8.7 represent the total payroll variable by the product of the two factors. The results show that holding constant the total number of employees as measured by log TE, the rate of return does rise with log AS. We can interpret this to mean that specific human capital per man rises with annual earnings. One should also note that in every case the coefficient of log AS is larger than the coefficient of log TE although in general the *t*-value of the log AS coefficient is lower.

A more demanding test of these findings requires each of the total payroll components to be split into the product of the number of employees and the annual earnings per employee. Thus, $PPA = PE \cdot PS$, where

PE = number of production worker employees,
PS = annual earnings per production worker,

and $MPA = ME \cdot MS$, where

ME = number of nonproduction worker employees,
MS = annual earnings per such employee.

It is worth noting that the annual earnings of the production workers in this sample are about 64 percent of the annual earnings of the nonproduction workers and that annual earnings of both tend to rise with concentration ratio. In fact the annual earnings of production workers in the most concentrated stratum are nearly $1,000 per year higher than in the least concentrated stratum, which is nearly 25 percent of the earnings in the latter group. The rise in the earnings of the nonproduction workers is less than $800, which is a bit more than 10 percent of the earnings in the least concentrated stratum. The results for table 8.8., then, are comparable to those in table 8.6 since each of the payroll components in the regressions of table 8.6 are split into the product of the number of employees and the annual earnings per employee. Moreover, one can argue that the larger coefficients of log MPA in table 8.6 as compared with the coefficients of log PPA are consistent with the present hypothesis insofar as nonproduction workers have larger annual earnings than production workers and seem to have more specific human capital per employee as well. Table 8.8 shows that in the sample as a whole both coefficients of log PS and of log MS are positive and larger in magnitude than the coefficients of log PE and of log ME. Indeed, the coefficient of log PS is more than twice as large as the coefficient of log PE while the coefficient of log MS is more than 10 times as high as the coefficient of log ME. However, all of the t-values of these coefficients are very small, which may be due to the fairly high correlation between log PS and log MS. In the substrata there is a less clear-cut picture. Thus in the upper concentration stratum the coefficient of log MS is negative while in the middle stratum the coefficient of log PS is negative. Moreover, in this stratum the coefficient of log ME is also negative.

Although one cannot always explain convincingly disappointing results, such is not the case for table 8.8. Multicollinearity among the explanatory variables when these are subject to measurement error is especially troublesome, and the more so, the larger the number of explanatory variables, and the smaller the size of the sample. Counting the dummy variables, the regressions in table 8.8 have 29 explanatory variables. This does not cause much trouble for the whole sample where we have nearly 400 observations, but it can be troublesome for the subsamples which are about a third as large. Moreover, the correlations among the explanatory variables are high. In the whole sample the correlation between log GBV and log INV is 0.81. Between log ME and log PE the correlation is 0.83, and between log MS and log PS the correlation is 0.65. Despite these high correlations we manage to obtain sensible results for the whole sample where we have 398 observations but not in all of the substrata. Surely it would be more disappointing to find coefficients with the "wrong" sign for the entire sample than for the subsamples. Therefore, it is safe to conclude that, despite a few disappointments in some

TABLE 8.8

Selected Statistics for Regressions with log OHA Dependent, 398 Four-digit SIC Manufacturing Industries, 1963, Stratified by Concentration, log PE, log ME, log PS, and log MS among the Explanatory Variables

			STRATA	
	TOTAL	[0, 25)	[25, 50)	[50, 100]
VARIABLES	COEFFICIENTS AND *t*-RATIOS			
C	0.3220	−0.0187	0.2346	1.0520
	2.32	0.04	0.59	2.68
log NO	0.1034	0.0757	0.0712	0.1837
	4.13	1.89	1.63	3.58
log INV	0.2031	0.1867	0.1654	0.3445
	6.36	4.65	3.32	3.76
log GBV	0.2536	0.1447	0.2837	0.3566
	5.78	2.83	3.89	3.08
log PE	0.1538	0.1784	0.0356	0.2045
	3.01	2.96	0.45	1.34
log PS	0.3306	0.2640	−0.2975	0.7900
	2.48	1.76	1.38	1.88
log ME	0.0306	0.2125	−0.3595	0.0796
	0.16	1.03	1.14	0.14
log MS	0.3189	0.2466	0.9384	−0.1061
	1.48	1.06	2.61	0.17
SPEC	−0.0004	−0.0377	−0.0390	−0.0025
	0.06	3.09	0.65	0.30
R²	0.9294	0.9716	0.9536	0.9153
S.E.	0.3187	0.2051	0.2804	0.3921

of the substrata, the evidence continues to support the contention that specific human capital per employee rises with the worker's annual earnings. Thus, firms invest relatively more in those of its employees who are more highly paid. The next section continues the exploration on the subject of specific human capital with fresh evidence from the employment turnover statistics of the Bureau of Labor Statistics.

After this detailed inspection of the regression results, let us consider the general pattern of the findings. We have seen that the coefficients of the components of tangible capital, *GBV* and *INV*, tend to rise with the concentration ratio while the coefficients of the components of specific human capital tend to vary inversely with the concentration ratio. To explain these results we begin by calculating the marginal contribution to overhead with respect to the various capital components. In this calculation we use equations of the same form as the regressions. Denote a generic capital component by K. Therefore,

(12)
$$\frac{\partial OHA}{\partial K} = a \frac{OHA}{K}.$$

As before let r denote the rate of return and let d_K denote the rate of depreciation of capital. If the amount of capital in an industry is at a level that maxi-

TABLE 8.9

Estimates of Marginal Contributions to Overhead Based on
Regressions in tables 8.5 and 8.6

	RATIO[a]	TABLE 8.5		TABLE 8.6	
		a_i	MP_i	a_i	MP_i
		TOTAL			
OHA/GBV	0.588	0.254	0.149	0.271	0.159
OHA/INV	1.558	0.205	0.320	0.192	0.299
OHA/TPA	0.911	0.480	0.437		
OHA/PPA	1.450			0.147	0.212
OHA/MPA	2.448			0.322	0.789
		[0, 25)			
OHA/GBV	0.744	0.127	0.094	0.151	0.112
OHA/INV	1.746	0.203	0.355	0.185	0.323
OHA/TPA	0.804	0.631	0.507		
OHA/PPA	1.231			0.164	0.202
OHA/MPA	2.314			0.432	1.000
		[25, 50)			
OHA/GBV	0.567	0.236	0.134	0.239	0.136
OHA/INV	1.347	0.216	0.291	0.201	0.271
OHA/TPA	0.920	0.502	0.462		
OHA/PPA	1.488			0.040	0.060
OHA/MPA	2.409			0.465	1.119
		[50, 100]			
OHA/GBV	0.503	0.446	0.224	0.439	0.221
OHA/INV	1.612	0.364	0.587	0.361	0.581
OHA/TPA	1.029	0.118	0.121		
OHA/PPA	1.698			0.119	0.202
OHA/MPA	2.614			0.009	0.023

a. These are weighted means by strata so that every industry receives a weight equal to its relative size as measured by the denominator of the ratio.

mizes the present value of the net return, then

(13) $$\frac{\partial OHA}{\partial K} = r + d_K = a\,\frac{OHA}{K}.$$

Let us apply this reasoning to the various forms of capital beginning with H, specific human capital.

(14) $$\frac{\partial OHA}{\partial H} = r + d_H.$$

However, H does not appear directly in the regressions, but is related to TPA as shown in (8). Therefore,

$$a\,(OHA)/H = [a\,(OHA)/(TPA)(r + d_H)]/[(\alpha - 1)(1 - \beta)]$$

Together with (14) it follows that

(15) $$OHA/TPA = (1 - \beta)(\alpha - 1)/a.$$

Consider table 8.9 which gives the ratios OHA/TPA by concentration ratio strata. The table shows that this ratio, measured at the mean values by stratum, increases as C increases. Hence the ratio $(1 - \beta)(\alpha - 1)/a$ should increase as C increases. This means that either $(\alpha - 1)(1 - \beta)$ should increase with C or a itself should decrease with C. I would regard the former possibility with more favor because it seems plausible to me that not only should specific human capital per man rise with the concentration ratio, but also that it should rise at an increasing rate while $(1 - \beta)$ itself is independent of C.

The argument that relates C and α is simple. In the high concentration industries an employee with skills specific to the *industry* has fewer employment alternatives. If it is also true, as seems likely, that these skills are correlated with the skills specific to the firm, then a firm would be more willing to invest in industry as well as firm-specific skills since the risk of losing its investment is lower if the employee is less likely to quit. Moreover, the quit rate is lower in the more concentrated industries, which is consistent with there being more specific human capital per employee in these industries. Therefore, one can expect a rising α with rising C. A similar argument applies for the two components of the total payroll, PPA and MPA. For these also we observe a rise in the ratios OHA/PPA and OHA/MPA with rising concentration ratios.

Equations (12) and (13) also yield implications for the two kinds of tangible capital, GBV and INV, where K stands for either GBV or INV. Table 8.9 shows that the marginal contribution to overhead rises with the concentration ratio. This is consistent with the evidence that the rate of return, r, rises with the concentration ratio if d_K, the rate of depreciation is independent of C, or approximately so, as seems plausible.

There are, however, some surprises in table 8.9. Thus one would not wish to argue that the marginal contribution of inventories always exceeds the marginal contribution of plant and equipment (represented in *GBV*) since it is likely that inventories have a higher rate of depreciation. These results have a statistical explanation. Gross book value is the original cost of plant and equipment while inventories include the stocks of finished goods, goods in process, and raw materials, all of which are valued on a more current basis. The dependent variable, value added less adjusted payrolls, is sensitive to the values of such current variables as prices and the physical quantity of sales. Hence one would expect a higher correlation between *OHA* and the more currently valued capital components such as *INV* than with items of longer average life valued at original cost. I would be reluctant to interpret the level of the calculated marginal contributions to overhead as accurate estimates of the true marginal contributions. More confidence can be given to the ranking of these estimates by concentration ratio strata since the rankings are less prone to these types of errors.

This completes the discussion of the regression analysis of the 398 industries in the sample of four-digit SIC census manufacturing industries for 1963. The next topic is a study of the employment turnover statistics in order to learn more about specific human capital.

4. An Analysis of Employment Turnover in Selected Manufacturing Industries

The Bureau of Labor Statistics collects and publishes detailed figures for about 100 four-digit SIC census manufacturing industries giving the monthly accession and separation rates of the labor force. Accessions consist of new

TABLE 8.10

Summary Statistics on the Turnover of Employment, Annual Averages of Monthly Rates, 1958–63 for 99 Four-digit SIC Manufacturing Industries

	VARIABLES[a]				
	New Hires	Other Accessions (Rehires)	Quits	Other Separations	Layoffs
Means percent	2.084	1.454	1.257	0.593	1.780
S.D.	0.915	0.982	0.669	0.181	1.219
	CORRELATION MATRIX				
New hire	1	0.3453	0.9062	0.5105	0.4829
Other acc.		1	0.1970	0.3783	0.9516
Quits			1	0.4248	0.2591
Other sep.				1	0.3827
Layoffs					1

NOTE: For a list of the industries in this sample see appendix 3.
a. The components of turnover include both production as well as nonproduction workers.

hires and a residual class. The latter includes, primarily, rehires of employees previously laidoff. Separations consist of three categories, quits, layoffs, and a residual class. The latter includes separations due to deaths, retirements, firings, the draft, and so forth. Seasonal variation of the demand for the final product is the cause of most layoffs and rehirings. The turnover figures for these five categories are available monthly for the period 1958 to the present. Table 8.10 shows the averages for the sample over the period 1958 to 1963. Note that these are the *monthly* rates. Thus, if the employees who left the work force of a given firm were a random sample of all of its employees, then the average duration of continuous employment would be about 27 months. In fact, turnover rates vary among different classes of workers, and the composition of the work force varies among different industries. Hence in order to isolate the effects of specific human capital on turnover, it is necessary to account for differences in the composition of employment by industry.

It is useful to regard the components of the turnover rates in two different ways according to whether the action is at the initiative of the worker or the employer and according to whether it represents a permanent or a temporary event. For example, new hires, layoffs, and rehires are at the initiative of the firm while quits are the workers' prerogative. Quits and new hires are permanent while layoffs are temporary. The multiple regressions shown below will attest to the usefulness of treating separately each of the turnover components. However, even the simple correlations of the turnover components reveal much about how they are related. Table 8.10 gives the simple correlations. We see that quits and new hires are closely correlated (the r is 0.91) and, similarly, other accessions (rehires) and layoffs are closely correlated ($r = 0.95$). Thus workers who quit are likely to be replaced and those who are laidoff are likely to be rehired. It is not as obvious a priori why layoffs and rehires should be positively correlated since one would think that the two should be negatively correlated. The positive correlation, however, is explained by the nature of the data. The figures are the monthly averages over a six year period, 1958 to 1963, and on average there is a positive correlation. With monthly figures we would not expect a positive correlation between layoffs and rehires if the average period of layoffs were longer than one month.

The quantity of labor services can change with a constant number of workers by varying the total hours worked per employee. As we shall see, this affects the components of turnover.

The existence of specific human capital has several implications for the relation between turnover and other factors. If a firm invests in the specific skills of one of its employees, then it will attempt to retain his services in order to prevent the loss of its investment. Hence the higher is the autonomous quit rate of a given class of workers, the more reluctant is the firm to make specific investments in the members of the class. Once a firm has invested in one of its employees, it has several means of persuading the employee to remain. These include the offer of higher wages, more regular employment, pensions, fringe benefits, deferred compensation, and so forth. Notwithstanding this, the firm recoups its investment in specific human capital by obtaining a marginal product from the worker that exceeds its payment to the worker. Moreover, the firm can increase the worker's incentive to remain in its employ by encouraging the worker to share in both the cost of the investment of specific human capital and in the return. Hence the firm and the worker have a mutual interest to protect their respective specific investments. However, a rational worker would not assume the whole investment in specific human capital because he would then bear the risk of losing the capital at the employer's whim. It follows that both parties have an incentive to share the cost and the returns. The return to the firm appears as part of

TABLE 8.11

Definitions, Mean Values and Standard Deviations of the Variables Used to Explain Components of Employment Turnover

DEFINITION OF VARIABLES[a]	MEAN	S.E.
$E\ 64/8$ = (total employment 1964)/(total employment 1958)	1.084	0.145
PRO/E = Weighted average ratio of production workers to total employees, 1958–1964[a]	0.782	0.221
FEM/E = Weighted average ratio of female workers to total employees, 1959–1964[a]	0.259	0.216
$DPRO/E = (PRO/E, 1958)-(PRO/E, 1964)$	−0.0298	0.334
$DFEM/E = (FEM/E, 1959)-(FEM/E, 1964)$		0.016
$AVHW$ = Average hourly earnings of production workers[a]	2.320	0.450
$THRS$ = Average weekly hours of production workers[a]	40.35	1.52
MS = Average annual earnings of nonproduction workers (thousands of dollars) in 1963	7.843	0.805
C = Concentration ratio in 1963	36.64	21.90
OHA/TE = Ratio of contribution to overhead to total employment (thousands of dollars) in 1963	5.358	4.053
$K/TE = (GBV+INV)/TE$ (thousands of dollars) in 1963	12.015	8.934
log $SIZE$ = log of total employment in the 4 Largest Companies in 1963	7.53	0.76

a. An average in this case refers to a weighted average by years. For instance, if P_t denotes the number of production workers in year t and E_t the total number of employees in year t, then

$$PRO/E = \sum_{1958}^{1964} P_t / \sum_{1958}^{1964} E_t.$$

Unless noted to the contrary all averages are for the period 1958 to 1964.

OHA while the return to the worker is part of his pay. Indeed, perhaps the dependence of OHA on the payroll results from the sharing of the cost and returns of specific human capital by the worker and the employer. The turnover figures show the effects of both kinds of specific human capital, that due to the firm's investment as well as that due to the worker's investment.

My intention is to explain the five components of turnover among manufacturing industries with a multiple regression analysis. There are five regressions, one for each of the five components of turnover. Let us now consider some of the reasons for differences in turnover among industries. According to the theory of specific human capital, the industries with lower turnover rates should be the ones with more specific human capital per man, other things equal. Among the "other things," is the composition of the labor force. The turnover statistics refer to the firm's total labor force at its establishments including both its production and its nonproduction workers. If it be true that turnover varies by kind of worker and that industries vary in the composition of their work force, then also the turnover rates will vary among industries.

The BLS data provide some pertinent information allowing us to calculate the proportion of women and the proportion of production workers by industry. Table 8.11 shows that 26 percent of all employees are women and that 78 percent of all employees are production workers. (The two classifications overlap in an unknown way.) If women or production workers have higher quit rates than men or nonproduction workers and if the proportion of women and of production workers varies by industry, then we would predict higher quit rates in those industries with a higher proportion of women and a higher proportion of production workers. Therefore, we are led to include the proportion of women and the proportion of production workers among the determinants of the turnover rates. Similarly, one would wish to include changes in these proportions. For example, the proportion of production workers has fallen on average from 1958 to 1963. If the decreases vary by industry, then this would explain some variation of the turnover rates by industry.

Similarly, the regressions include changes in total employment. Obviously, new hirings necessarily rise with a rise in the work force. In addition, if some of the newly hired workers are dissatisfied with their new jobs, they will quit and seek other employment. Hence both the quit rate and the new hiring rate should vary directly with the change in the work force. This is why the regressions include the ratio of 1964 to 1958 employment. On average, industry employment in the sample rose by 8 percent during this period.

To describe how specific human capital affects the components of turnover, it is helpful to adopt a simple algebraic framework. The components of human capital can be classified by the degree of their specificity. The most specific is the cost of acquiring and maintaining the worker's skill and knowledge that is only of use to his current employer while the least specific has value to the widest variety of firms. Although there is a continuum of specificity, let us distinguish only two kinds, S and G; S denotes the specific human capital and G denotes all other forms of human capital. The total human capital, H, is the sum of S and G,

(1) $$H = S + G.$$

Specific human capital may be further divided into the cost borne by the firm and that borne by the worker. Let

SF = firm's investment in specific human capital,
SW = worker's investment in specific human capital.

Thus,

(2) $$S = SF + SW.$$

We are now ready to determine the implications for the components of turnover of these forms of human capital beginning with the quit rate Q and the layoff rate L. To a linear approximation

(3) $$Q = a_0 + a_1 G + a_2 SF + a_3 SW.$$

(4) $$L = b_0 + b_1 G + b_2 SF + b_3 SW.$$

Direct tests of the theory require data giving the values of the three explanatory variables in these two equations; but, since, the requisite data are unavailable, we must perforce proceed indirectly by relying on postulated relations between the observed variables and the unobserved human capital components. In this exercise we shall derive the implications of the theory for the coefficients of (3) and (4).

The theory of specific human capital implies that

$$a_2 \leqslant 0, \quad a_3 < 0, \quad |a_3| > |a_2|,$$
$$b_2 < 0, \quad b_3 \leqslant 0, \quad |b_2| > |b_3|.$$

The larger is the worker's investment in specific capital, the smaller his quitting propensity so that $a_3 < 0$. Similarly, the larger the firm's investment in specific human capital, the more reluctant is the firm to layoff the worker. This assumes that the firm incurs a risk of being unable subsequently to rehire the laidoff worker so that the firm is least likely to part with those workers in whom it has the largest investment. At first blush one would think that both a_2 and b_3 would be zero because it is of no concern to one party that the other may lose by its action. Though there may be some truth in this, another factor must be admitted. It is argued above that both parties have an incentive to share in the cost of specific human capital, which requires some implicit agreement on how to share the return. In particular the firm must protect the worker's investment in specific human capital or it would lose its reputation for fair dealing and thereby fail to attract the kind of employee that it desires. Put differently, if the firm did not protect the worker's share of the specific human capital, then it would find that its labor cost is increased for a given quantity and quality of labor input. Hence b_3 is likely to be negative though of smaller size than b_2. The same considerations tend to make the worker solicitous of the firm's welfare though not to the same degree because individual workers are less well known than individual firm's labor policies.

However, high executives who are well known in the industry are likely to be as scrupulous in these matters as are firms, and for the same reasons. Hence although a_2 is probably negative, it is closer to zero than is b_3.

It is not clear how G should affect quits and layoffs. In so far as G affects the quit rate it is perhaps more likely to raise than to lower it if the more highly educated employees keep in closer touch with the market and shift jobs more frequently. Pending further research, pronouncements on this relation seem hazardous.

The larger the amount of human capital in the form of G and of SW the higher is the worker's wage rate, W. To a linear approximation

$$(5) \qquad W = \alpha_1 G + \alpha_2 SW$$

with $\alpha_1, \alpha_2 \geqslant 0$. It follows that in a regression with the quit rate as the dependent variable and the wage rate as an explanatory variable, the coefficient of W would be negative unless there is a strong inverse relation between G and SW. However, G and SF are probably complementary so that on balance the quit rate varies inversely with the wage rate; the higher the wage rate, the larger is the worker's specific investment, and the smaller his quitting propensity. The layoff rate may be independent of the wage rate if the wage rate is independent of the firm's investment in specific human capital SF.

Next, we derive some implications about the observed regression coefficients assuming complementarity between nonhuman and human capital. Suppose that

$$(6) \qquad K = \beta_1 G + \beta_2 SW + \beta_3 SF,$$

where K denotes the tangible capital and

$$\beta_1, \beta_2, \beta_3 \geqslant 0.[7]$$

Let Q and L be regressed on W and K giving

$$(7) \qquad Q = c_0 + c_1 W + c_2 K,$$

$$(8) \qquad L = d_0 + d_1 W + d_2 K.$$

Together with (5) and (6), these two equations yield some inferences about the relations between the observable regression coefficients and the unobserved human capital coefficients. Thus, ignoring the constant terms,

$$Q = (c_1\alpha_1 + c_2\beta_1) G + (c_1\alpha_2 + c_2\beta_2) SW + c_2\beta_3 SF,$$

and

$$L = (d_1\alpha_1 + d_2\beta_1) G + (d_1\alpha_2 + d_2\beta_2) SW + d_2\beta_3 SF,$$

---

7. Griliches (1969) gives some results consistent with this assumption. Rosen's data for the railroad industry also supports this assumption (1965).

so that in terms of (3) and (4) we obtain the following relations:

$$a_1 = c_1\alpha_1 + c_2\beta_1 \qquad \text{of indeterminate sign};$$

(9) $$a_2 = c_1\alpha_2 + c_2\beta_2 < 0;$$

$$a_3 = c_2\beta_3 \leqslant 0;$$

and

$$b_1 = d_1\alpha_1 + d_2\beta_1 \qquad \text{of indeterminate sign};$$

(10) $$b_2 = d_1\alpha_2 + d_2\beta_2 \leqslant 0;$$

$$b_3 = d_2\beta_3 < 0.$$

These relations imply a negative coefficient of capital, K, in the layoff regression and a negative or zero K coefficient in the quit-rate regression. We may also conclude that the more strongly the worker's share of specific human capital affects the wage rate, the greater the likelihood of an inverse relation between the quit rate and the wage rate. However, it is difficult to reach any definite conclusions about the sign of the wage rate coefficient in the layoff rate regressions; any sign would be consistent with the preceding analysis. Plainly, $b_1 = 0$ implies $d_1 > 0$, if $d_2 < 0$.[8]

Table 8.12 gives some pertinent empirical results. In both the quit and the layoff regressions the coefficient of capital per worker is negative as predicted

---

8. The text illustrates a special case of a general approach. Let x denote an m-vector of observed variables and let y denote an n-vector of the unobserved or latent variables. Thus the components of y correspond to the components of various kinds of human capital while the components of x refer to those variables that are related to human capital such as the wage rate. Equation

(i) $$x = Ay$$

gives the relation between x and y corresponding to (5) and (6) in the text. The regression is given by

(ii) $$t = b'x,$$

where t denotes a scalar-valued component of turnover and b is the m-vector of regression coefficients. Therefore,

(iii) $$t = b'Ay,$$

Finally, define the n-vector a according to

(iv) $$a = A'b.$$

The theory determines the signs of the elements of both A and a, while b is the estimated regression coefficients. Hence tests of the theory require predictions of the values of b given the elements of a and A in (iv). A favorable case is when A is nonsingular so that $m = n$, and, even so, it is not always possible to predict the sign of b.

One can sometimes solve for b by the following procedure.

(v) $$Aa = AA'b.$$

If A has rank m so that AA' is invertible, then from (v) we obtain

(vi) $$b = (AA')^{-1}Aa.$$

This relation is a starting point for a systematic study of the identifiability properties of the system. A different procedure is required if $m > n$ so that the rank of A is $n < m$. In the latter case it is more difficult to identify the model. These brief remarks must suffice for a large and difficult subject.

TABLE 8.12

Regressions with Components of Employment Turnover as Dependent Variables, 99 Four-digit SIC Manufacturing Industries, 1958–64

EXPLANATORY VARIABLES	DEPENDENT VARIABLES				
	New Hires	Rehires Other Accessions	Quits	Other Separations	Layoffs
	COEFFICIENTS, *t*-RATIOS, ELASTICITIES				
Constant	10.15	−7.612	6.353	1.524	−5.389
	4.03	3.92	5.08	2.12	2.69
$E64/8$	0.5536[a]	−1.3247	0.2570	0.1633	−1.8649
	1.25[b]	2.05	1.08	1.35	2.30
	0.29[c]	−0.99	0.22	0.30	−1.14
PRO/E	0.4303[a]	6.0059	0.8195	0.1556	6.5275
	1.51[b]	5.35	1.97	1.90	4.61
	0.16[c]	3.23	0.51	0.20	2.87
FEM/E	−1.204	0.794	...	−0.300	...
	2.15	1.18		2.09	
	−0.15	0.14		−0.13	
$AVHW$	−0.9084	1.2473	−0.5636	−0.0748	1.2342
	3.14	2.72	4.33	1.26	2.87
	−1.01	1.99	−1.04	−0.29	1.61
C	−0.0120	−0.00766	−0.00602	...	−0.01161
	3.67	1.39	3.62		2.07
	−0.21	−0.19	−0.17		−0.24
$D(PRO/E)$[d]	...	3.2423	0.389	...	3.6764
		4.74	1.50		4.14
MS	−0.1190	0.2332	−0.1825	...	0.2751
	1.03	1.36	2.95		1.30
	−0.45	1.26	−1.14		1.21
$THRS$	−0.1234	...	−0.07245	−0.0226	...
	2.20		2.77	1.41	
	−2.39		−2.32	−1.54	
OHA/TE	0.0218	0.04152	...	...	0.06683
	1.12	1.36			1.79
	0.06	0.15			0.20
K/TE	−0.0289	−0.0402	−0.0100	−0.0056	−0.0564
	3.05	2.73	2.27	2.55	3.12
	−0.17	−0.33	−0.096	−0.11	−0.38
$D(FEM/E)$[d]	...	−10.65	...	...	−16.94
		1.92			2.42
log $SIZE$	...	0.2082	...	...	...
		1.38			
R^2	0.6377	0.3479	0.7870	0.1954	0.3088
S.E.	0.5779	0.8417	0.3223	0.1681	1.0637

a. Coefficients   b. *t*-ratios   c. Elasticities

d. Since the explanatory variable is a difference, the elasticities calculated at the mean values would be misleading.

by this analysis. Moreover, the elasticity of the layoff rate with respect to capital is −0.38 while it is −0.10 for the quit rate so that the relative magnitudes are consistent with the theory. The quit rate varies inversely with both the average wage rate of production workers as well as the average annual salary of nonproduction workers, *MS*. For both, the elasticity is about −1. The coefficients of both of these variables are positive in the layoff regressions.[9]

One explanatory variable in these regressions is the production workers' weekly hours which has a negative coefficient in the quit rate regression and a nearly zero coefficient in the layoff rate regression. This result is also consistent with the theory because of the typical nature of the wage structure. Overtime wage rates are about 50 percent higher than straight-time wage rates, making the worker's marginal wage rate rise with average weekly hours, which raises their return on their investment in specific human capital. Hence total hours reflect an aspect of the wage structure which should affect the quit rate in the same way as the average hourly wage rate. Moreover, just as wages exert a stronger influence on the quit than on the layoff rate, so too for total hours, which is precisely what the regression shows.

Both the quit rate and the layoff rate vary inversely with the concentration ratio. In some other regressions not presented herein the log of the number of companies appears as an explanatory variable and has a positive coefficient. This means that the quit and layoff rates are lower the larger is the relative size of the four leading companies and the smaller is the total number of companies in the industry. These findings are consistent with the view that companies in the more concentrated industries invest more heavily in their employees and that the workers increase their own specific human capital investment in step with the company's investment. Hence total specific human capital per worker rises with the concentration ratio.

There is another way of presenting the argument for using the concentration ratio as an explanatory variable in terms of the competitive state of the industry. In the preceding section we see that the rate of return is a rising function of the concentration ratio. Therefore, it is consistent with this finding to suppose that competition for workers varies inversely with the

---

9. Because the dependent variable in these regressions has an estimate of employment in the denominator while some of the variables used to represent the effects of the composition of the labor force also contain an estimate of this variable in their denominators, there is a possibility of some upward bias in the estimates of the coefficients of these variables. One can try to avoid this by choosing the same regression form as in section 3, namely, one in which the dependent variable would be the absolute number of quits, layoffs, and so on, while among the explanatory variables there is total industry employment. However, this procedure may well introduce more measurement error than it is designed to prevent because the absolute numbers of the components of accessions and separations are not published, only the rates are. Hence one cannot be sure of what measure of total employment to use as a multiplier for converting the ratios to absolute numbers. This can introduce measurement error and also bias the estimates of the coefficients.

concentration ratio making firms in the more concentrated industries more willing to invest in specific human capital. This might result from an implicit understanding among the competitors not to raid each other's work force. Nor is this all. Under these conditions a company may even be willing to invest in industry specific human capital as well as in company specific human capital. For example, suppose a company has a 30 percent share of industry sales. If there is tacit collusion among the firms in the industry, then a company may expect to receive 30 percent of the return on its investment in industry specific human capital. Consequently, the investment in both industry and company specific skills may well be a rising function of the concentration ratio.

Before examining the coefficient of the gross return on capital per employee, OHA/TE, let us pause to consider how the variables representing the labor force composition affect quits and layoffs. For the most part the coefficients of these variables have the expected signs. Thus the quit rate is higher, the higher is the proportion of production workers and the larger is the rise in this proportion $D(PRO/E)$. Industries with larger increases of employment have higher quit rates and such industries also have lower layoff rates. Also, the layoff rate varies directly with the proportion of production workers and with rises in this proportion. Surprisingly, an increase in the proportion of females tends to *lower* the layoff rate although the average duration of female employment is less than that of males. Neither FEM/E nor $D(FEM/E)$ has a statistically significant coefficient in the quit rate regression nor does the former variable significantly affect layoffs.

Table 8.12 includes figures for two regressions for which the dependent variables are new hires and rehires, components of the accession rates. We know that the simple correlation between new hires and quits is 0.91 and that between rehires and layoffs is 0.95. Do the accession components give any new information beyond that contained in the corresponding regressions for quits and layoffs? If the corresponding accession and separation components were perfectly correlated, then with the same explanatory variables in both regressions, the regression coefficients could only differ by a scale factor. Hence it is better to look at the elasticities, which are independent of the absolute units, than to look at the coefficients themselves. First, consider the rehires and the layoffs regressions. A glance at table 8.12 is enough to show that the same variables have nearly the same elasticities in the two regressions. Therefore, there is no new information in the rehire regression beyond that given in the layoff regression. The elasticities of the explanatory variables in the quits and new hires regressions are also nearly equal with only two exceptions, the proportion of production workers, PRO/E, and average salaries of nonproduction workers, MS. The elasticities for these two variables are substantially lower in the new hires than in the quits regression. Moreover, the R^2's differ more for these two regressions than for layoffs and rehires. It follows that new hires are less like quits than rehires are like layoffs, which is hardly surprising since, unlike rehires, one cannot replace a person who has quit by a literally perfect substitute. Moreover, the theory of human capital herein seems of slight relevance for explaining variations in new hiring rates among industries. Nevertheless, the striking similarity between new hires and quits, notwithstanding the differences we have noted, supports the belief that an effort is made to replace the quits with close substitutes.

The last regression in table 8.12 has "other separations," a small and heterogeneous category as the dependent variable. Because of its diverse character, it would be imprudent to explain these regression results using the theory of specific human capital. The only explanatory variables of any importance are capital per worker with a negative coefficient and a t-ratio of 3.1 and the proportion of females with a negative coefficient and a 2.1 t-ratio.

One variable remains to be discussed before showing how labor force turnover affects the gross rate of return, log $SIZE$, where $SIZE$ is the number of employees in the four largest companies. It is sometimes argued that the larger the company, the more easily it can rely on its own resources to shift its employees internally without much loss of firm specific human capital. The variable log $SIZE$ was introduced to test this notion, but it only manages to enter one regression where it has a small positive coefficient.

Gross profit per employee, OHA/TE, is an explanatory variable in one of the two regressions of our special interest, layoffs, where it has a positive t-value of nearly 1.8, but it fails to have a significantly nonzero coefficient in the quit regression. We shall systematically explore the relation between profits and the two components of separations, layoffs and quits, and thereby obtain strong confirmation of the theory of specific human capital.

How should quits and layoffs affect industry returns according to the theory of specific human capital? We begin with (3) and (4) which assert that quits vary inversely with SW and that layoffs vary inversely with SF. That is, the larger is the worker's specific human capital, the smaller is his quit propensity and the larger is the firm's specific human capital, SF, the less likely is he to be laidoff. To show these effects on the industry return we need a regression giving as nearly as possible a correspondence between L and SF and one between Q and SW. With these goals in view, consider the following equation:

$$(11) \quad \log OHA = a_0 + a_1 \log K + a_2 \log NO + a_3 C + a_4 \log TE + \\ a_5 \log AS + a_6 Q + a_7 L.$$

Except for the addition of the variables Q and L, this is the same as the regressions given in the preceding section. In (11) a rise in Q implies a rise in the depreciation of the firm's specific human capital for the following reasons. First, a rise in Q implies a rise in the rate of depreciation of SF, and second the regression holds TE and L constant so that it holds constant the total stock of firm-owned specific human capital. Therefore, under these conditions the gross return to total SF must rise in equilibrium to compensate for the rise in the rate of depreciation of *SF*. Therefore, the coefficient of *Q* should be positive.

Holding *Q* and *TE* constant, a rise in *L* implies a smaller total stock of *SF*. Therefore, since the stock of tangible capital is held constant by the presence of log *K* in (11), the total return must be lower and the coefficient of *L* is accordingly negative. In other words when *L* is lower, the accounting measure of the tangible capital is closer to the true capital so that the measured rate of return is closer to the true rate.

The estimates of (11) are as follows:

(12) $\log OHA = 0.696 + 0.685 \log K + 0.1204 \log NO + 0.006749\ C$
         0.68   6.20   2.16   2.18

   $+ 0.1017 \log TE + 0.2582 \log AS + 0.1822\ Q - 0.03803\ L.$
       0.72   0.68   1.29   1.08

   $R^2 = 0.712,\quad \text{S.E.} = 0.400$

Thus the signs of the coefficients of both *Q* and *L* conform to the predictions made by the theory of specific human capital.

A more demanding test of this theory distinguishes between the permanent and the transitory components of quits and layoffs. Both *Q* and *L* are averages for the period 1958–63 while the dependent variable refers to the single year 1963. Perhaps the components of *Q* and *L* for that year happen to be out of line and have a different impact on *OHA* than in other years. Therefore, we should separate *Q* and *L* into their permanent and transitory elements. Let

   Q^* = predicted average quit rate from a regression equation,
   Q = actual average quit rate for 1958–63,
   $Q63$ = observed quit rate in 1963,
   L^* = predicted average layoff rate from a regression equation,
   L = actual average layoff rate for 1958–63,
   $L63$ = observed layoff rate in 1963.

Instead of (11) we shall estimate a regression equation including the variables as follows:

$$Q^*,\ L^*,\ Q63 - Q,\ L63 - L,\ Q - Q^* \quad \text{and} \quad L - L^*.[10]$$

---

10. $Q^*, L^*, Q - Q^*$, and $L - L^*$ are computed from the regressions as follows:
(i) $Q^* = 4.3330 + 0.3021\ E\ 64/8 + 0.07768 \log NO - 0.005702\ K/E - 2.1890 \log AS;$
         11.69   1.39   2.93   1.36   14.98
   $R^2 = 0.7974,\qquad \text{S.E.} = 0.3028.$
(ii) $L^* = -6.0708 - 1.5363\ E\ 64/8 + 0.3627 \log NO - 0.02751\ K/E - 2.3699 \log AS$
         0.99   1.91   2.60   1.61   1.08
   $+ 5.0372\ PRO/E + 0.3585\ FEM/E + 1.8036\ AVHW + 2.7958\ D(PRO/E)$
       2.46   0.32   2.39   2.26
   $+ 12.897\ D(FEM/E) - 0.01924\ THRS + 0.4639\ MS + 0.2324 \log SIZE;$
       1.80   0.16   1.79   1.01
   $R^2 = 0.3306,\qquad \text{S.E.} = 1.0598.$

4. *Analysis of Employment Turnover*

Log *OHA* should vary directly with Q^* and inversely with L^* for precisely the reasons given above. The transitory components due to year effects are represented by $Q63 - Q$ and $L63 - L$, and log *OHA* should vary inversely with these variables. Thus, if either the quit rate or the layoff rate happens to be below normal, then the rate of return should be above normal. The third component, which is the residual of the regression equation used to predict Q and L, can be used to test the accuracy of the regression used to predict Q^* and L^*. We would expect $Q - Q^*$ and $L - L^*$ not to affect log *OHA* with perfect regressions for predicting Q and L ($R^2 = 1$). The estimated regression is as follows:

$$(13) \quad \log OHA = -1.291 + 0.694 \log K + 0.0917 \log NO + 0.005559\, C$$
$$ 0.32 \quad\;\; 4.80 \qquad\quad 1.05 \qquad\qquad 1.96$$

$$+ 0.1639 \log TE + 0.9736 \log AS + 0.5886\, Q^*$$
$$ 1.00 \qquad\qquad 0.56 \qquad\qquad 0.71$$

$$- 0.4830\,(Q63 - Q) - 0.2315\, L^* - 0.04194\,(L63 - L)$$
$$ 1.37 \qquad\qquad 3.71 \qquad\quad 0.31$$

$$+ 0.06531\,(L - L^*)$$
$$ 1.74$$

$$R^2 = 0.768, \quad \text{S.E.} = 0.3650$$

This regression gives a better fit than (12). The coefficients of Q^* and L^* have the signs predicted by the theory of specific human capital. Moreover, both of the transitory components, $Q63 - Q$ and $L63 - L$, have the predicted negative signs. The regression residual for quits, $Q - Q^*$, fails to enter (13) because it would make too small a contribution toward reducing the standard error of the residual. However, the layoff regression residual, $L - L^*$, does enter (13) with a positive coefficient and a *t*-ratio of 1.74. Since the quit rate regression fits the data better than does the layoff regression, the difference between the behavior of their residuals in (13) is not unexpected. Nevertheless L^* has a stronger inverse effect on log *OHA* in (13) than has L in (12).

The coefficients of the other variables in (12) and (13) are consistent with the results for the comparable regressions shown in table 8.7. It will be recalled that the table 8.7 regressions include dummy variables for two-digit industry effects but that neither (12) nor (13) include such variables. This is because with a smaller sample size made necessary by the limitations of available data it would not be wise to use up so many precious degrees of freedom by inclusion of the two-digit industry dummies. In addition since the present sample

These two regressions should be compared with those in table 8.12. Observe that unlike the regressions in table 8.12, *OHA/E* is not used as an explanatory variable in either (i) or (ii). This is because Q^* and L^* are themselves to be used as explanatory variables of log *OHA* as is shown in (12) and (13).

is not a random one from all four-digit SIC manufacturing industries, the standard statistical tests for determining the closeness of (12) and (13) to the comparable regressions in table 8.7 are not applicable.

The results in the two regressions (12) and (13) together with the findings given in table 8.12 support the theory of specific human capital. They also suggest new avenues of research. First, it is desirable to know more about the labor force composition by industry. In particular more information about the occupational mix and the workers' levels of schooling would resolve some ambiguities about specific human capital. It would also be helpful to know more about other labor force characteristics that might affect labor turnover by industry such as the age distribution of the workers, the racial composition, the occupational mix, and so on. Separate turnover data for the major groups of workers would also help resolve some ambiguities about the nature of specific human capital.

It would seem that a direct research objective should be the estimation of the stock of specific human capital but this may encounter almost insurmountable obstacles. Although some specific human capital results from direct outlays by the firm in training its workers, probably much firm specific human capital results from the firm paying the workers a wage rate above their marginal product initially. While this occurs, the firm accumulates a stock of specific human capital upon which it subsequently expects a return. In addition there is the complication that the worker may himself bear some of the cost of creating firm specific human capital by accepting a wage rate below his opportunity cost. Indeed this is the mechanism by which a worker acquires general on-the-job training that has value to companies other than his current employer. This all suggests the desirability of estimating the determinants of the worker's marginal product, but this is especially difficult because the theory of specific human capital implies that the wage rate will be below the marginal product after the creation of a stock of specific human capital. Hence the wage rate is no simple guide to the marginal product. Nor is this all. To the extent that the worker and the firm share the specific human capital, they also share the return. Therefore, the worker's wage rate exceeds what he could get elsewhere. It follows that we cannot use the wage rate of comparable workers in other companies to measure the given worker's marginal product. Of course, although the worker receives less than his marginal product, he is not being exploited. The difference between the marginal product and the wage rate gives the return to the firm's outlay on specific human capital including an allowance for its depreciation. This return constitutes part of the firm's total return on all of its capital.

5. A Brief Survey of Findings by Other Investigators

Since Collins and Preston (1968) contains an excellent and detailed survey of the researches on the relation between the concentration ratio and rates of return, my survey of these findings may be brief. Indeed it is only necessary for me to discuss the especially relevant ones and the more recent findings not given in Collins and Preston.

Most studies show a positive relation between concentration and rates of return. Although most of the correlations are "statistically significant," the r^2 seldom exceeds 10 percent in agreement with my findings.

Bain's 1951 study is among the first. It is of special interest because in some respects my results using a fresh and different kind of sample confirm his. His sample includes 42 out of the 340 manufacturing industries in the 1935 Census of Manufactures. The sampling unit is a company instead of an establishment as in my study. This restricts the admissible industries to those in which the companies are specialized. The rate of return is measured by the ratio of net profit after taxes to net worth at the beginning of the year. Bain reports a correlation of 0.33 between the concentration ratio and the estimated rate of return, which is about the same as in my sample. Bain's most interesting result, however, is his finding of a nonlinear relation. This is because "there is a rather distinct break in average profit-rate showing at the 70 percent concentration line, and that there is a significant difference in the average of industry average profit rates above and below this line." Since the 70 percent concentration ratio refers to the share of the *8* largest firms, this finding agrees tolerably well with my results for the *4* firm concentration ratio in the stratum for $C \geqslant 50$ percent.

Although in an earlier study, Stigler (1963) declared that his findings did not corroborate Bain's for the post World War II period, in a more recent study (1968) he finds a significant and positive rank correlation between the rate of return and the concentration ratio for a sample of 17 highly concentrated manufacturing industries. As an alternative measure of profitability Stigler uses the ratio of the market value of the shareholders' equity to the book value of their equity. For this measure as well he finds a significantly positive rank correlation with the concentration ratio. In addition he finds a higher correlation with the various measures of profitability using the Herfindahl index to measure relative size. Since the Herfindahl index includes the effect of the number of companies, it combines both C and NO. Therefore, my findings are consistent with Stigler's most recent study although mine deal with virtually all of the four-digit SIC manufacturing industries.

The Collins-Preston study (1968) measures the rate of return by the ratio of OH to the value of shipments, which they call the price-cost margin. Their sample is taken from the 1958 Census of Manufactures four-digit industries. They relate the price-cost margin to the four firm concentration ratio, the capital output ratio, and a variable to discriminate between local and national markets. They find a positive correlation between the concentration ratio and the price-cost margin.

There have also been a few studies of specific human capital in terms of the turnover rates of the labor force. One of the first of these and, to date, still the most important is Walter Oi's doctoral dissertation (1961) partly summarized in a journal article (1962). In addition to presenting a theoretical analysis of specific human capital, Oi gives an empirical analysis. He measures turnover as the lower of the accession or the separation rate and does not study their components separately. Not only is this undesirable in theory but it is also undesirable in practice as is shown by the evidence in section 4.

However, Oi does give a separate analysis of the quit rate. His data are for 64 three-digit SIC census manufacturing industries over the period 1951–58. This sample differs from mine in two respects, it precedes it in time and it uses more aggregative industries. The dependent variables in the two pertinent regressions in Oi's study are the log of his turnover measure and the log of the quit rate. The explanatory variables in both regressions are as follows: relative wage, seasonal variability, average firm size measured in terms of employment, percentage of nonproduction workers, cyclical variability, and the secular change in relative wage. The coefficients of the relative wage and of the percentage of nonproduction workers are both negative in agreement with my results. The coefficients of seasonal and cyclical variability are both positive as expected. However, in contrast to my regressions, Oi finds a statistically significant negative effect of firm size on turnover. The coefficient of the secular wage change in Oi's regression is negative in the turnover and positive in the quit rate regression, albeit with very low t-ratios in both regressions. This variable was tried in my analysis and found to have too low a t-ratio to warrant inclusion in the final results. In addition to the difference in samples and the measurement of the dependent variables, there are several other important differences between Oi's regressions and those reported herein. The concentration ratio, capital per employee, the salaries of nonproduction workers including those at central offices and auxiliaries, total weekly hours, and the percentage of women workers all appear as important explanatory variables of the components of employment turnover in my findings but are not present in Oi's study.

A second and nearly contemporaneous study of employment turnover using the theory of specific human capital is Mincer (1962). This work challenges the interpretation of the inverse relation between turnover and wage rates as the result of specific human capital. Mincer points out that general training is also likely to have a positive effect on wages and a negative effect on turnover. It follows that the negative wage coefficient in the turnover regression reflects both general and specific training. To meet this objection Mincer presents new evidence. This consists of a regression to explain the proportion of workers by occupation who had full-time work in a given year. The sample consists of 87 male occupations for the year 1949. The explanatory variables are the full-time mean incomes by occupation, median years of schooling, proportion of workers less than twenty-five years old, and the proportion employed in durables and construction. The latter two variables are intended to correct for differences in the composition of occupations not germane to the specific human capital theory under investigation.

Mincer claims that if wage rates are held constant, then years of schooling and specific training vary inversely. He goes on to conclude that since the regression does hold wages constant, the specific human capital theory implies a negative years-of-schooling coefficient. In fact, he finds a statistically significant positive coefficient which he finds "puzzling" on the basis of the theory of specific human capital.

Puzzlement is unwarranted because the theory of specific human capital does not have the implication given by Mincer. It is necessary to make two distinctions in bringing the theory to bear on the regression results. First, one must distinguish between the cost of the human capital incurred by the firm and that incurred by the individual. The latter raises the worker's marginal product, not his wage rate. The cost of the skills borne by the worker raises both his wage and the marginal product. Hence holding the wage rate constant while the level of education rises means that the cost of the human capital borne by the *worker* must decrease, but it does not mean that the cost of the specific human capital borne by the *firm* must decrease. Second, one must distinguish between voluntary and involuntary unemployment. Thus, quits represent the former while layoffs represent the latter. The measure of turnover used by Mincer, percentage of workers with full-time employment, includes voluntary as well as involuntary unemployment. Yet the theory of specific human capital implies different effects on these. Holding the wage rate constant while increasing the level of education does imply a decrease in that portion of the specific human capital the cost of which is borne by the worker. Hence in a regression with the quit rate as the dependent variable, the coefficient of years of schooling should be positive if wages are held constant. However, with the layoff rate as the dependent variable and wages held constant, the sign of the years of schooling coefficient may be either negative or positive although a negative sign is more plausible if the firm invests more specific human capital in those workers having more formal schooling. Thus holding the wage rate constant does not hold the worker's marginal product constant. It follows that in a regression with layoffs dependent, the coefficient of years of schooling would be negative, given the wage rate, if the firm's specific human capital is complementary with formal schooling. Since Mincer's dependent variable combines the effects of both quits and layoffs, it follows that the theory of specific human capital is consistent with either a positive or a negative coefficient of years of schooling with wages constant.

In fact Parsons' doctoral dissertation (1970) shows that the theory of specific human capital receives strong confirmation if the quit and layoff rates are studied separately. He finds that holding the wage rate constant, the quit rate varies directly with years of schooling while the layoff rate varies inversely with years of schooling as predicted by the theory. This assumes, in agreement with Mincer, that the firm's specific capital per man is higher the more formal schooling per worker. Parsons explores many other theoretical and empirical implications of the theory of specific human capital in his work including a theory of optimal search for better jobs by workers and analysis of annual time series figures on quit rates by industry.

6. Conclusions

This research reveals a number of characteristics of the high-concentration industries. Rates of return rise with concentration. Tangible capital per employee rises with concentration. Specific human capital per employee rises with concentration. Highly concentrated industries contain the establishments of more diversified companies. In addition we know from other studies that concentration changes very slowly over time. Hence the belief that competition and concentration vary inversely is consistent with the evidence.

There is also evidence in this research to show that in the highly concentrated industries a large number of companies seem to be attracted by the high rates of return. Hence we should expect a positive association between rates of return and concentration by industry over time. The evidence for the first assertion is derived from the regressions in section 3 which shows that the coefficient of the number of companies in an industry is positive and rising with the concentration ratio. This is consistent with the view that competition ultimately prevails so that concentration and rates of return decline.

This research also suggests that several hitherto neglected capital components are among the important determinants of a company's return. Among these is the firm's investment in the specific skills and training of its employees that are of special benefit to the firm. It makes such investments in the expectation of obtaining a return comparable to the return on other forms of capital. However, unlike tangible capital, specific human capital is embodied in the worker. Hence the firm can lose its specific human capital whenever the worker quits. This fact creates a mutual interest between the worker and the company to share the cost and the returns of specific human capital, but the peculiar nature of human capital, that the company does not own it, makes it possible to study the implications of the existence of such capital by examining the components of labor force turnover. Despite the indirect nature of the evidence for the existence of specific human capital, there can be little doubt that it yields the firm a return. Therefore, omission of this capital from the calculation of the rate of return distorts the relation between concentration and the rate of return.

Other kinds of intangible capital should be included in order to calculate correctly the rate of return, but this is difficult because the necessary figures are lacking. Among the more important and neglected components of a firm's intangible capital are its investments in research and in marketing.

The emphasis on the importance of measuring intangible capital does not imply that we should neglect the task of improving the measures of tangible capital. For instance, gross book value is based on the original cost of plant and equipment which causes error in times of changing prices. We have seen some consequences in the empirical results where the price changes bias upward the coefficient of inventories, which are valued on a more current basis, and bias downward the coefficient of gross book value.

Even with good measures of tangible and intangible capital there are still some difficult conceptual problems due to the nature of the organization of some companies. Somehow we must learn how to relate the activities at the company headquarters to operations at the plant. This is not easily done with the present collection of census statistics. Although much progress has been made in the 1963 census, as witness the analysis in appendix 1, it is still impossible to relate the company's financial capital to its plants in order to avoid the present underestimates of total capital. For this we shall need a better-integrated set of company and establishment statistics.

All of this notwithstanding, it remains true that relative size, the salient aspect of industry structure considered herein, is positively related to rates of return, the measure of industry performance.

Appendix 1: Estimation of Payrolls, Annual Earnings, and Employment of Nonproduction Workers

According to the Census of Manufactures, a manufacturing company consists of establishments, central offices, and auxiliaries. Establishments are the sites of the manufacturing activities and are classified into four-digit SIC census manufacturing industries according to the primary products of the establishment. This means that establishments generally make several related products. The central offices and auxiliaries (CAO's) include the company headquarters, and other facilities not located at establishments such as research laboratories, testing grounds, warehouses, and manufacturer's sales branches. Simple companies combine at one location all of these activities while complex companies have one or more establishments and one or more CAO's. In addition to the four-digit SIC industries which classify the establishments, there are enterprise industries which classify companies. An enterprise consists of one or more four-digit SIC industries. Companies are classified into enterprise industries according to the primary activities of their establishments. There are several types of companies as follows:

1. Single-unit (one establishment);
2. Multiple-unit,
    a. Single-enterprise industry,
    b. Multi-enterprise industry.

Most of the complications are due to the companies in 2.b. Fortunately, the 1963 census provides, for the first time, the means to relate certain statistics at the establishment and company levels. For our purposes the pertinent statistics are employment and payrolls. In this endeavor we are guided by tables 4B and 5A, *1963 Enterprise Statistics Part 1*. Table 4B gives the total employment for every enterprise industry and its four-digit SIC components. For example, the enterprise industry 20B, prepared meats and dressed poultry, consists of two four-digit industries—2013, meat processing, and 2015, poultry dressing plants. The goal is to estimate the employment and payroll at the CAO's corresponding to each of the four-digit establishments operated by the company. Thus, we wish to impute the employment and payrolls at the CAO's to the establishments. Since the employment and the payrolls of the nonproduction workers at the establishments are already known, these figures can be added to the ones attributed from the CAO's to give an estimate of total payrolls and employment of all nonproduction workers by four-digit class.

Let

$$e_{ij} = \text{employment in enterprise industry } i \text{ and four-digit class } j.$$

The unknowns to be estimated are as follows:

$$e_{ij}^1 = \text{employment of single-enterprise-industry companies,}$$

$$e_{ij}^2 = \text{employment of multi-enterprise-industry companies.}$$

(1) $e_{ij} = e_{ij}^1 + e_{ij}^2.$

Let

$$e_{i.} = \text{total employment of all establishments in enterprise industry } i$$

$$E_i = \text{total employment in enterprise industry } i.$$

It is important to realize that $e_{i.}$ does not necessarily equal E_i. This is because some establishments in a given four-digit class belong to companies whose primary activities place them in an enterprise industry composed of other four-digit industries. For example, suppose that a cigarette company owns an establishment that makes soft drinks. This employment is included in the enterprise industry 21A, tobacco products, and not in the enterprise industry 20P, bottled soft drinks and flavorings, because the soft drink activity is secondary to the cigarette company. However, the four-digit classification of establishments does include the employment in the establishment belonging to the cigarette company. Hence $e_{i.}$ on an establishment basis for soft drinks is less than E_i for the enterprise industry while $e_{i.}$ on an establishment basis for cigarettes is below E_i for the enterprise industry tobacco products. These considerations lead to the definition of the specialization ratio as follows:

(2) $s_{ij} = e_{i.}/(E_i - CAO \text{ employment}).$

An example and interpretation of this ratio is given in the text in section 3.
Table 5A of *Enterprise Statistics* gives us

E_i^1 = total employment in enterprise industry i of single-unit companies,

E_i^2 = total enterprise industry i's employment of multi-unit companies.

Since single-unit companies have one establishment, such companies correspond to 1.0 in the classification above while the multi-unit companies include
both 2.a and 2.b. Single-unit companies are obviously specialized to a single
four-digit SIC industry. It is assumed that the ratio e_{ij}^1/e_{ij} for the four-digit
SIC industry is the same as for the enterprise industry i that contains this
four-digit category.

(3) $e_{ij}^1 = e_{ij}(E_i^1/E_i),$

therefore, is the estimate of the total employment of single-unit companies in
the four-digit SIC industry (i, j). A similar assumption yields an estimate of
e_{ij}^2. Thus,

(4) $e_{ij}^2 = s_{ij}e_{ij}(E_i^2/E_i).$

Notice, however, that this estimate is adjusted to take into account the enterprise industry specialization ratio.

The only payroll component that it is necessary to estimate is the payroll
at the *CAO*'s corresponding to the four-digit SIC industries since the 1963
census already provides the employment and payrolls of nonproduction
workers located at the establishment. Thus we need only estimate the corresponding payrolls of the nonproduction workers who are *not* at the establishments, namely the nonproduction workers at the *CAO*'s. The *Enterprise
Statistics* Table 5A gives us the total *CAO* payrolls of the multi-unit companies. Denote this figure by X_i^2.

Let

x_{ij}^2 = *CAO* payrolls of multi-unit companies in enterprise industry i and
four-digit SIC industry j.

The estimate of x_{ij}^2 is given by

(5) $x_{ij}^2 = X_i^2(e_{ij}^2/e_i.).$

This estimate, x_{ij}^2, is added to the four-digit SIC estimate of the payrolls of
nonproduction workers to give the variable, *MPA*, that enters the multiple
regressions described in sections 3 and 4.

Appendix 2: Description of the Samples

1. THE 1958 SAMPLE

The data are taken from a special census tabulation which gives figures on
gross and net book value together with related data. This tabulation includes

data for 277 four-digit SIC manufacturing industries based on the 1945 classifications as revised in 1954. The regression analysis uses 265 observations and omits the 12 industries given as follows:

2084   wines and brandy (a)
2251   full-fashioned hosiery (a)
2271   wool carpets and rugs (a)
3312   blast furnaces and steel mills (a)
3331   primary copper (a, b)
3332   primary lead (a, b)
3334   primary aluminum (a, b)
3424   files (b)
3583   sewing machines (b)
3585   refrigeration machinery (a)
3661   radios and related products (a)
3717   motor vehicles and parts (a)

a. Value of shipments are not given because there are substantial interplant transfers of the same company.
b. Disclosure, concentration figures are not given because this would violate the census disclosure rules.

2. THE 1963 SAMPLE

There is a total of 417 four-digit SIC manufacturing industries. The classification is based on the 1957 manual. The sample omits the 19 industries as follows:

2391   curtains and draperies, error in the number of companies and establishments
2661   building paper and board mills, no inventory figures
2711   newspapers, especially misleading concentration figure because of the local nature of the industry
2814   cyclic coal tar, disclosure
2911   petroleum refining, unreliable estimate of OHA
3151   leather gloves, no concentration ratio
3297   nonclay refractories, disclosure for the 20th to 50th largest firms
3332   primary lead, disclosure
3334   primary aluminum, disclosure
3411   metal cans, unreliable estimate of OHA
3492   safes and vaults, disclosure
3636   sewing machines, disclosure
3641   electric lamps, census error in classifying establishments
3662   radio, TV, communications, excluded by accident
3721   aircraft, unreliable estimate of OHA
3723   aircraft propellers and parts, disclosure
3871   watches and clocks, unreliable estimate of OHA
3872   watchcases, same reason as 3297
3942   dolls, disclosure

3. SAMPLE OF FOUR-DIGIT SIC MANUFACTURING INDUSTRIES FOR LABOR FORCE TURNOVER

2011	2221*	2431	28234**	3321
2015	2231*	2432	2834	3322
2041	2241*	24412**	2841	3323
2042	2254	2511	2844	3351
2051	2311*	2512	3211*	3352
2052	2321	2515	3221	3357
2071	2327	2631*	3229	3361
2082	2328	2643	3241*	33629**
2111*	2341	26512**	3251	3391
2121*	2342	2653	3291	3429
2211*	2421	2821	3312	3433
3441	3533	3585	3651*	39413**
3443	35356**	3611	3661	3949
34469**	3541	3612	3694	
3452	35428**	3613	3713	
3461	3545	3621	3722	
3481	3551	3622	3731	
34948**	3552	3633	3811*	
3511	3561	3634	3821	
3519	3562	3642	3822	
3522*	3566	36434**	3861*	

* Three-digit SIC industry.
** combination of two four-digit industries, e.g., 24412 means 2441 and 2442. In these cases the appropriate explanatory variables are weighted averages of the four-digit components.

The source for the employment turnover data: U.S. Dept. of Labor, Bureau of Labor Statistics, 1965, *Employment and Earnings Statistics for the United States, 1909–1965*, Bull. no. 1312–13. For a description of the methods of collecting the data and related information see U.S. Dept. of Labor, Bureau of Labor Statistics, 1966, *Handbook of Methods for Surveys and Studies*, Bull. no. 1458, pp. 34–40.

Appendix 3: The Two-Digit Industry Effects

Table 13 gives the coefficients of the dummy variables that represent the two-digit industry effects for the regressions which divide total payrolls into log *PPA*, production workers, and log *MPA*, nonproduction workers. The other regression statistics are given in table 8.6. A blank indicates that the given two-digit industry is not represented by any four-digit constituent in the given concentration ratio stratum. For example, there are no four-digit industries in the two-digit group, lumber, that have concentration ratios above 50 percent. Perhaps the most interesting finding is that in the most· concentrated stratum the industries are more nearly alike than in the other two less concentrated strata. This is indicated by the generally lower values for the *t*-ratios in the stratum [50–100]. Finally, it would be possible to represent common industry effects in a more detailed way by using a dummy variable

TABLE 8.13

Regression Coefficients for Two-digit SIC Industry Dummy Variables for the Regressions in Table 8.6

Two-digit SIC Industry	Total	Concentration Ratio Strata [0, 25)	[25, 50)	[50–100]
20, food	0.782	1.018	0.871	−0.111
	3.57	3.33	2.43	0.19
21, tobacco	0.839	...	−0.422	0.146
	3.03		0.86	0.23
22, textile mills	0.357	0.812	0.608	−0.896
	1.66	2.69	1.77	1.55
23, apparel	0.716	0.911	0.870	−0.391
	3.35	3.23	2.44	0.54
24, lumber	0.198	0.572	0.227	...
	0.85	1.83	0.57	
25, furniture	0.443	0.758	0.464	...
	1.99	2.50	1.29	
26, paper	0.473	0.488	0.577	−0.412
	2.08	1.57	1.70	0.70
27, printing	0.535	0.756	0.783	−0.491
	2.39	2.51	2.16	0.71
28, chemicals	0.678	1.141	0.736	−0.379
	3.05	3.66	2.08	0.64
29, petroleum	0.717	1.408	0.791	−1.044
	2.82	4.22	2.19	1.57
30, rubber	0.507	0.816	0.890	−0.601
	1.87	2.32	2.26	0.94
31, leather	0.573	0.790	0.732	−0.307
	2.52	2.77	1.94	0.47
32, stone, clay and glass	0.370	0.681	0.672	−0.635
	1.68	2.27	1.87	1.10
33, primary metals	0.316	0.836	0.482	−0.984
	1.38	2.81	1.27	1.59
34, fabricated metals	0.378	0.645	0.540	−0.440
	1.72	2.19	1.47	0.79
35, machinery	0.241	0.536	0.454	−0.777
	1.10	1.78	1.28	1.34
36, electrical machinery	0.429	0.738	0.747	−0.554
	1.89	2.18	2.03	0.93
37, transportation equipment	0.212	0.806	0.296	−0.864
	0.86	2.58	0.69	1.34
38, instruments	0.386	0.625	0.578	−0.522
	1.60	1.72	1.55	0.85
39, miscellaneous	0.468	0.718	0.611	−0.487
	2.28	2.54	1.84	0.83

for every three-digit SIC industry. However, this would be unwise both because it would use up too many degrees of freedom and because it verges on the untestable doctrine that every industry is unique; and, therefore, the set of four-digit SIC industries cannot be studied as a whole. The results shown in table 8.13 are substantially the same as the coefficients of the industry dummy variables for the other regressions reported in tables 8.5–8.8.

Appendix 4: The Relative Size Distribution of Firms

In the preceding analysis the relative size distribution of firms is represented by a single figure, the four-firm concentration ratio, which raises the question of whether other measures of relative size would give different results. Some evidence suggests that the results are probably not very sensitive to the particular concentration ratio, be it the four-, eight-, or twenty-firm concentration ratio, because of a tendency for the logs of the shares by firm to be positively correlated. It appears that one may be able to predict with some success the log of the shares by firm in an industry as a function of the log of the share of the leading firm in the industry.

Table 8.14 gives some evidence to support these views. In this table we find the regression statistics for three regressions relating the logs of the shares of nonoverlapping groups of firms. Thus the first regression gives the log of the

TABLE 8.14
Regression Equations for Concentration Ratios, 1963[a]

(1)	$\log (C8 - C4) = 0.9262 + 0.4310 \log (C4)$, $\qquad\qquad\qquad$ 8.10 $\quad$ 13.13	$R^2 = 0.3013, n = 402$
(2)	$\log (C20-C8) = 1.4091 + 0.4933 \log (C8-C4)$, $\qquad\qquad\qquad$ 11.85 $\quad$ 10.19	$R^2 = 0.2087, n = 396$[b]
(3)	$\log (C50-C20) = 0.4594 + 0.7568 \log (C20-C8)$, $\qquad\qquad\qquad$ 2.62 $\quad$ 11.67	$R^2 = 0.2669, n = 376$[b]

a. $CJ =$ share of value added accounted for by the J largest firms in the industry.
The number below the regression coefficient is its t-ratio.
b. The reduction in sample size is made necessary by the omission of those industries which attain 100 percent of the market for the 8 largest firms so that $C20 - C8 = 0$ which implies a value of the log equal to minus infinity. In additional cases, the combined share of the 20 largest firms reaches 100 for the first time so that although $C20 - C8 > 0$, $C50 - C20 = 0$. Hence additional industries must be excluded from the third regression and this accounts for the reduction in sample size.

combined share of the 5th to the 8th largest firm as a linear function of the log of the 1st to the 4th largest firm. The R^2 is 0.30. The other two regressions give similar results for the other two rankings. Hence there is reason to believe that the eight- or twenty-firm concentration ratio instead of the four-firm concentration ratio would not materially change the regression results in sections 3 and 4.

There is another interesting and important fact about the relative size distribution of firms implicit in the results shown in table 8.14. This refers to the use of the log of the shares instead of the shares themselves in describing the relative size distribution. Development of this point requires some algebra. Let

c_{ij} = share of the value added by the ith largest firm in industry j

n_j = number of firms in industry j.

Consequently,

(1)
$$\sum_{i=1}^{n_j} c_{ij} = 1.$$

The first regression in table 8.14 relates

$$\log\left[\sum_{i=1}^{4} c_{ij}\right] \quad \text{and} \quad \log\left[\sum_{i=5}^{8} c_{ij}\right]$$

Instead one can calculate regressions that would relate

$$\sum_{i=1}^{4} c_{ij} \quad \text{and} \quad \sum_{i=5}^{8} c_{ij}.$$

However, to do the latter introduces a bias tending to lower the correlation among the shares because the shares must sum to one. Therefore, the covariance between the shares of nonoverlapping sets of companies tends to be negative. To prove this, let

$$\sigma_{ik} = \text{cov}\,(c_{ij}, c_{kj}).$$

Thus σ_{ik} is the covariance between the shares of companies of rank i and k by industry. By the definition of a covariance (E denoting the expected value),

(2)
$$\sigma_{ik} = E c_{ij}\, c_{kj} - E c_{ij}\, E c_{kj}.$$

Since

$$\sum_{i=1}^{n_j} c_{ij} = 1, \qquad\qquad \text{for all } j,$$

$$E \sum_i c_{ij} c_{kj} = E c_{kj}$$

so that

(3)
$$\sum_i E c_{ij} c_{kj} = E c_{kj}.$$

It follows that

$$\sum_i \sigma_{ik} = \sum_i E c_{ij} c_{kj} - \sum_i E c_{ij}\, E c_{kj}$$

$$= E c_{kj} - E c_{kj}$$

(4)
$$\sum_i \sigma_{ik} = 0.$$

Therefore, since $\sigma_{ii} > 0$, it follows that

(5) $$\sum_{i, i \neq k} \sigma_{ik} < 0.$$

Next let us derive the results for the covariance among the logs of the shares. Let G_j = geometric mean share of firms in industry j. Therefore,

(6) $$\sum_{i=1}^{n_j} \log c_{ij} = n_j \log G_j.$$

Equation (6) implies that

(7) $$\sum_{i=1}^{n_j} E\left[(\log c_{kj}, \log c_{ij})\right] = E\left[(n_j \log G_j)(\log c_{kj})\right].$$

Let

$$m_{ik} = \text{cov}(\log c_{ij}, \log c_{kj}).$$

Thus, m_{ik} is the covariance between the log of the shares of companies of ranks i and k across industries so that it corresponds to σ_{ik}. Now

$$\sum_i m_{ik} = \sum_i E\left[(\log c_{ij})(\log c_{kj})\right] - \sum_i (E \log c_{ij})(E \log c_{kj})$$

$$= E\left[(n_j \log G_j)(\log c_{kj})\right] - E \log c_{kj} \sum_i E(\log c_{ij})$$

$$= E\left[(n_j \log G_j)(\log c_{kj})\right] - E \log c_{kj}(En_j \log G_j)$$

(8) $$= \text{cov}(n_j \log G_j, \log c_{kj}).$$

In contrast with (4), the right-hand side of (8) is not necessarily zero so that there is not necessarily a built-in bias towards a negative covariance between the logs of the shares as there is between the shares themselves. Hence the regression coefficients in table 8.14 are not biased downward because of the arithmetic that forces the sums of the shares to be one. In general the relations among the logs of the shares have several interesting properties in terms of entropy that are worth analyzing in their own right.[11]

---

11. Gort (1963) reaches the erroneous conclusion that the shares of the firms in the same industry are uncorrelated because he fails to take into account the bias resulting from regressing the shares of nonoverlapping groups of firms. For a discussion of concentration in terms of entropy, see Stigler (1968, chap. 4) and Theil (1967, chap. 8).

References

Ahlfors, Lars V. 1953. *Complex Analysis.* New York: McGraw-Hill.

Aumann, Robert. 1961. The Core of a Cooperative Game without Side Payments. *Trans. Am. Math. Soc.* 98: 539–52.

————. 1962. Cooperative Games without Side Payments. *Recent Advances in Game Theory*, pp. 83–99. Princeton: Princeton Univ. Conference (private printing for members of the conference).

————. 1964. Markets with a Continuum of Traders. *Econometrica* 32: 39–50.

————. 1966. Existence of Competitive Equilibria in Markets with a Continuum of Traders. *Econometrica* 34: 1–17.

Bain, Joe S. 1951. Relation of Profit Rate to Industry Concentration: American Manufacturing, 1936–1940. *Q.J.E.* 65: 293–324 (erratum, p. 602).

Becker, Gary S. 1962. Investment in Human Capital: A Theoretical Analysis. *Investment in Human Beings, J.P.E. Supplement* 70: 9–49.

Bochner, Solomon. 1959. *Lectures on Fourier Integrals*, trans. Morris Tenenbaum and Harry Pollard. Annals of Mathematical Studies 42. Princeton: Princeton Univ. Press.

Böhm-Bawerk, Eugen von. 1930. *The Positive Theory of Capital*, trans. William A. Smart (photo-offset of 1891 ed.). New York: Stechert.

Cagan, Phillip. 1956. The Monetary Dynamics of Hyperinflation. *Studies in the Quantity Theory of Money*, ed. Milton Friedman. Chicago: Univ. of Chicago Press.

Chamberlin, E. H. 1933. *The Theory of Monopolistic Competition.* Cambridge, Mass.: Harvard Univ. Press.

Collins, Norman R., and Lee E. Preston. 1968. *Concentration and Price-Cost Margins in Manufacturing Industries.* Berkeley and Los Angeles: Univ. of Calif. Press.

Cournot, Augustin. 1960. *Researches into the Mathematical Principles of the Theory of Wealth*, trans. Nathaniel Bacon from 1838 French ed., introductory essay by Irving Fisher. New York: Kelley.

Debreu, G., and I. Herstein. 1953. Nonnegative Square Matrices. *Econometrica* 21: 597–607.

Debreu, G., and H. Scarf. 1963. A Limit Theorem on the Core of an Economy. *Int. Econ. Rev.* 4: 235–46.

Demsetz, Harold. 1962. The Effect of Consumer Experience on Brand Loyalty and the Structure of Market Demand. *Econometrica* 30: 22–33.

Dunford, Nelson, and Jacob T. Schwartz. 1958. *Linear Operators Part I: General Theory.* New York: Interscience Publishers.

Edgeworth, Francis Y. 1881. *Mathematical Psychics.* London: Kegan-Paul.

Farrell, M. J. 1959. The Convexity Assumption in the Theory of Competitive Markets. *J.P.E.* 67: 377–91.

————. 1961. A Reply. *J.P.E.* 69: 484–9.

Federal Trade Commission. 1969. *Economic Report on the Influence of Market Structure on the Profit Performance of Food Manufacturing Companies.* Staff Report by Willard F. Mueller; William H. Kelly; and Russell C. Parker. Washington, D.C.: U.S. Govt. Printing Office.

Feller, William. 1962. *An Introduction to Probability Theory and Its Applications.* 2d ed., vol. 1. New York: Wiley.

————. 1966. *An Introduction to Probability Theory and Its Applications*, vol. 2. New York: Wiley.

Fellner, William. 1949. *Competition among the Few.* New York: Knopf.

Friedman, Milton. 1953. *Essays in Positive Economics.* Chicago: Univ. of Chicago Press.

Fourt, Louis A., and J. W. Woodlock. 1960. Early Prediction of Success for New Grocery Products. *J. Marketing* 25: 31–8.

367

Gale, David. 1960. *The Theory of Linear Economic Models.* New York: McGraw-Hill.

Gnedenko, B. V., and A. N. Kolmogorov. 1954. *Limit Distributions for Sums of Independent Random Variables,* trans. K. L. Chung. Reading, Mass.: Addison-Wesley.

Gort, Michael. 1963. Analysis of Stability and Change in Market Shares. *J.P.E.* 71: 51–63.

Graves, L. M. 1946. *The Theory of Functions of Real Variables.* New York: McGraw-Hill.

Griliches, Z. 1961. A Note on the Serial Correlation Bias in Estimates of Distributed Lags. *Econometrica* 29: 65–73.

———. 1967. Distributed Lags: A Survey. *Econometrica* 35: 16–49.

———. 1969. Capital-Skill Complementarity. *Rev. Econ. Stat.* 51: 465–68.

Halmos, Paul R. 1957. *Introduction to Hilbert Space.* 2d ed. New York: Chelsea Publishing.

———. 1958. *Finite-Dimensional Vector Spaces.* 2d ed. Princeton: Van Nostrand.

Hardy, G. H. 1949. *Divergent Series.* Oxford: Oxford Univ. Press.

Hardy, G. H.; J. E. Littlewood; and G. Polyà. 1952. *Inequalities.* 2d ed. Cambridge: Cambridge Univ. Press.

Helson, Henry. 1964. *Lectures on Invariant Subspaces.* New York: Academia Press.

Hicks, J. R. 1946. *Value and Capital.* 2d ed. Oxford: Oxford Univ. Press.

———. 1956. *A Revision of Demand Theory.* Oxford: Oxford Univ. Press.

Hotelling, Harold. 1938. The General Welfare in Relation to Problems of Taxation and of Railway and Utility Rates. *Econometrica* 6: 242–69.

Karlin, Samuel. 1959. *Mathematical Methods and Theory in Games, Programming and Economics,* vol. 1. Reading, Mass.: Addison-Wesley.

Kessel, Reuben. 1971. A Study of the Effects of Competition in the Tax-Exempt Bond Market. *J.P.E.* 79: 706–38.

Knight, Frank H. 1921. *Risk, Uncertainty and Profit.* Boston: Houghton Mifflin.

Knopp, Konrad. 1951. *Theory and Application of Infinite Series,* trans. R. C. H. Young. 2d Eng. ed. London: Blackie.

Koopmans, Tjalling C., and Martin Beckmann. 1957. Assignment Problems and the Location of Economic Activity. *Econometrica* 25: 53–76.

Kruskal, William H., and Lester G. Telser. 1960. Food Prices and the Bureau of Labor Statistics. *J. Bus.* 33: 258–79.

Kuhn, H. W., and A. W. Tucker. 1950. Nonlinear Programming. *Proceedings of the Second Berkeley Symposium on Mathematical Statistics and Probability,* ed. Jerzy Neyman. Berkeley and Los Angeles: Univ. of Calif. Press.

Levitin, E. S., and B. T. Polyak. 1966. Constrained Minimization Methods, trans. ed. *Zh. Vychisl. Mat. i. Mat. Fiz.* 6, 5: 787–823 (*U.S.S.R. Computational Math. and Math. Physics,* 6: 1–50).

Liusternik, L., and V. Sobolev. 1961. *Elements of Functional Analysis,* trans. A. E. Labarre, Jr.; H. Izbicki; and H. W. Crowley. New York: Ungar.

Loève, Michael. 1955. *Probability Theory.* 2d ed. Princeton: van Nostrand.

Lucas, W. F. 1967. A Counterexample in Game Theory. *Management Science* 13: 766–67.

———. 1968. A Game with No Solution. *Bull. Am. Math. Soc.* 74: 237–39.

Luce, R. Duncan, and Howard Raiffa. 1957. *Games and Decisions.* New York: Wiley.

Marshall, Alfred. 1920. *Principles of Economics.* 8th ed. London: Macmillan.

Mood, A. M., and F. Graybill. 1963. *Introduction to the Theory of Statistics.* New York: McGraw-Hill.

Mincer, Jacob. 1962. On-the-Job Training: Costs, Returns and Some Implications. *Investment in Human Beings, J.P.E.* Supplement 70: 50–79.

Nash John. F. 1950a. The Bargaining Problem. *Econometrica* 18: 155–62.

Nash, John F. 1950b. Equilibrium Points in *N*-Person Games. *Proc. Nat. Acad. Sci., U.S.A.* 36: 48–49.

———. 1951. Non-cooperative Games. *Ann. Math.* 54: 286–95.

———. 1953. Two-Person Cooperative Games. *Econometrica* 21: 128–40.

Natanson, I. P. 1961. *Theory of Functions of a Real Variable*, trans. Leo F. Boron, ed. in collaboration with Edwin Hewitt, rev. ed., 2 vols. New York: Ungar.

Nerlove, Marc. 1958. *The Dynamics of Supply: Estimation of Farmers' Response to Price*. Baltimore: Johns Hopkins Press.

Neumann, John von, and Oskar Morgenstern. 1944, 1947 (2d ed.). *The Theory of Games and Economic Behavior*. Princeton: Princeton Univ. Press.

Oi, Walter Y. 1961. Labor as a Quasi-Fixed Factor of Production. Ph.D. dissertation, Univ. of Chicago.

———. 1962. Labor as a Quasi-Fixed Factor. *J.P.E.* 70: 538–55.

Okuguchi, K. 1970. Adaptive Expectations in an Oligopoly Model. *Rev. Econ. Stud.* 37: 233–37.

Parsons, Donald. 1970. Specific Human Capital, Layoffs and Quits. Ph.D. dissertation, Univ. of Chicago.

Rosen, Sherwin. 1965. Short Run Employment Variation on Class-1 Railroads in the United States, 1947–1963. Ph.D. dissertation, Univ. of Chicago.

Rothenberg, Jerome. 1960. Non-Convexity, Aggregation and Pareto Optimality. *J.P.E.* 58: 435–68.

Samuelson, Paul A. 1953. *Foundations of Economic Analysis*. Cambridge, Mass.: Harvard Univ. Press.

———. 1966. Constancy of the Marginal Utility of Income. *The Collected Scientific Papers of Paul A. Samuelson*, ed. Joseph E. Stiglitz, vol. 1. Cambridge, Mass.: M.I.T. Press.

Scarf, Herbert. 1962. An Analysis of Markets with a Large Number of Participants. *Recent Advances in Game Theory*, pp. 127–55. Princeton: Princeton Univ. Conference (private printing for members of the conference).

———. 1967. The Core of an *N* Person Game. *Econometrica* 35: 50–69.

———. 1970. On the Existence of a Cooperative Solution for a General Class of *N*-Person Games. Unpublished Cowles Foundation Discussion Paper no. 293.

Shapley, Lloyd S. 1967. On Balanced Sets and Cores. *Naval Research Logistics Quarterly*, 14: 453–60.

Shapley, L. S., and M. Shubik. 1966. Quasi-Cores in a Monetary Economy with Nonconvex Preferences. *Econometrica* 34: 805–27.

———. 1968. On Market Games. RAND Corporation unpublished memorandum RM-5671-PR.

———. 1969. On the Core of an Economic System with Externalities. *Am. Econ. Rev.* 59: 678–84.

———. Two-Sided Markets: The Assignment Game. Chap. 6 in forthcoming monograph.

Shubik, M. 1959a. Edgeworth Market Games. *Contributions to the Theory of Games*, ed. A. W. Tucker and R. D. Luce, vol. 4. Princeton: Princeton Univ. Press.

———. 1959b. *Strategy and Market Structure*. New York: Wiley.

Stigler, George J. 1963. *Capital and Rates of Return in Manufacturing Industries*. Princeton: Princeton Univ. Press.

———. 1968. *The Organization of Industry*. Homewood, Ill.: Irwin.

Sudman, Seymour. 1962. On the Accuracy of Consumer Panels. Ph.D. dissertation, Grad. School of Business, Univ. of Chicago.

Telser, L. G. 1961. How Much Does It Pay Whom to Advertise? *Am. Econ. Rev.* 51: 194–205.

Telser, L. G. 1962a. The Demand for Branded Goods as Estimated from Consumer Panel Data. *Rev. Econ. Stat.* 44: 300–24.

———. 1962b. Advertising and Cigarettes. *J.P.E.* 70: 471–99.

———. 1964. Advertising and Competition. *J.P.E.* 72: 537–62.

———. 1966. Supply and Demand for Advertising Messages. *Am. Econ. Rev.* 56: 457–66.

———. 1968. Some Aspects of the Economics of Advertising. *J. Bus.* 41: 166–73.

———. 1969a. On the Regulation of Industry: A Note. *J.P.E.* 77: 937–52. 1971. Correction. *J.P.E.* 79: 932.

———. 1969b. Another Look at Advertising and Concentration. *J. Ind. Econ.* 18: 85–94.

———. 1971. The Core of a Market, Money and the Role of Prices. Unpublished Center Report, Center for Mathematical Studies in Business and Economics, Univ. of Chicago.

Telser, L. G., and R. L. Graves. 1971. *Functional Analysis in Mathematical Economics: Optimization over Infinite Horizons.* Chicago: Univ. of Chicago Press.

Theil, H. 1961. *Economic Forecasts and Policy.* 2d rev. ed. Amsterdam: North-Holland Publishing.

———. 1967. *Economics and Information Theory.* Amsterdam: North-Holland Publishing.

U.S. Senate Committee on the Judiciary. 1961. *Administered Prices Drugs.* 87th Congress: First Session, Report no. 448.

Uzawa, H. 1960. Preference and Rational Choice in the Theory of Consumption. *Mathematical Methods in the Social Sciences 1959*, ed. K. J. Arrow; S. Karlin; and P. Suppes, Stanford: Stanford Univ. Press.

Vainberg, M. M. 1964. *Variational Methods for the Study of Non-linear Operators*, trans. A. Feinstein. San Francisco: Holden-Day.

Varga, Richard S. 1962. *Matrix Iterative Analysis.* Englewood Cliffs, N.J.: Prentice-Hall.

Viner, Jacob. 1952. Cost Curve and Supply Curves. Reprinted in *Readings in Price Theory*, ed. G. J. Stigler and K. Boulding. Chicago: Irwin.

Weiss, Leonard W. 1966. Concentration and Labor Earnings. *Am. Econ. Rev.* 56: 96–117.

Whittle, Peter. 1963. *Prediction and Regulation.* London: English Univs. Press.

Widom, Harold. 1965. Toeplitz Matrices. *Studies in Real and Complex Analysis*, ed. I. I. Hirshman, Jr. Mathematical Assoc. of America Studies in Mathematics, vol. 3. Englewood Cliffs, N.J.: Prentice-Hall.

Wilks, Samuel S. 1962. *Mathematical Statistics.* New York: Wiley.

Name Index

Aumann, Robert, 58, 207–8

Bain, Joe S., 353
Becker, Gary S., 316 n
Beckmann, Martin, 97
Böhm-Bawerk, von Eugen, 18, 26, 38, 57

Cournot, Augustin, 132, 147
Collins, Norman, 352–53

Debreu, Gerard, 58, 270
Demsetz, Harold, 304
Director, Aaron, 305

Edgeworth, Francis, Y., 57–58

Farrell, Michael, 58
Friedman, Milton, 62 n, 63

Gale, David, 92
Graves, Robert L., 240, 256, 258, 263
Griliches, Zvi, 289 n, 344 n
Gort, Michael, 315 n, 365 n

Halmos, Paul R., 247–48
Hardy, G. H., 228
Helson, Henry, 256
Herstein, Israel, 270
Hicks, J. R., 62 n

Kessell, Reuben, 131 n
Knight, Frank H., 282
Koopmans, Tjalling, 97
Kruskal, William H., 277

Liusternik, L., 257
Loève, Michael, 258

Lucas, W. F., 194 n
Luce, R. Duncan, 141, 194 n, 209

Marshall, Alfred, 2 n, 63, 316
Mincer, Jacob, 2 n, 316, 354–55
Morgenstern, Oskar, 38, 57–58, 125 n, 194, 208

Nash, J. F., Jr., 132, 133 n, 215 n
Natanson, I. P., 232
Neumann, von, John, 38, 57–58, 125 n, 194, 208

Oi, Walter Y., 316 n, 354

Parsons, Donald, 355–56
Preston, Lee, 352–53
Polyak, B. T., 240

Raiffa, Howard, 141 n, 194 n, 209
Rothenberg, Jerome, 58

Scarf, Herbert, 58, 60 n, 209
Shapley, Lloyd S., 59–61
Shubik, Martin, 58–59, 61 n
Sobolev, V., 257
Starret, David, 61 n
Stigler, George J., 204 n, 353
Sudman, Seymour, 276

Telser, Lester G., 240, 256, 258, 263, 276, 277, 305

Varga, Richard S., 258, 259, 269–70
Viner, Jacob, 183

Widom, Harold, 249

371

Subject Index

Transitory components of quits and layoffs,
    350–52
Transferable utility, 4–9
Transformation
    linear, 233, 243
    positive definite representation of, 256
True capital, estimates of, 329
Turnover of work force, and specific human
    capital, 315

Utility
    maximization consistency with
      maximum consumer surplus, 64
    measurability of, 5

VA (value added), 318, 320, 322
VS (value of shipments), 318
Variables excluded, effect of 283–85

Wage rate
    during training period, 328
    and marginal product of labor according
      to specific human capital theory, 352
    relation to G and SW, 344
Wealth, interpretations of, 5–6